Changing Climates, Earth Systems and Society

International Year of Planet Earth

Series Editors:

Eduardo F.J. de Mulder
Executive Director International Secretariat
International Year of Planet Earth

Edward Derbyshire
Goodwill Ambassador
International Year of Planet Earth

The book series is dedicated to the United Nations International Year of Planet Earth. The aim of the Year is to raise worldwide public and political awareness of the vast (but often under-used) potential of Earth sciences for improving the quality of life and safeguarding the planet. Geoscientific knowledge can save lives and protect property if threatened by natural disasters. Such knowledge is also needed to sustainably satisfy the growing need for Earth's resources by more people. Earths scientists are ready to contribute to a safer, healthier and more prosperous society. IYPE aims to develop a new generation of such experts to find new resources and to develop land more sustainably.

For further volumes:
http://www.springer.com/series/8096

John Dodson
Editor

Changing Climates, Earth Systems and Society

Springer

Editor
Prof. John Dodson
Australian Nuclear Science &
Technology Organisation
Institute for Environmental Research
Kirrawee NSW 2232
Australia
jdd@ansto.gov.au

ISBN 978-90-481-8715-7 e-ISBN 978-90-481-8716-4
DOI 10.1007/978-90-481-8716-4
Springer Dordrecht Heidelberg London New York

Library of Congress Control Number: 2010932611

Printed on acid-free paper

Springer is part of Springer Science+Business Media (www.springer.com)

Foreword

The International Year of Planet Earth (IYPE) was established as a means of raising worldwide public and political awareness of the vast, though frequently under-used, potential the Earth Sciences possess for improving the quality of life of the peoples of the world and safeguarding Earth's rich and diverse environments.

The International Year project was jointly initiated in 2000 by the International Union of Geological Sciences (IUGS) and the Earth Science Division of the United Nations Educational, Scientific and Cultural Organisation (UNESCO). IUGS, which is a Non-Governmental Organisation, and UNESCO, an Inter-Governmental Organisation, already shared a long record of productive cooperation in the natural sciences and their application to societal problems, including the International Geoscience Programme (IGCP) now in its fourth decade.

With its main goals of raising public awareness of, and enhancing research in the Earth sciences on a global scale in both the developed and less-developed countries of the world, two operational programmes were demanded. In 2002 and 2003, the Series Editors together with Dr. Ted Nield and Dr. Henk Schalke (all four being core members of the Management Team at that time) drew up outlines of a Science and an Outreach Programme. In 2005, following the UN proclamation of 2008 as the United Nations International Year of Planet Earth, the "Year" grew into a triennium (2007–2009).

The Outreach Programme, targeting all levels of human society from decision makers to the general public, achieved considerable success in the hands of member states representing over 80% of the global population. The Science Programme concentrated on bringing together like-minded scientists from around the world to advance collaborative science in a number of areas of global concern. A strong emphasis on enhancing the role of the Earth sciences in building a healthier, safer and wealthier society was adopted – as declared in the Year's logo strap-line "Earth Sciences *for* Society".

The organisational approach adopted by the Science Programme involved recognition of ten global themes that embrace a broad range of problems of widespread national and international concern, as follows.

- Human health: this theme involves improving understanding of the processes by which geological materials affect human health as a means identifying and reducing a range of pathological effects.
- Climate: particularly emphasises improved detail and understanding of the nonhuman factor in climate change.

- Groundwater: considers the occurrence, quantity and quality of this vital resource for all living things against a background that includes potential political tension between competing neighbour-nations.
- Ocean: aims to improve understanding of the processes and environment of the ocean floors with relevance to the history of planet Earth and the potential for improved understanding of life and resources.
- Soils: this thin "skin" on Earth's surface is the vital source of nutrients that sustain life on the world's landmasses, but this living skin is vulnerable to degradation if not used wisely. This theme emphasizes greater use of soil science information in the selection, use and ensuring sustainability of agricultural soils so as to enhance production and diminish soil loss.
- Deep Earth: in view of the fundamental importance of deep the Earth in supplying basic needs, including mitigating the impact of certain natural hazards and controlling environmental degradation, this theme concentrates on developing scientific models that assist in the reconstruction of past processes and the forecasting of future processes that take place in the solid Earth.
- Megacities: this theme is concerned with means of building safer structures and expanding urban areas, including utilization of subsurface space.
- Geohazards: aims to reduce the risks posed to human communities by both natural and human-induced hazards using current knowledge and new information derived from research.
- Resources: involves advancing our knowledge of Earth's natural resources and their sustainable extraction.
- Earth and Life: it is over two and half billion years since the first effects of life began to affect Earth's atmosphere, oceans and landmasses. Earth's biological "cloak", known as the biosphere, makes our planet unique but it needs to be better known and protected. This theme aims to advance understanding of the dynamic processes of the biosphere and to use that understanding to help keep this global life-support system in good health for the benefit of all living things.

The first task of the leading Earth scientists appointed as Theme Leaders was the production of a set of theme brochures. Some 3,500 of these were published, initially in English only but later translated into Portuguese, Chinese, Hungarian, Vietnamese, Italian, Spanish, Turkish, Lithuanian, Polish, Arabic, Japanese and Greek. Most of these were published in hard copy and all are listed on the IYPE website.

It is fitting that, as the International Year's triennium terminates at the end of 2009, the more than 100 scientists who participated in the ten science themes should bring together the results of their wide ranging international deliberations in a series of state-of-the-art volumes that will stand as a legacy of the International Year of Planet Earth. The book series was a direct result of interaction between the International Year and the Springer Verlag Company, a partnership which was formalised in 2008 during the acme of the triennium.

This IYPE-Springer book series contains the latest thinking on the chosen themes by a large number of Earth science professionals from around the world. The books are written at the advanced level demanded by a potential readership consisting of Earth science professionals and students. Thus, the series is a legacy of the Science Programme, but it is also a counterweight to the Earth science information in several media formats already delivered by the numerous National Committees of the

International Year in their pursuit of world-wide popularization under the Outreach Programme.

The discerning reader will recognise that the books in this series provide not only a comprehensive account of the individual themes but also share much common ground that makes the series greater than the sum of the individual volumes. It is to be hoped that the scientific perspective thus provided will enhance the reader's appreciation of the nature and scale of Earth science as well as the guidance it can offer to governments, decision-makers and others seeking solutions to national and global problems, thereby improving everyday life for present and future residents of Planet Earth.

Eduardo F.J. de Mulder
Executive Director International Secretariat
International Year of Planet Earth

Edward Derbyshire
Goodwill Ambassador
International Year of Planet Earth

Preface

This book series is one of the many important results of the International Year of Planet Earth (IYPE), a joint initiative of UNESCO and the International Union of Geological Sciences (IUGS), launched with the aim of ensuring greater and more effective use by society of the knowledge and skills provided by the Earth Sciences.

It was originally intended that the IYPE would run from the beginning of 2007 until the end of 2009, with the core year of the triennium (2008) being proclaimed as a UN Year by the United Nations General Assembly. During all three years, a series of activities included in the IYPE's science and outreach programmes had a strong mobilizing effect around the globe, not only among Earth Scientists but also within the general public and, especially, among children and young people.

The Outreach Programme has served to enhance cooperation among earth scientists, administrators, politicians and civil society and to generate public awareness of the wide ranging importance of the geosciences for human life and prosperity. It has also helped to develop a better understanding of Planet Earth and the importance of this knowledge in the building of a safer, healthier and wealthier society.

The Scientific Programme, focused upon ten themes of relevance to society, has successfully raised geoscientists' awareness of the need to develop further the international coordination of their activities. The Programme has also led to some important updating of the main challenges the geosciences are, and will be confronting within an agenda closely focused on societal benefit.

An important outcome of the work of the IYPE's scientific themes includes this thematic book as one of the volumes making up the IYPE-Springer Series, which was designed to provide an important element of the legacy of the International Year of Planet Earth. Many prestigious scientists, drawn from different disciplines and with a wide range of nationalities, are warmly thanked for their contributions to a series of books that epitomize the most advanced, up-to-date and useful information on evolution and life, water resources, soils, changing climate, deep earth, oceans, non-renewable resources, earth and health, natural hazards, megacities.

This legacy opens a bridge to the future. It is published in the hope that the core message and the concerted actions of the International Year of Planet Earth throughout the triennium will continue and, ultimately, go some way towards helping to establish an improved equilibrium between human society and its home planet. As stated by the Director General of UNESCO, Koichiro Matsuura, "Our knowledge of the Earth system is our insurance policy for the future of our planet". This book series is an important step in that direction.

R. Missotten
Chief, Global Earth Observation Section
UNESCO

Alberto C. Riccardi
President
IUGS

Contents

Contributors

Peter J. Barrett Antarctic Research Centre, Victoria University of Wellington, Wellington 6012, New Zealand; Climate Change Institute, Victoria University of Wellington, Wellington 6012, New Zealand, Peter.Barrett@vuw.ac.nz

P.J. Bartlein Department of Geography, University of Oregon, Eugene, Oregon, USA 97403-1257, bartlein@uoregon.edu

Nancy A.N. Bertler Antarctic Research Centre, Victoria University of Wellington, Wellington 6012, New Zealand; GNS Science, National Isotope Centre, Lower Hutt 5040, New Zealand, Nancy.Bertler@vuw.ac.nz

Maarten Blaauw School of Geography, Archaeology and Palaeoecology, Queen's University Belfast, Belfast, BT7 1NN, Northern Ireland, maarten.blaauw@qub.ac.uk

Frank M. Chambers Department of Natural and Social Sciences, Centre for Environmental Change and Quaternary Research, University of Gloucestershire, Cheltenham, GL50 4AZ, UK, fchambers@glos.ac.uk

Torben R. Christensen Department of Earth and Ecosystem Sciences, Lund University, SE 223 62, Lund, Sweden, torben.christensen@nateko.lu.se

Rixt de Jong Institute of Geography, University of Bern, CH-3012 Bern, Switzerland, dejong@giub.unibe.ch

François de Vleeschouwer Department of Ecology and Environmental Sciences, Umeå University, SE 901 87, Umeå, Sweden, fdevleeschouwer@gmail.com

David Dodman Human Settlements and Climate Change Groups, International Institute for Environment and Development, London WC1H 0DD, UK, david.dodman@iied.org

John Dodson Institute for Environmental Research, Australian Nuclear Science and Technology Organisation, Sydney, NSW, 2234 Australia, John.Dodson@ansto.gov.au

Emily Fearnley School of Population Health, University of Queensland, Brisbane, Australia, e.fearnley@uq.edu.au

Walter Finsinger Centre for Bio-Archaeology and Ecology (UMR 5059 CNRS), University of Montpellier 2, F-34090 Montpellier, France, walter.finsinger@univ-montp2.fr

Timothy J. Foxon Sustainability Research Institute, Centre for Climate Change Economics and Policy, School of Earth and Environment, University of Leeds, Leeds LS2 9JT, UK, t.j.foxon@leeds.ac.uk

Stefan Fronzek Finnish Environment Institute, 00251 Helsinki, Finland, stefan.fronzek@ymparisto.fi

Emmanuel Gandouin Institut Méditerranéen d'Ecologie et de Paléoécologie, IMEP-CNRS, UMR 6116, Bâtiment Villemin, Domaine du Petit Arbois, Université Paul Cézanne, 13545 Aix-en-Provence Cedex 04, France, e.gandouin@univ-cezanne.fr

Patrick Gonzalez Center for Forestry, University of California, Berkeley, CA 94720-3114, USA, pgonzalez@cal.berkeley.edu

S.P. Harrison Department of Biological Sciences, Macquarie University, North Ryde, NSW, 2109 Australia, sharriso@science.mq.edu

Saleemul Huq Climate Change Group, International Institute for Environment and Development, London WC1H 0DD, UK, saleemul.huq@iied.org

Rod Janssen Energy and Environment Consultant, London W4 1DX, UK, rod.janssen@gmail.com

Margareta Johansson Department of Earth and Ecosystem Sciences, Lund University, SE 223 62, Lund, Sweden, margareta.johansson@nateko.lu.se

Ulla Kokfelt Department of Earth and Ecosystem Sciences, Lund University, SE 223 62, Lund, Sweden, ulla.kokfelt@geol.lu.se

Mariusz Lamentowicz Department of Biogeography and Palaeoecology, Faculty of Geographical and Geological Sciences, Adam Mickiewicz University, PL-61-680 Poznan, Poland; Ecosystem Boundaries Research Unit, Wetlands Research Group, Swiss Federal Research Institute WSL, CH-1015 Lausanne, Switzerland, mariuszl@amu.edu.pl

Isabelle Larocque-Tobler Institute of Geography, University of Bern, 3012 Bern, Switzerland; Institut National de Recherche Scientifique, Centre Eau, Terre et Environnement, Québec, Québec, Canada G1K 9A9, larocque@giub.unibe.ch, isabelle_larocque@ete.inrs.ca

Isabelle Laurion Institut National de Recherche Scientifique, Centre Eau, Terre et Environnement, Québec, Québec, Canada G1K 9A9, isabelle_laurion@ete.inrs.ca

Gaël Le Roux EcoLab UMR CNRS 5245, Toulouse and EcoLab/Campus Ensat, 31326 Castanet-Tolosan, France, gael.leroux@ulg.ac.be

J.R. Marlon Department of Geography, University of Oregon, Eugene, Oregon, USA 97403-1257, jmarlon@uoregon.edu

Dmitri Mauquoy School of Geosciences, University of Aberdeen, Aberdeen AB24 3UF, UK, d.mauquoy@abdn.ac.uk

Edward A.D. Mitchell Laboratory of Soil Biology, Institute of Biology, University of Neuchâtel, CH-2009 Neuchâtel, Switzerland; Ecosystem Boundaries Research Unit, Wetlands Research Group, Swiss Federal Research Institute WSL, CH-1015

Lausanne, Switzerland; Laboratoire des Systèmes écologiques, Ecole Polytechnique Fédérale de Lausanne (EPFL), CH-1015 Lausanne, Switzerland, edward.mitchell@unine.ch

Robert Moschen Institute of Chemistry and Dynamics of the Geosphere 4: Agrosphere Isotope Geochemistry and Climate Dynamics Research Centre, 52425 Jülich, Germany, r.moschen@fz-juelich.de

Jonathan E. Nichols NASA Goddard Institute for Space Studies, 2880 Broadway, New York, NY, 10025 USA, jnichols@giss.nasa.gov

Philippe Ponel Institut Méditerranéen d'Ecologie et de Paléoécologie, IMEP-CNRS, UMR 6116, Bâtiment Villemin, Domaine du Petit Arbois, Université Paul Cézanne, 13545 Aix-en-Provence Cedex 04, France, philipe.ponel@univ-cezanne.fr

Hannah Reid Climate Change Group, International Institute for Environment and Development, London WC1H 0DD, UK, hannah.reid@iied.org

Emanuela Samaritani Laboratory of Soil Biology, Institute of Biology, University of Neuchâtel, CH-2009 Neuchâtel, Switzerland; Ecosystem Boundaries Research Unit, Wetlands Research Group, Swiss Federal Research Institute WSL, CH-1015 Lausanne, Switzerland; Laboratoire des Systèmes écologiques, Ecole Polytechnique Fédérale de Lausanne (EPFL), CH-1015 Lausanne, Switzerland, emanuela.samaritani@gmail.com

Elisabeth Simelton Centre for Climate Change Economics and Policy, University of Leeds, Leeds LS2 9JT, UK, E.S.Simelton@leeds.ac.uk

Monique Stewart Institute of Geography, University of Bern, 3012 Bern, Switzerland, monique.stewart@giub.unibe.ch

Bas van Geel Department of Paleoecology and Landscape Ecology, Institute for Biodiversity and Ecosystem Dynamics, Universiteit van Amsterdam, 1098 XH Amsterdam, The Netherlands, B.vanGeel@nva.nl

Philip Weinstein School of Population Health, University of Queensland, Brisbane, Australia, p.weinstein@uq.edu.au

Xiaoping Yang Institute of Geology and Geophysics, Chinese Academy of Sciences, Beijing 100029, China, xyang@mail.igcas.ac.cn, xpyang@263.net.cn

Reviewers

A. Challinor University of Leeds, Leeds, UK

John Dodson Institute for Environmental Research, Australian Nuclear Science and Technology Organisation, Sydney, NSW 2234, Australia

E. Fraser University of Leeds, Leeds, UK

M. Gagen University of Wales, Swansea, UK

A. Gouldsen University of Leeds, Leeds, UK

I. Larocque Institut National de Recherche Scientifique, Québec, Canada

H.W. Li Institute of Geology and Geophysics, Beijing, China

D. McCarroll University of Wales, Swansea, UK

Colin Prentice Macquarie University, Sydney, Australia

Elisabeth Simelton Centre for Climate Change Economics and Policy, University of Leeds, Leeds LS2 9JT, UK

F. Wagner-Cremer University of Utrecht, Utrecht, The Netherlands

Introduction

Changing climates can bring benefits and threats to societies and the range of these will have a disproportional spread. The negatives often befall those societies that have the lowest capacity to cope. The upward global temperature trend shows that each recent decade has been warmer than its predecessor and there is abundant evidence that major components of the climate system are shifting. As a consequence there are clear and abundant signs of change in patterns of vegetation, fire, ice, water and land cover. These are all linked as part of the Earth System, and there are strong feedbacks between them all. Humans utterly depend on Earth System resources and services for economic and cultural well-being. It follows that changes in climate will ultimately drive significant changes in the life chances of people, well-being of societies and the future of nations.

The science is clear that a significant part of recent global warming is driven by human activities. One of the main components of this is the accumulation of anthropogenically derived greenhouse active gases in the Earth's atmosphere. This gives rise to an example of a major feedback loop. In it, human economic activity (energy and resource use, agriculture, forestry and more) leads to changes in Earth system elements (climate, land cover, sea level change and more) which feedback into the distribution and kinds of human activities which are economic. Societies have no option but to mitigate the causes to slow the effect, or adapt to the new order. The scale of the problem is so large that no single nation can solve the problem alone. The problem is compounded by the fact that the complexity of the total change is a combination of natural, solar and anthropogenic forcings. We are faced with the need to develop new knowledge, a better understanding of how biophysical and socio-economic systems interact, develop new strategies and actions, and new international co-operative arrangements in order to solve what is the defining challenge of the twenty-first century.

Recently a large number of books have been published on the theme of climate change. In this volume we have opted for a different approach. We have endeavoured to identify a selection of major themes and to cover these well. Chapters on vegetation, fire, deserts and ice sheets consider the past from palaeo-data sets and outline likely trajectories based on the implications of future climate change. Embedded in each of these chapters is information on the methods employed to give state-of-art analyses.

The chapters on peatlands, lakes and rivers focus more on aquatic and semi-aquatic environments. These settings contain some of the most detailed palaeo-data sets and the authors give up-to-date overviews on the types of records these systems contain, how they are obtained, and developing generalized syntheses of regional variation and the kinds of changes at risk within them that might result from climate change.

The final set of chapters (9–12) focus on the human dimensions of climate change. The two main thematic topics covered are human health and food security; the final two chapters outline the role of policy and organisational structures that are likely to have some success for humans to make progress on the enormous challenge ahead. This is likely to be the largest challenge of all as Copenhagen 2009 demonstrated.

Australian Nuclear Science and Technology Organisation John Dodson
Australia

Setting the Scene: How Do We Get to a Fitting Future?

John Dodson

Abstract Climate is a term referring to long-term patterns of weather conditions, and it is sometimes difficult to say where weather ends and climate begins. There is abundant evidence that climates have changed significantly in the past and that these were driven by a great complex of natural and usually interlinked processes with strong feedback mechanisms. It is a complex science. There is now clear evidence that human impact on land cover and environmental systems, through economic activity, have added a new driver of climate trends and variability. The best science we have informs us that certain kinds of extremes will occur more frequently and generally upward trends in temperature will disrupt many environmental and socio-economic systems. Records from the past show that many societies failed to recognize the onset of new trends in climate and lived in denial of change until it was too late. Humans now live in a globalized world of linked economies, resources, information, and agriculture and we are vulnerable together. The scale of the projected changes requires global cooperation and action for solutions. The cost of abatement or adaptation increases greatly as action is deferred. Only national governments can direct the scale of changes required, and few have done this. Are we being forced to follow the old behavioral pattern of denial on a global scale?

J. Dodson (✉)
Institute for Environmental Research, Australian Nuclear Science and Technology Organisation, Sydney, NSW, 2234 Australia
e-mail: John.Dodson@ansto.gov.au

Keywords Climate change · Paleoclimates · Societal response · Economics · International action

Introduction

It seems that hardly a week goes by without news of loss of life or damage caused by extremes in weather somewhere in the world. Bushfires, cyclones, floods, droughts, heat waves, and snowstorms are generated by weather systems, and this is the way it has been for millions of years. Some of the earliest written records of societies contain references to extreme events, and large magnitude weather events can place burdens on societies for many years.

For most of us weather refers to daily, weekly, or seasonal variations, and we use forecasts to help plan our day or weekend. For others it helps predict opportunity or difficulty for the day or days ahead. Many societies have sophisticated weather prediction systems. These are based on models and comparative records of past observations. They require a sophisticated understanding of how weather systems form, and the relationships between air masses, topography, moisture sources, and sea conditions. We are often concerned with how these compare to normal or average conditions at any particular time of the year.

Climate refers to long-term patterns, with the main drivers set by latitude, altitude, ocean currents, distribution of land, sea and ice, and patterns of prevailing winds and the mean locations of high or low pressure systems. These combine in usually different ways from place to place and give us what we recognize as regional climates.

J. Dodson (ed.), *Changing Climates, Earth Systems and Society*, International Year of Planet Earth,
DOI 10.1007/978-90-481-8716-4_1, © Springer Science+Business Media B.V. 2010

It is knowledge of climate patterns that sets the scene for the kinds of environmental processes, vegetation, fire patterns, lifestyles, agriculture, forestry, and building codes. Most of these have long-term currency. Good climate science also reveals information about variability and risk, which can be used to develop policy and build systems to mitigate extreme events and to identify change in climate patterns.

Climate Change

Trends in climate which alter mean states, increase variability, or the frequency of extreme events can greatly alter biophysical, economic, and social systems. It is therefore fundamentally important to accurately predict any likelihood or degree of change. By its very nature, since climate refers to long-term averages and adaptation to them, change can lead to lack of fitness in physical and social systems to new climates. And this is THE big risk for societies. Investment for the future requires identification of what the future might look like.

Advances in the earth and biological sciences have recognized that climates have changed substantially in the past. Sediments in oceans and lakes, land form types especially those created by ice sheets and deserts, fossil soils, dust accumulations on land, ice and ocean floors, and tree ring, and coral and speleothem records among other records contain features of past climates. The science of climate change in the past has revealed quite clearly that the measured records that have good global coverage over the last 50–100 years show only a small portion of the variability that Earth's climate systems may undergo. Recent advances in technology in drilling, geochemical analyses, and nuclear techniques in dating and isotope analyses show that not only have there been great changes in the past but in many cases these can be sudden, with one climate state switching to another within a year or two. Some ancient societies have had to deal with this.

The measured records of temperature change (NASA, 2010) reveal the year-to-year variability we all experience and know about. Aggregated temperature records, when compared to the 1951–1980 taken as a mean, show that each decade since that time has shown warming at a rate of about 0.2°C per decade. Each decade has been warmer that the one before. The Northern Hemisphere has recorded a greater rise over all but for the Southern Hemisphere 2009 is the warmest since temperature records began (about the 1880s). Global temperature in 2009 is virtually tied with 1998, 2002, 2003, 2006 and 2007, and these are the warmest on record. Despite the objectivity of this kind of data we still hear climate skeptics telling the press that the world is cooling.

The many thousands of records of past climates allow regional and inter-regional comparisons to be made. They reveal that some climate shifts are global in scale, as in the case of the ice ages themselves, while others are strongly regionalized in their expression or impact. This has provided a sound basis to examine how climate phenomena are linked and in some cases to identify the key driving or forcing factors. This provides the key elements in understanding the physics of climate systems, and the basis for modeling and forecasting of climate. The ability to do the latter well on regional and relevant time scales may be just the key toward producing information relevant for the future of societies. More on this later.

Climate Impacts on Societies

Human nature has probably been relatively constant for many tens of thousands of years. At least we know that brain capacity and physiological systems are constant on such time scales. People have chosen to live in social groups, develop hierarchies, complex decision-making processes, and strive to improve life chances, security, and dominance over foes and the environment. This has been a great driver of technological innovation, and from this social change. Not all changes have been positive for all groups.

Much human behavior is driven by knowledge, experience, and inbuilt expectations. In the case of experience, societies have tended to think that climates of the future will look much like those of their recent experience. If the current year is colder, drier, or stormier than recent years, then not to worry, the usual pattern will return next year or the year after.

One way of testing how humans respond to major climate change is to compare evidence from places where there are good archaeological and well-dated and clearly associated paleoclimate records. There are now many documented stories of this kind.

The classic Mayan civilization ruled Mesoamerica for 600 years from 250 AD. It was a very sophisticated society in that it was highly stratified and had widespread and spectacular urban centres connected by vast trade networks. They used mathematics, had their own sporting contests, and had a complex calendar to record seasons and events and to help in forecasting. However, around 850–900 AD the Mayan empire collapsed dramatically. Until 1995 many hypotheses had been raised to account for the collapse. The first solid evidence that this might be due to climate change came from sediments in closed lake basins of the Yucatan Peninsula (Hodell et al. 1995). These studies give a record of climate variability of climate over the last 7,000 years, and crucially right through the Mayan period. The sediments indicate an unprecedented dry period coincided with the collapse of the Maya. Pollen analyses show (Rosenmeier et al. 2002) that rainforest was also severely disrupted, and marine sediments off the coast of Venezuela (Haug et al. 2003) had indications of major soil erosion at the same time. The drought probably lasted for about 160 years, which was likely to be an impossible length of time for any society to weather.

While climate itself may not have been the main cause of the collapse it must have been a contributing factor, even for a sophisticated society like the Maya. One can imagine the first year or so of drought being ridden out by drawing on resources, and a general feeling that order would be restored once the weather systems returned to "normal." Of course in this case they did not. Societal resilience was gradually "broken down" and it appears that human sacrifices to the Sun God were made in an attempt to break the pattern. Eventually the classic Mayan civilization collapsed.

One may ask how modern societies might cope with new conditions that are outside the norm of experience. Do we learn from experience? Will climate change mean more frequent hurricanes in the Caribbean? Many believe so. If so is it wise to reconstruct the city of New Orleans in the same way it was before? The recent severe bushfires in Southern Australia have caused massive loss of life and property. Such fires are expected to occur more frequently under future climate scenarios. Already people are thinking of rebuilding in these devastated regions. Is this wise?

Climate Models

Forecasting future climates depends on the ability to model the science of climate systems. This essentially involves writing equations to describe how wind patterns, and temperature and precipitation will vary over regions. These are constructed into models that incorporate three-dimensional representations of the ocean and sea ice, and they couple these with the carbon cycle and atmospheric chemistry. All the components have feedbacks, which the various systems must take account of. The most realistic models are very hungry for computing power and the solutions of many embedded equations are approximations. Models are not flawless, as many are willing to point out, but they are the best we have at present. And great strides have been made in improving the underlying science and parameterizations, and the various models have been tested against each other by hindcasting to simulate past patterns of climate established from geological records. Progress in improving models has been substantial over the last decade or so.

In general climate models simulate the broad patterns of climate quite well. Precipitation patterns are less well-modeled than temperature, and for the latter models simulate slightly higher values for 1979–1998, for example, than has been measured. These are evidence that models are on the right track but also show there are other factors that have not been identified or included just yet.

These limitations are probably small compared to the uncertainty of future economic and population growth, and political decisions that will substantially affect greenhouse gas and aerosol composition of the atmosphere.

Considerable effort is now being made to improve climate models and to link them to other kinds of models such as economic models, ecosystem models, and agricultural crop models, for example, for wheat, rice, and soya production. These are usually set up to account for feedbacks between the various components of model types and can guide policy and company decision makers as they ponder future impacts and business opportunities or possible problems in profit futures.

The Intergovernmental Panel on Climate Change (2007) is the world's largest co-operative effort to consider the state of climate science, and extensive use of models is applied to help identify adaptation and mitigation strategies to minimize future impacts on biophysical and economic systems. It is economic and technological systems that have caused the human impacts on the climate system. We can already measure the feedbacks to society from human induced climate change. Those who choose to ignore or deny these will eventually have to concede that to do nothing will ultimately find the feedbacks so strong that they cannot be denied for longer. Stern (2006) and Garnaut (2008) provide two recent analyses and judgements on the economic impacts of future climate. The gist of their conclusions is that future societies will have a softer economic landing if wise decisions on emissions are made soon. The alternative will be very expensive in both monetary and social and political terms.

Two weeks of negotiation in December 2009 have just concluded in Copenhagen. Over 190 of the world's leaders came together to negotiate an agreement for a global approach to reduce the quantum of future anthropogenic driven climate change. Nothing on this scale had ever been attempted before and the General Secretary of the United Nations, Ban Ki-Moon, described the outcome as the beginning of hope. The spirit is clear – there is agreement that warming should be less than 2°C globally, to minimize risk to environment and economic systems, and there is to be a fund of $100 billion to help undeveloped economies adapt.

This is a remarkable achievement given the huge disparity between the various vested interests in the countries involved. It carries significant political risk for those who helped craft the agreement. In my own country (Australia, and a nation particularly vulnerable to future climate change) there is a significant minority in favor of limiting action on climate change as it will have a net cost to the economy. They choose to ignore the Stern and Garnault arguments that the cost to future generations grows greatly as the environmental impacts impinge ever more on economic and social systems.

However, the Copenhagen Agreement brought despair to a significant proportion of environmentalists, and many people from low lying nations such as the Maldives and Bangladesh for not going far enough. There is a growing body of voices calling the Copenhagen Agreement a fiasco, and expressing the view that politics will prevent such meetings ever achieving united effort to mitigate anthropogenic driven climate change.

Where to now? The Copenhagen Agreement needs to be suitably hardened to ensure the main agreements are honored. The world heads to Mexico City later in 2010. Will we move forward? Systems need to be put in place to encourage development of the new technologies to confront movement to low carbon economies. These technologies need to be shared generously with those less fortunate. Who will show leadership? What a wonderful world this can be if we all share a common future of increased opportunity and life chances for all. This is something that armed conflicts can never achieve.

This Book

In recent years a large number of books have appeared on the climate change theme. Here we bring together a number of world authorities on 13 themes. They offer their perspective on how climates of the past and present shape key themes; and how in future climates may change the world as we know it. Forewarning and action will better prepare our children and grandchildren for the future.

References

Garnaut R (2008) The Garnaut Climate Change Review. Cambridge: Cambridge University Press.

Haug GH, Gunther D, Peterson LC, Sigman DM, Hughen KA, Aeschlimann B (2003) Climate and the collapse of Maya civilization. Science 299:1731–1735.

Hodell DA, Curtis JH, Brenner M (1995) Possible role of climate in the collapse of classic Maya civilization. Nature 375: 391–394.

Intergovernmental Panel on Climate Change (2007) Climate Change 2007. Cambridge: Cambridge University Press.

NASA (2010) GISS (Goddard Institute for Space Studies) surface temperature analysis. http://data.giss.nasa.gov/gistemp

Rosenmeier MF, Hodell DA, Brenner M, Curtis JH, Guilderson TP (2002) A 4000-year lacustrine record of environmental change in the southern Maya lowlands, Petén, Guatemala. Quat Res 57:183–190.

Stern N (2006) The Economics of Climate Change. Cambridge: Cambridge University Press.

Impacts of Climate Change on Terrestrial Ecosystems and Adaptation Measures for Natural Resource Management

Patrick Gonzalez

Abstract Emissions from motor vehicles, power plants, deforestation, and other human sources are warming the Earth and damaging ecosystems and human well-being. Field observations from around the world have detected significant changes in terrestrial ecosystems and attributed them to climate change rather than other factors. Climate change has shifted the ranges of plants, animals, and biomes, altered the timing of life events such as plant flowering and animal migration, increased wildfires, and driven 75 frog and other amphibian species to extinction. Projections of future climate change and analyses of vulnerability indicate that unless we substantially reduce greenhouse gas emissions, further warming may overwhelm the adaptive capacity of many species and ecosystems. Climate change could convert extensive land areas from one biome to another, alter global biogeochemical cycles, and isolate or drive numerous species to extinction. Natural resource managers are developing adaptation measures to help species and ecosystems cope with the impacts of climate change.

Keywords Adaptation · Climate change · Ecological impacts · Natural resource management · Vegetation shifts

P. Gonzalez (✉)
Center for Forestry, University of California, Berkeley, CA 94720-3114, USA
e-mail: pgonzalez@cal.berkeley.edu

Introduction

Emissions from motor vehicles, power plants, deforestation, and other human sources are warming the Earth and damaging ecosystems and human well-being. Climate change increased global average surface temperature 0.7 ± 0.2°C from 1906 to 2005 (IPCC, 2007a). Field observations show that this warming has shifted the geographic ranges of plants, animals, and biomes (major vegetation formations) around the world (IPCC, 2007b; Rosenzweig et al., 2008; Gonzalez et al., 2010). Climate change has also altered the phenology (timing of life events such as plant flowering and animal migration) of numerous species on all continents (IPCC, 2007b; Rosenzweig et al., 2008). Climate change has lifted the cloud deck in Costa Rica cloud forest, driving 75 frog and other amphibian species to extinction (Pounds et al., 2006).

Projections of future climate change using general circulation models, dynamic global vegetation models, and climate envelope models indicate that unless we substantially reduce greenhouse gas emissions, increased temperatures and other changes in climate could exceed the resilience of many ecosystems (IPCC, 2007b). Climate change could convert extensive land areas from one biome to another, increase wildfire, and isolate or drive to extinction numerous plant and animal species. Approximately 20–30% of species assessed so far are at high risk of extinction if global mean temperatures increase 2–3°C above preindustrial levels (Thomas et al., 2004; IPCC, 2007b). Greenhouse gas emissions and modified evaporation and runoff due to deforestation and forest degradation could substantially change global

J. Dodson (ed.), *Changing Climates, Earth Systems and Society*, International Year of Planet Earth, DOI 10.1007/978-90-481-8716-4_2, © Springer Science+Business Media B.V. 2010 biogeochemical cycles. All of these ecological impacts of climate change threaten to reduce ecosystem services, including the provision of erosion control, firewood, flood control, food, forest carbon sequestration, freshwater, medicine, timber, and other necessities for survival (Millennium Ecosystem Assessment, 2006).

The world can avoid the worst impacts of climate change by improving energy efficiency, expanding

Table 1 Major observed and projected impacts of climate change on terrestrial ecosystems and potential adaptation measures

	Observed impacts in the twentieth century with global increase of 0.7ºC	Projected impacts in the twenty-first century with global increase of 2–4ºC	Adaptation measures
Biomes	Biome shifts toward polar or equatorial regions and upslope, in Africa, Europe, and North America (Gonzalez, 2001; Peñuelas and Boada, 2003; Beckage et al., 2008; Kullman and Öberg, 2009, Gonzalez et al., 2010)	Biome shifts toward polar or equatorial regions and upslope across extensive areas (IPCC, 2007b; Sitch et al., 2008; Gonzalez et al., 2010)	Managing for habitat type, management of climate change refugia
Species ranges	Range shifts in the direction expected with climate change for 80% of 434 plant and animal species from around the world (Parmesan and Yohe, 2003)	Range disappearance for 15–37% of plant and animal species examined from around the world (Thomas et al., 2004)	Protection of climate change refugia, habitat corridors
Phenology	Earlier flowering, leafing, and migration for 62% of 677 plant and animal species from around the world (Parmesan and Yohe, 2003)	Continued changes in flowering, leafing, and animal migration (IPCC, 2007b)	Food propagation for mistimed migrants
Species extinctions	Extinction of 75 frog and other amphibian species in Costa Rica (Pounds et al., 2006), significant declines in Antarctic penguins (Emslie et al., 1998; Fraser et al., 1992; Barbraud and Weimerskirch, 2001; Wilson et al., 2001)	High risk of extinction for 20–30% of assessed species from around the world (Thomas et al., 2004; IPCC, 2007b)	Managed relocation
Wildfire	Fire frequency increase of 400% in western United States (Westerling et al., 2006), increase of burned forest area in Canada (Gillett et al., 2004)	High fire forest dieback risk for half of the area of the Amazon (Golding and Betts, 2008), increased fire in boreal forest (Balshi et al., 2009)	Prescribed burning
Pests	Extensive forest dieback from bark beetles in western North America (Breshears et al., 2005; Raffa et al., 2008)	Continued bark beetle infestations in western North America (Kurz et al., 2008)	Prescribed burning
Global biogeochemistry	Increase in global net primary productivity from CO_2 fertilization and longer growing seasons (Nemani et al., 2003), decrease in tropical forest biomass in Costa Rica from increased respiration at night (Clark et al., 2003)	Enhanced vegetation growth slows, then reverses by the end of the twenty-first century, so that the terrestrial biosphere converts from a carbon sink to a carbon source (IPCC, 2007a)	Natural regeneration and enrichment planting of adapted plant species

public transit, installing wind, solar, and other renewable energy systems, conserving forests, and using other currently available measures to reduce greenhouse gas emissions (Pacala and Socolow, 2004; IPCC, 2007c). Nevertheless, a lag between emissions to the atmosphere and warming of the land and oceans commits the world to another 0.3–0.9°C of warming from 2000 to 2100, due to cumulative emissions since the beginning of the Industrial Revolution (Wigley, 2005; IPCC, 2007a). Consequently, human communities and ecosystems will need to adapt to a certain amount of warming. Natural resource managers are developing adaptation measures to help species and ecosystems cope with the impacts of climate change (IPCC, 2007b; US CCSP, 2008).

This chapter discusses the impacts of climate change on terrestrial ecosystems and adaptation measures for natural resource management. Terrestrial ecosystems include tundra, forests, woodlands, grasslands, and deserts. Other chapters in the book cover agricultural, coastal, freshwater, marine, and wetland ecosystems. This chapter first presents information on ecological impacts that field observations have already detected and attributed to twentieth century climate change. It then presents projections of potential impacts of continued climate change in the twenty-first century. The sections on observed and projected impacts each examine three broad areas: vegetation, fauna, and global biogeochemical cycles of carbon and water. Finally, the chapter presents adaptation solutions that could help species and ecosystems cope with climate change. Table 1 summarizes the major impacts and adaptation options. This chapter depends especially on the findings of the Intergovernmental Panel on Climate Change (IPCC).

Observed Impacts

Observed Changes in Climate

Human activities have raised carbon dioxide (CO_2), the principal greenhouse gas, to its highest level in the atmosphere in 800,000 years (Lüthi et al., 2008). Analyses reveal that the added gas bears the unique chemical signature of burned coal and oil and not the sign of gases from volcanoes or geysers (IPCC, 2007a). The accumulation of greenhouse gases has raised global temperatures to their warmest levels in 1,300–1,700 years (Mann et al., 2008). Orbital cycles and other natural factors account for only 7% of observed warming (IPCC, 2007a).

In 2008, motor vehicles, power plants, and other fossil fuel-burning industrial sources emitted greenhouse gases to the atmosphere at a rate (mean ± 66% confidence interval) of 8.7 ± 0.5 Gt C/y and deforestation contributed emissions of 1.2 ± 0.75 Gt C/y, while global vegetation and soils removed greenhouse gases from the atmosphere at a rate of 4.7 ± 1.2 Gt C/y (IPCC, 2007a; Le Quéré et al., 2009). While the biosphere is currently a net carbon sink, human activities emit twice the amount of greenhouse gases than vegetation, soils, and the oceans can naturally absorb. That is the fundamental imbalance that causes climate change.

Climate change warmed global temperatures 0.7 ± 0.2°C from 1906 to 2005 (IPCC, 2007a). Temperatures in the Arctic have been warming at almost twice the rate of the rest of the world (ACIA, 2004; IPCC, 2007a). Winter temperatures in some parts of the temperate zone have warmed almost as quickly as annual temperatures in polar areas (USGCRP, 2009).

In the twentieth century, climate change reduced Northern Hemisphere snow cover 7% and accelerated the melting of glaciers around the world to its greatest rate in 5,000 years (IPCC, 2007a). From 1890 to 1986, warm temperatures melted 160 m of ice thickness from Tasman Glacier, New Zealand (Kirkbride, 1995). In the western United States, more precipitation has been falling as rain than as snow since 1949 (Knowles et al., 2006). Climate change is increasing the proportion of Atlantic hurricanes in the most intense categories (Mann et al., 2009).

Climate change has increased the intensity and length of droughts since 1970, especially in subtropical and tropical areas (IPCC, 2007a). Warmer ocean temperatures and the reduction of continental vegetation cover significantly reduced rainfall in the African Sahel during the nineteenth and twentieth centuries in the most severe drought in the instrumental record in the world (Zeng et al., 1999; Giannini et al., 2003; Dai et al., 2004).

Observed Impacts on Vegetation

Field observations demonstrate that climate change has altered the distribution and condition of vegetation around the world. Warmer temperatures and changing patterns of precipitation have shifted the geographic range of plants and biomes, altered plant phenology, increased wildfire, and exacerbated pest outbreaks (IPCC, 2001b; IPCC, 2007b). Climate change affects ecosystems at the same time as other potential stresses, including acid rain, agricultural expansion, air and water pollution, dams, deforestation, desertification, increased livestock herding, invasive species, ozone, urbanization, and water withdrawals. Although ecosystems are not static, climate change and these other stresses are pushing some ecosystems out of historic ranges of variability.

Determination of impacts of climate change on ecosystems involves two distinct research procedures: detection and attribution. Detection is measurement of historical changes that are statistically significantly different from natural variability (IPCC, 2001a). Attribution is determination of the relative importance of different factors in causing observed change. If statistical analysis and multiple lines of evidence demonstrate that observed changes are (1) unlikely to be due entirely to natural variability, (2) consistent with estimated or modeled responses, and (3) inconsistent with alternative plausible explanations, then the analysis and evidence can reasonably attribute the cause of the observed change to climate change (IPCC, 2001a). Attribution of ecological impacts to the human activities that cause climate change requires a two-step 'joint attribution': attribute ecological changes to changes in climate factors, then attribute changes in climate factors to human emissions of greenhouse gases (IPCC, 2007b; Rosenzweig et al., 2008).

Building on IPCC (2001b, 2007b) detection of numerous ecological changes and their attribution to human-caused climate change, Rosenzweig et al. (2008) assembled over 29,500 time series of statistically significant temperature-related changes detected in physical and ecological systems and statistically attributed over 90% of those changes to observed human-caused increases in temperature. The database of observed changes included any published statistically significant trend in a physical or ecological system related to temperature, occurring between 1970 and 2004 and documented with at least 20 years of data. Over 28,000 of the cases examined terrestrial ecosystems.

Biomes are the major vegetation formations of the world, including tundra, forests, woodlands, grasslands, and desert. Spatial patterns of temperature and precipitation determine the global distribution of biomes. When climate change exceeds plant physiological thresholds, alters mortality and recruitment, and modifies wildfire and other disturbances, it can shift the location of biomes latitudinally (toward polar and equatorial regions) and elevationally (up mountain slopes).

Observed changes in temperature or precipitation that fall one-half to two standard deviations outside of historical mean values have caused biome changes in the twentieth century (Gonzalez, 2001; Peñuelas and Boada, 2003; Beckage et al., 2008; Kullman and Öberg, 2009, Gonzalez et al., 2010). In Africa, climate change and desertification have caused a long-term decline in rainfall that has caused extensive forest dieback (Fig. 1) and shifted the Sahel (savanna), Sudan (woodland), and Guinea (tropical forest) ecological

Fig. 1 Yir (*Prosopis africana*) tree that died in Senegal as part of a vegetation shift in the African Sahel driven by a rainfall decline caused by climate change (Gonzalez, 2001). Photo by P. Gonzalez

zones 25–30 km toward the Equator from 1945 to 1993 (Gonzalez, 2001). In Spain, temperate broadleaf forest has shifted upslope into montane heathland (Peñuelas and Boada, 2003). In the northeast United States, temperate broadleaf forest has shifted upslope to replace boreal conifer forest (Beckage et al., 2008). In Scandinavia, boreal conifer forest has shifted upslope to replace alpine grassland (Kullman and Öberg, 2009).

Field observations have detected range shifts of many individual plant species. Of 434 plant and animal species examined with distribution data spanning at least 20 years, Parmesan and Yohe (2003) found that 80% shifted in the direction expected with climate change. For the 99 Northern Hemisphere plant and animal species with suitable range data, Parmesan and Yohe (2003) calculated average shifts of 6.1 km per decade northward or 6.1 m per decade upslope. Examining over 1,400 plant and animal species with suitable distribution data spanning at least 10 years, Root et al. (2003) found that 80% have shifted in the direction expected with climate change. For 171 forest plant species in France, Lenoir et al. (2008) found that the average optimum elevation (elevation with maximum probability of finding a species) shifted upslope by 29 m per decade between the periods 1905–1985 and 1986–2005. Numerous cases of drought-induced forest dieback around the world demonstrate that climate change has increased tree mortality in many ecosystems (Allen et al., 2010).

Plant species and biomes also shifted extensively across the globe during the 11,000–21,000 years from the Last Glacial Maximum to the present (Overpeck et al., 2003). Tree lines shifted 1,000 km away from the poles during those long millennia (ACIA, 2004). In contrast, climate change in the late twentieth century has shifted some plant ranges by that same magnitude in less than a century (NAS, 2008).

Phenology is the timing of life events, including, for plants, leaf unfolding, spring flowering, fruit ripening, leaf coloring, and leaf fall. Climate change has altered the phenology of numerous plant species (Parmesan and Yohe, 2003; Root et al., 2003; IPCC, 2007b). Meta-analysis of published research on 677 plant and animal species examined with data spanning at least 20 years found 62% of the species exhibited spring advance (Parmesan and Yohe, 2003).

In 21 European countries, climate change advanced flowering times earlier in the spring for 78% of 542 plant species from 1971 to 2000 (Menzel et al., 2006). In England, climate change has advanced spring flowering for 385 plant species (Fitter and Fitter, 2002). In the eastern United States, climate change advanced the date of spring flowering of 100 tree and forb species by an average of 2.4 days from 1970 to 1999 (Abu-Asab et al., 2001). Examination of the longest series of direct phenology observations in the world, the records of cherry (*Prunus* spp.) blossoming in Japan, shows no clear trend from 1400 to 1900, but significant advance after 1952 (Menzel and Dose, 2005). Records of tree leaf unfolding in England since 1736 show an average advance of 2.5 ± 1.7 days/century for oaks (*Quercus* spp.; Thompson and Clark, 2008). Analysis of 172 plant and animal species showed an average advance of spring phenology events of 2.3 days per decade, primarily in the twentieth century (Parmesan and Yohe, 2003).

Fire forms a natural component of forest and grassland ecosystems (Bowman et al., 2009). Many plant species depend on fire to initiate germination, remove competing plant species, or control insects and pathogens. Fire-dependent vegetation covers much of the world, especially in the tropics and subtropics. Climate change is altering key factors that control fire: temperature, precipitation, humidity, wind, biomass, vegetation species composition and structure, and soil moisture. Consequently, wildfire frequency and extent has increased in some ecosystems. In mid-elevation conifer forests of the western United States, an increase in spring and summer temperatures of 1°C from 1970 to 2003, earlier snowmelt, and longer summers increased fire frequency 400% and burned area 650% (Westerling et al., 2006). An increase of burned area in forests across Canada from 1920 to 1999 is consistent with climate change and not natural variability (Gillett et al., 2004). Across North American boreal forest, total burned area increased by a factor of 2.5 from 1959 to 1999, whereas burned area of human-ignited fires remained constant (Kasischke and Turetsky, 2006).

Climate warming is changing the abundance and range of pests and pathogens (IPCC, 2007b). In the United States, climate change extended the range of at least two species of damaging bark beetles from 1960 to 1994 (Williams and Liebhold, 2002). An epidemic of mountain pine beetle and spruce bark beetle now spreads across western North America, damaging conifer tree species across at least 47,000 km^2 (Raffa et al., 2008). Intense drought and beetle damage have caused massive dieback of pinyon pine (*Pinus edulis*) across the southwest United States (Breshears et al., 2005).

Observed Impacts on Fauna

Climate change has lifted the cloud deck in the Monteverde cloud forest, Costa Rica, causing a fungus infection that has driven the golden toad (*Bufo periglenes*) and 74 other amphibian species to extinction (Pounds et al., 2006). These comprise the only documented species extinctions to date caused by climate change. Climate change has also driven other amphibian population declines in Latin America and the Caribbean (Alexander and Eischeid, 2001; Ron et al., 2003; Burrowes et al., 2004).

Polar bears (*Ursus maritimus*) inhabit sea ice over the continental shelves and inter-island archipelagos of the Arctic, depending for food on seals that breed on the ice. On Hudson Bay, Canada, break-up of sea ice advanced three weeks from the 1970s to the 1990s, driving polar bears ashore earlier with reduced fat reserves and causing them to fast for longer periods of time (Stirling et al., 1999). Preliminary estimates indicate that the western Hudson Bay population has declined from 1,200 bears in 1987 to 950 bears in 2004 (Stirling et al., 1999).

Melting of Antarctic sea ice has caused significant population declines of Adélie penguins (*Pygoscelis adeliae*) and emperor penguins (*Aptenodytes forsteri*) because they depend on sea ice to feed on marine species (Emslie et al., 1998; Fraser et al., 1992; Wilson et al., 2001). At Terre Adélie (66°S), emperor penguin populations declined 50% from 1952 to 2000 (Barbraud and Weimerskirch, 2001). On Anvers Island (64–65°S), Adélie penguin populations have declined 70% (Emslie et al., 1998; Fraser et al., 1992), although populations are thriving further south at Ross Island (77°S), an effective poleward range shift (Wilson et al., 2001).

Climate change has shifted the range of many types of animals (IPCC, 2007b). Field observations document upslope range shifts of 16 butterfly species in Spain (Wilson et al., 2005), 37 dragonfly and damselfly species in Great Britain (Hickling et al., 2005), the white stork (*Ciconia ciconia*) in Poland (Tryjanowski et al., 2005), and the grey-headed flying fox (*Pteropus poliocephalus*) in Australia (Tidemann et al., 1999). From 1914 to 2006, half of 28 small mammal species monitored in Yosemite National Park, USA, shifted upslope an average of ~500 m, consistent with an observed 3°C increase in minimum temperatures (Moritz et al., 2008). From 1900 to 1998, two-thirds of 35 nonmigratory butterfly species examined in Europe shifted their range north, while only two species ranges shifted south (Parmesan et al., 1999).

As described in the previous section, meta-analyses of time series of at least 10 or 20 years have demonstrated twentieth century changes in the phenology of numerous plant and animal species. Animal life events that are occurring earlier include emergence from hibernation, amphibian calling and mating, spring bird migration, egg-laying, and appearance of butterflies (Parmesan and Yohe, 2003; Root et al., 2003; Rosenzweig et al., 2008). Short-range bird migration has advanced for nine bird species in Australia (Green and Pickering, 2002), 36 bird species in Sweden (Stervander et al., 2005), and 52 bird species in the eastern United States (Butler, 2003). In the Rocky Mountains, USA, emergence of yellow-bellied marmots (*Marmota flaviventris*) from hibernation advanced 23 days from 1975 to 1999, consistent with a local temperature increase of 1.4°C (Inouye et al., 2000). During the same period, snowmelt and plant flowering did not change, generating a possible phenology mismatch between marmots and their food plants.

Observed Impacts on Global Biogeochemistry

Biogeochemical cycles are the circulation of carbon, water, and other chemical compounds essential to life through the atmosphere, oceans, land, vegetation, and animals of the Earth. Satellite data, field sampling, and computer modeling indicate that climate change is particularly altering the global carbon and water cycles (IPCC, 2001a; IPCC, 2007a). This section describes some of the major connections between observed impacts of climate change on terrestrial ecosystems and global biogeochemical cycles.

Increased atmospheric CO_2 concentrations can enhance vegetation growth though the 'CO_2 fertilization' effect. CO_2 fertilization and lengthening

Fig. 2 Tropical rainforest at La Selva Biological Station, Costa Rica, viewed from the carbon flux tower that furnished data to help detect a decrease in forest biomass caused by climate change (Clark et al., 2003). Photo by P. Gonzalez

of the growing season together may be increasing global carbon sequestration by vegetation. Analysis of the satellite-derived Normalized Difference Vegetation Index (NDVI); (Tucker, 1979) showed a 6% increase in global net primary productivity (NPP) from 1982 to 1999, with substantial increases in tropical ecosystems (Nemani et al., 2003). A review of observations from various forest ecosystems also suggest global increases in forest productivity (Boisvenue and Running, 2006). In the Amazon, CO_2 fertilization and faster forest turnover rates may be causing an increase in the density of lianas (Phillips et al., 2002). On the other hand, warmer night temperatures reduced forest biomass in Costa Rica tropical rainforest (Fig. 2) from 1984 to 2000, due to increased respiration at night (Clark et al., 2003).

In a self-reinforcing cycle, climate change is increasing wildfire in some forest ecosystems (Gillett et al., 2004; Westerling et al., 2006), releasing more CO_2, which causes climate change. Wildfire currently emits 2–4 Pg C/y, up to half the amount of greenhouse gases of fossil fuel burning (Schultz et al., 2008; Bowman et al., 2009). Of the 2–4 Pg C/y of wildfire emissions, burning to deforest land may account for 0.6 Pg C/y (Bowman et al., 2009).

The increase in the proportion of Atlantic hurricanes in the most intense categories due to climate change (Mann et al., 2009) may also be increasing physical disturbance of forest ecosystems, windthrow, and carbon emissions from dead trees. One storm, Hurricane Katrina, felled trees containing ~0.1 Pg C (Chambers et al., 2007).

Due to increased inputs of heat energy into the atmosphere, land, and oceans, climate change is increasing convection and precipitation globally (IPCC, 2007a). In a self-reinforcing cycle, increased precipitation contributes to increased growth in forest ecosystems, increasing evapotranspiration inputs that contribute to cloud formation and precipitation. On the other hand, warmer temperatures in some ecosystems may be reducing soil moisture, increasing evapotranspiration, and decreasing moisture available for plant growth (NAS, 2008).

Projected Impacts

Projected Changes in Climate and Uncertainties

Climate change is a function of two sets of factors: greenhouse gas emissions and the complex responses of the atmosphere, land, and oceans. Consequently, projections of future climate depend on emissions scenarios (derived from trends in population, resource use per person, and emissions per unit of resource use) and general circulation models (GCMs; computer simulations of the atmosphere, land, and oceans). The relatively straightforward physics of the greenhouse effect governs global temperatures. In contrast, changes in precipitation, cloudiness, cyclones, the El Niño-Southern Oscillation, and other climate phenomena depend on more complex processes.

IPCC has developed a standard set of six emissions scenarios, ranging from a high-energy efficiency future to continuation of business-as-usual, on which research groups in 10 countries have run 23 GCMs (IPCC, 2007a). Uncertainties in projections of future climate change derive mainly from the range in emissions estimates from the scenarios and accuracy differences among the GCMs.

IPCC projects global average temperature increases of 1.8–4°C between the periods 1980–1999 and 2090–2099 for the six scenarios (IPCC, 2007a). Analyses of potential mitigating effects of emissions reductions policies project warming of 0.8–7.8°C with no emissions reductions and 0.5–4.4°C with a range of emissions reduction policies (Van Vuuren et al., 2008). Actual greenhouse gas emissions through 2004 have exceeded the highest IPCC emissions scenario, suggesting warming in the upper part of the range of projections (Raupach et al., 2007).

Global average temperatures in the Last Glacial Maximum were approximately 4–7°C cooler than present (IPCC, 2007a; NAS, 2008). The warming after the ice ages occurred over 10–20 millennia, in contrast to projections of future warming of a similar magnitude in just one century.

GCMs project an increase in global average precipitation. Boreal and polar areas may receive large increases while subtropical areas may experience decreases in rainfall (IPCC, 2007a). Most projections indicate more frequent extremes in temperature and precipitation, leading to more frequent droughts and flooding.

Projected Impacts on Vegetation

The magnitude of projected climate changes would render ecosystems vulnerable to biome shifts, phenology changes, wildfire increases, species extinctions, and other impacts. Assessment of potential impacts requires analysis of the vulnerability of ecosystems. Vulnerability to climate change is the degree to which a system is susceptible to, and unable to cope with, adverse effects (IPCC, 2007b). Vulnerability is a function of three components: exposure, sensitivity, and adaptive capacity. Exposure consists of the climate that a species or ecosystem experiences. Sensitivity is the degree to which a species or ecosystem changes due to climate change. Adaptive capacity is the ability of a species or ecosystem to adjust, moderate potential damage, take advantage of new conditions, or cope with warming and other impacts (IPCC, 2007b).

Climate projections provide information on exposure. To assess sensitivity and adaptive capacity, dynamic global vegetation models (DGVMs) (Daly et al., 2000; Sitch et al., 2008) simulate the spatial distribution of vegetation, biomass, and wildfire based on climate, soil, and observed characteristics of plant functional types. Climate envelope, bioclimatic, or niche models simulate the geographic range of individual species based on areas that seem to fall within the climate tolerance of a plant or animal (Elith et al., 2006). Uncertainties in DGVMs and climate envelope models arise from incomplete information on relationships of organisms to climate parameters, dispersal capabilities, inter-specific interactions, and evolutionary adaptations.

Vulnerability analyses indicate that projected climate change, combined with deforestation, pollution, and other stresses, could overwhelm the adaptive capacity of many species and ecosystems by 2100 (IPCC, 2007b). Climate change and other stresses may elicit threshold-type responses, in which species or ecosystems experience sudden changes at threshold levels where pressures overwhelm adaptive capacities (Burkett et al., 2005). Some potential changes, such as species extinctions, will be irreversible.

Spatial analyses of observed twentieth century climate change and projected twenty-first century vegetation indicates that one-tenth to one-half of global land may be highly (confidence ~0.80) to very highly (confidence ~0.95) vulnerable to vegetation shifts (Gonzalez et al., 2010). DGVMs project extensive latitudinal and elevational biome shifts (Fig. 3) around the world (IPCC, 2007b; Sitch et al., 2008; Gonzalez et al., 2010). The most vulnerable biomes may be alpine grassland, tundra, and boreal forest, replaced in many areas by temperate conifer forest, boreal forest, and temperate conifer forest, respectively. Conditions favorable to alpine ecosystems may completely disappear from the tops of mountains.

Increased drought and fire threaten to cause extensive dieback of Amazon rainforest. High fire risks are projected for half of the area of the Amazon (Golding and Betts, 2008). A global mean temperature rise >2°C could convert 20–90% of Amazon tropical evergreen broadleaf forest to grassland (Jones et al., 2009).

Projected global increases in drought due to climate change (Burke et al., 2006) threaten to exacerbate desertification and cause biome changes in

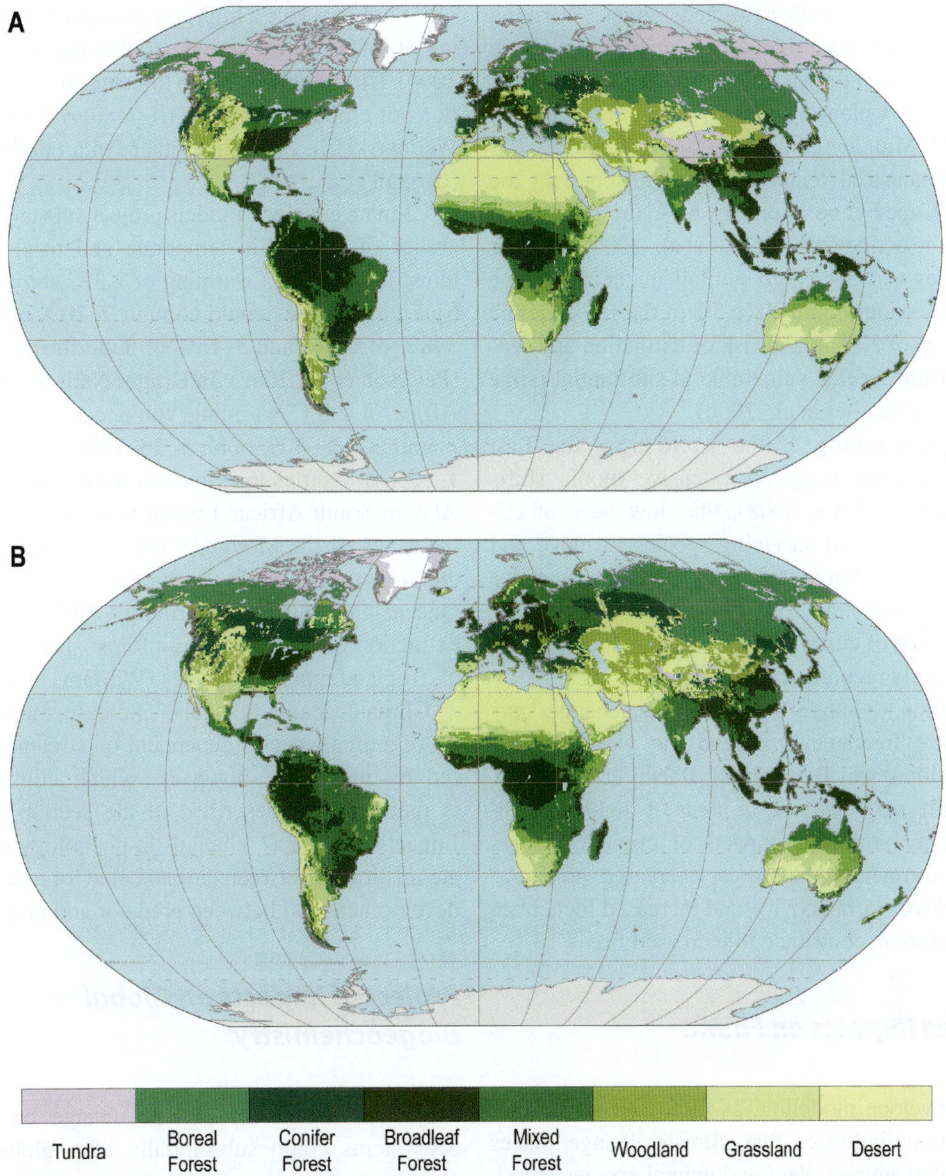

Fig. 3 Projected biome changes shown by modeling of global vegetation under (**a**) observed climate in the period 1961–1990 and (**b**) projected climate in the period 2071–2100, showing the worst case of nine combinations of three IPCC (2007a) emissions scenarios and three general circulation models, with global average temperature increases of 2.4–4.0°C (Gonzalez et al., 2010)

the African Sahel and other arid, semi-arid, and dry subhumid areas (IPCC, 2001b). Drought in Southern Africa could decrease vegetation cover in the Kalahari enough to remobilize sand dunes (Thomas and Leason, 2005).

Climate change may also shift geographic ranges of numerous individual plant species (IPCC, 2007b). Contracting patches of habitat, dispersal limitations, and novel climates that species have never before encountered (Williams et al., 2007) may overwhelm the adaptive capacities of some species. Climate envelope modeling of the ranges of 1,103 plant and animal species under projected warming 2–3°C above preindustrial temperatures indicate that climate change places 15–37% of examined species at risk of extinction (Thomas et al., 2004).

Climate change threatens the unique flora of South Africa and Namibia, characterized by high numbers of endemic species with restricted ranges. Warming of 1.5–2.7°C above pre-industrial temperatures could contract the area of the Succulent Karoo biome by 80% and push 2,800 plant species to extinction (Hannah et al., 2002; Midgley et al., 2002). Warming of 2.7°C above pre-industrial temperatures could contract the area of the Cape Fynbos biome by 65%, driving 23% of the species to extinction (Thomas et al., 2004). Warmer temperatures and decreased rainfall under mid-range climate projections could place 5% of the 159 endemic plant species of Namibia at risk of extinction and render half of the species vulnerable to substantial range contractions (Thuiller et al., 2006).

Although warming of the world after the Last Glacial Maximum triggered extensive biome shifts and extinctions (NAS, 2008), the slow pace of climate change allowed surviving species to move and reassemble into functional ecosystems as the ice retreated (Pitelka et al., 1997; Overpeck et al., 2003). The rapid pace of current climate change may prevent such an orderly reconfiguration.

In addition to range shifts, climate change may increase fire frequencies around the world (IPCC, 2007b; Golding and Betts, 2008; Balshi et al., 2009), although fire may decrease in areas of higher precipitation. Many projected impacts of climate change, including warmer temperatures, decreased precipitation, increased storm activity, and increased fuels from dying vegetation, contribute to increased fire.

Projected Impacts on Fauna

Climate envelope modeling, as discussed in the previous section, indicates that climate change places 15–37% of examined plant and animal species at risk of extinction (Thomas et al., 2004). The species most at risk are limited-range endemics and alpine and polar species. Unique characteristics of polar animals render them more vulnerable to climate change. Many polar animal species depend on habitats with extensive snow and ice. Polar ecosystems are relatively simple, with low densities of predators and competitors. In the Arctic, most animals have evolved fewer traits for competition, predator avoidance, and resistance to pathogens not found in cold regions (ACIA, 2004). Due to their high mobility, the dominant response of many arctic animals to climate change may be relocation rather than adaptation (ACIA, 2004).

Polar bears face a high risk of extinction with warming of 2.8°C above preindustrial temperatures (ACIA, 2004). Climate change also threatens to reduce populations of emperor penguins (*A. forsteri*) (Barbraud and Weimerskirch, 2001) and other Antarctic bird species (Croxall et al., 2002).

Climate envelope models project substantial vulnerability of species in temperate and tropical ecosystems. In Mexico, a warming of 2.3ºC above preindustrial temperatures could commit 2–20% of mammals, 3–8% of birds, and 3–15% of butterflies to extinction (Peterson et al., 2002). In Kruger National Park, South Africa, a 2.3ºC warming could commit 24–59% of mammals, 28–40% of birds, 13–70% of butterflies, and 1–45% of reptiles to extinction (Erasmus et al., 2002). Also in South Africa, a warming of 4ºC could reduce the range of the Mountain Wheat-ear bird (*Oenanthe monticola*) by half (Simmons et al., 2004). In Queensland, Australia, warming > 4ºC could cause the extinction of all endemic rainforest species, including 57 frog and mammal species (Williams et al., 2003).

Climate change can also increase the vulnerability of animal species dependent on streams and rivers fed by mountain snowpack, which climate change is reducing. As described in the section "Observed Impacts on Fauna", changes in phenology could initiate mistiming between animal behavior and food plant development and between predator and prey activity.

Projected Impacts on Global Biogeochemistry

Projected impacts of climate change on terrestrial ecosystems could substantially alter global biogeochemical cycles. Experimental enrichment of four forest sites to the CO_2 concentrations that would accompany a 3°C warming above preindustrial temperatures increased NPP 23 ± 2% (Norby et al., 2005). At higher CO_2 concentrations, plants would not need to open their stomata as much, which, in other experiments, yielded plant water savings of 5–15% (Wullschleger and Norby, 2001; Cech et al., 2003). Projections indicate that the combination of CO_2 fertilization, improved water use efficiency, and the expansion of global forest area due to biome shifts will continue to increase global NPP and total sequestration of carbon in vegetation through much of the twenty-first century (IPC, 2007a; Sitch et al., 2008).

Yet, saturation of CO_2 fertilization by mid-century, nitrogen and phosphorus limitations to plant growth, heat stress on plants, increased soil respiration, increased fire and other disturbances, and methane emissions from melting of tundra and permafrost may convert terrestrial ecosystems into a net emitter of greenhouse gases by 2100 (IPCC, 2007a). Continued tropical deforestation would add even more emissions.

In polar and boreal areas, the replacement of snow-covered ground by boreal conifer forest with dark foliage can reduce albedo (reflectance) and increase local temperature (Bala et al., 2007). This local warming, in addition to forest decline at the southern edge of boreal forest and potential increases in wildfire, may offset cooling effects of increased carbon sequestration in the new boreal forest areas (ACIA, 2004; IPCC, 2007a).

Projected increases in wildfire (Sitch et al., 2008; Balshi et al., 2009) may create a positive feedback for climate warming through significant emissions of greenhouse gases that would further increase temperatures (Randerson et al., 2006). In another possible positive feedback cycle, climate change may exacerbate outbreaks of bark beetles, causing extensive forest dieback, increasing greenhouse gas emissions, and further increasing global temperatures (Kurz et al., 2008).

In addition to increasing greenhouse gas emissions to the atmosphere, forest dieback and degradation could substantially alter segments of the hydrologic cycle (IPCC, 2007a). Forest dieback in the Amazon and other tropical rainforests could reduce precipitation regionally. Projected reductions of vegetation cover in many areas could increase runoff, decrease soil moisture, and decrease precipitation in affected areas. Decreased precipitation could create positive feedback cycles in affected areas by further reducing vegetation cover and increasing greenhouse gas emissions.

Adaptation

Types of Adaptation

Adaptation is an adjustment in natural or human systems in response to climate change, to moderate harm or exploit new conditions (IPCC, 2007b). Adaptation to climate change falls into three broad types. First, natural selection of individual plants and animals with resilient characteristics will, as these individuals pass their genes to offspring, drive the evolution of species more adapted to changed climate conditions. Second, natural resource management agencies and individual people can adjust land and water management practices at specific sites to help individual plant and animal species cope with climate change. Third, natural resource management agencies and other organizations can adjust management plans across broad landscapes to facilitate adaptation of species and ecosystems. In the first type of adaptation, plant and animal species are adapting. In the second and third types, agencies and people are adapting.

Evolutionary Species Adaptation

Observations indicate that some species are evolving to adapt to climate change. In Europe, the blackcap warbler (*Sylvia atricapilla*) evolved so that the direction of its migration route extends its winter range northward (Berthold et al., 2003). In England, the speckled wood butterfly (*Pararge aegeria*) has evolved in a way that dispersal morphology and life history traits have allowed the species to expand its geographic range (Hill et al., 1999; Hughes et al., 2003). The North American red squirrel (*Tamiasciurus hudsonicus*) has also evolved so breeding occurs slightly earlier as climate change advances the beginning of spring (Berteaux et al., 2004).

Species-Specific Natural Resource Management Adaptation

This section provides examples of adaptation measures in which natural resource agencies and individual people adjust land and water management practices at specific sites to help individual plant and animal species cope with climate change. Scott et al. (2008) describe some of these examples.

Prescribed burning is the planned ignition of fire to simulate the natural effects of fire in an ecosystem adapted to fire. Fire forms a natural component of forest and grassland ecosystems. In areas projected to experience an increase in fire frequency due to climate change, the preemptive use of fire can reduce the amount of litter and woody debris that might cause catastrophic stand-replacement fires and damage tree species adapted to less intense fire regimes. Although prescribed burning may release greenhouse gases in the short term, it can create conditions favorable for the growth of large trees, increasing carbon sequestration in the long term.

Natural regeneration and enrichment planting of adapted plant species starts with identification of native species that already grow in an area and that possess characteristics adapted to projected climate conditions. Natural regeneration would involve protection of existing small trees of the species of interest to increase their survival. Enrichment planting would involve planting new seedlings where the existing density of the species is sparse. High genetic diversity of species at the low elevation edge of their range may require special protection of those areas to conserve and propagate their seeds (Hampe and Petit, 2005).

Food propagation for mistimed migrants would involve the planting of food plants in situations where climate change has decoupled the phenology of animals and their food plants. This may be necessary, for example, with migratory birds that arrive earlier in the spring at breeding grounds where food plants are not developing earlier.

Riparian reforestation of native riparian tree species along river and stream banks could provide shade to keep water temperatures from warming excessively during summer months. This could create thermal refugia for fish and other species.

Managed relocation is an intervention technique that involves the intentional movement of populations or species from current areas of occupancy to locations where the probability of future persistence is projected to be higher (Richardson et al., 2009). It may become necessary in extreme cases where climate change threatens to strand limited-range endemic species, polar species, or alpine species on mountain peaks or other locations where warming may eventually eliminate all suitable habitat. The release of species that could become invasive in the areas of relocation presents a challenge to this adaptation measure.

Monitoring species abundance and distributions in permanent ecological plots will provide essential data to track the effectiveness of adaptation measures.

Landscape-Scale Natural Resource Management Adaptation

Biome shifts and other impacts of climate change threaten to reduce the effectiveness of the existing network of national parks, forests, reserves, and other natural resource management areas because managers generally designed those networks without considering the dynamics of climate change. This section provides examples of adaptation measures at the landscape scale, in which natural resource management agencies and other organizations adjust landscape management plans to facilitate adaptation of species and ecosystems. Generally, landscape adaptation measures concern the structured configuration of a network of existing natural resource areas combined with the targeted acquisition of new areas. Scott et al. (2008) describe some of these examples.

Analysis of vulnerability of geographic areas to climate change would involve spatial analyses to categorize areas within a landscape into three classes: areas of high, medium, or low vulnerability (Gonzalez et al., 2010). Analyses would require spatial data on exposure (climate projections), sensitivity (ecological models), and adaptive capacity (field data). The results would guide prioritization of geographic areas and planning of management actions (Hannah et al., 2002). In British Columbia, Canada, one proposed adaptation system classifies forest areas based on four combinations of exposure (low or high) and adaptive capacity (low or high) and assigns natural regeneration, mixed-species planting, enrichment planting, fire fuel reduction, and strict protection measures based on the classification (Nitschke and Innes, 2008). A global analysis of observed climate and projected vegetation identifies vulnerable areas and potential refugia around the world (Gonzalez et al., 2010).

In existing management areas, prioritization of places of higher vulnerability for adaptation measures would channel resources to those areas that may require more intensive management. Potentially greater disturbances and species turnover in vulnerable areas would require costly adaptation measures such as prescribed burning and invasive species removal. Areas of unique ecological or cultural value would continue to merit high priority.

Acquisition of new areas in climate change refugia would take advantage of less demanding management needs of those areas. Climate change refugia are locations more resistant to climate change due to wide climate tolerances of individual species, the presence of resilient assemblages of species, and local topographic and environmental factors. Because of the lower probability of drastic change, refugia will likely require less intense management interventions to maintain viable habitat and cost less than management of vulnerable areas (Griffith et al., 2009). Acquisition of new areas should continue to adhere to the principles of representation, the protection of examples of different ecosystem types practiced, for example, by Parks Canada, and replication, the protection of several separate areas of the same type of habitat to provide insurance of loss of any single replicate (US CCSP, 2008).

Establishment and maintenance of corridors will facilitate species dispersal and migration as climate change shifts the location of habitats over time. Fragmentation of habitat will impair the ability of species to adapt to climate change (IPCC, 2007b). Because fundamental ecosystem functions, including fire, food webs, and nutrient cycling, often require land areas of thousands of km^2, corridors can economically increase range sizes by linking up existing, but often small, management areas. Projects to reduce deforestation and degradation (REDD) can also serve to reduce fragmentation and to restore forest ecosystem services, including carbon sequestration.

Managing for habitat type rather than managing for specific species would involve the identification and conservation of functional groups (e.g., perennial grasses in a grassland ecosystem) or habitat types (e.g., tropical rainforest) instead of specific species. This potential adaptation measure recognizes that assuring the vibrant functioning of an ecosystem could more effectively conserve more species than dedicating scarce resources to the conservation of a few individual endangered species.

Monitoring ecosystem-level indicators, such as species richness, biomass, and densities of plants and animals, in permanent ecological plots will provide essential information to track impacts of climate change, ecosystem function, and the effectiveness of adaptation measures.

References

Abu-Asab MS, Peterson PM, Shetler SG, Orli SS (2001) Earlier plant flowering in spring as a response to global warming in the Washington, DC, area. Biodivers Conserv 10:597–612.

Allen CD, Macalady AK, Chenchouni H, Bachelet D, McDowell N, Vennetier M, Kitzberger T, Rigling A, Breshears DD, Hogg EH, et al. (2010) A global overview of drought and heat-induced tree mortality reveals emerging climate change risks for forests. For Ecol Manage 259:660–684.

Arctic Climate Impact Assessment (ACIA) (2004) Impacts of a Warming Arctic: Arctic Climate Impact Assessment. Cambridge: Cambridge University Press.

Alexander MA, Eischeid JK (2001) Climate variability in regions of amphibian declines. Conserv Biol 15: 930–942.

Bala G, Caldeira K, Wickett M, Phillips TJ, Lobell DB, Delire C, Mirin A (2007) Combined climate and carbon-cycle effects of large-scale deforestation. Proc Natl Acad Sci USA 104:6550–6555.

Balshi MS, McGuire AD, Duffy P, Flannigan M, Kicklighter DW, Melillo J (2009) Vulnerability of carbon storage in North American boreal forests to wildfires during the 21st century. Glob Chang Biol 15:1491–1510.

Barbraud C, Weimerskirch H (2001) Emperor penguins and climate change. Nature 411:183–186.

Beckage B, Osborne B, Gavin DG, Pucko C, Siccama T, Perkins T (2008) A rapid upward shift of a forest ecotone during 40 years of warming in the Green Mountains of Vermont. Proc Natl Acad Sci USA 105:4197–4202.

Berteaux D, Réale D, McAdam AG, Boutin S (2004) Keeping pace with fast climate change: Can arctic life count on evolution? Integr Comp Biol 44:140–151.

Berthold P, Gwinner E, Sonnenschein E (2003) Avian Migration. Berlin, Germany: Springer.

Boisvenue C, Running SW (2006) Impacts of climate change on natural forest productivity—Evidence since the middle of the 20th century. Glob Chang Biol 12:862–882.

Bowman DMJS, Balch JK, Artaxo P, Bond WJ, Carlson JM, Cochrane MA, D'Antonio CM, DeFries RS, Doyle JC, Harrison SP, et al. (2009) Fire in the Earth system. Science 324:481–484.

Breshears DD, Cobb NS, Rich PM, Price KP, Allen CD, Balice RG, Romme WH, Kastens JH, Floyd ML, Belnap J, Anderson JJ, Myers OB, Meyer CW (2005) Regional vegetation die-off in response to global-change-type drought. Proc Natl Acad Sci USA 102:15144–15148.

Burke EJ, Brown SJ, Christidis N (2006) Modelling the recent evolution of global drought and projections for the 21st century with the Hadley Centre climate model. J Hydrometeorol 7:113–1125.

Burkett VR, Wilcox DA, Stottlemyer R, Barrow W, Fagre D, Baron J, Price J, Nielsen JL, Allen CD, Peterson DL, et al. (2005) Nonlinear dynamics in ecosystem response to climatic change: Case studies and policy implications. Ecol Complex 2:357–394.

Burrowes PA, Joglar RL, Green DE (2004) Potential causes for amphibian declines in Puerto Rico. Herpetologica 60:141–154.

Butler CJ (2003) The disproportionate effect of global warming on the arrival dates of short-distance migratory birds in North America. Ibis 145:484–495.

Cech P, Pepin S, Körner C (2003) Elevated CO_2 reduces sap flux in mature deciduous forest trees. Oecologia 137:258–268.

Chambers JQ, Fisher JI, Zeng H, Chapman EL, Baker DB, Hurtt GC (2007) Hurricane Katrina's carbon footprint on U.S. Gulf Coast forests. Science 318:1107.

Clark DA, Piper SC, Keeling CD, Clark DB (2003) Tropical rain forest tree growth and atmospheric carbon dynamics linked to interannual temperature variation during 1984–2000. Proc Natl Acad Sci USA 100:5852–5857.

Croxall JP, Trathan PN, Murphy EJ (2002) Environmental change and Antarctic seabird populations. Science 297:1510–1514.

Dai A, Lamb PJ, Trenberth KE, Hulme M, Jones PD, Xie P (2004) The recent Sahel drought is real. Int J Climatol 24:1323–1331.

Daly C, Bachelet D, Lenihan JM, Neilson RP, Parton W, Ojima D (2000) Dynamic simulation of tree–grass interactions for global change studies. Ecol Appl 10:449–469.

Elith J, Graham CH, Anderson RP, Dudík M, Ferrier S, Guisan A, Hijmans RJ, Huettmann F, Leathwick JR, Lehmann A, Li J, et al. (2006) Novel methods improve prediction of species' distributions from occurrence data. Ecography 29:129–151.

Emslie SD, Fraser W, Smith RC, Walker W (1998) Abandoned penguin colonies and environmental change in the Palmer Station area, Anvers Island, Antarctic Peninsula. Antarct Sci 10:257–268.

Erasmus BFN, Van Jaarsveld AS, Chown SL, Kshatriya M, Wessels KJ (2002) Vulnerability of South African animal taxa to climate change. Glob Chang Biol 8:679–693.

Fitter AH, Fitter RSR (2002) Rapid changes in flowering time in British plants. Science 296:1689–1691.

Fraser WR, Trivelpiece WZ, Ainley DG, Trivelpiece SG (1992) Increases in Antarctic penguin populations: Reduced competition with whales or a loss of sea ice due to environmental warming. Polar Biol 11:525–531.

Giannini A, Saravanan R, Chang P (2003) Oceanic forcing of Sahel rainfall on interannual to interdecadal time scales. Science 302:1027–1030.

Gillett NP, Weaver AJ, Zwiers FW, Flannigan MD (2004) Detecting the effect of climate change on Canadian forest fires. Geophys Res Lett 31:L18211. doi:10.1029/2004GL020876.

Golding N, Betts R (2008) Fire risk in Amazonia due to climate change in the HadCM3 climate model: Potential interactions with deforestation. Glob Biogeochem Cycles 22:GB4007. doi:10.1029/2007GB003166.

Gonzalez P (2001) Desertification and a shift of forest species in the West African Sahel. Clim Res 17:217–228.

Gonzalez P, Neilson RP, Lenihan JM, Drapek RJ (2010) Global patterns in the vulnerability of ecosystems to vegetation shifts due to climate change. Glob Ecol Biogeogr.

Green K, Pickering CM (2002) A potential scenario for mammal and bird diversity in the Snowy Mountains of Australia in relation to climate change. In C Körner, E Spehn (Eds), Mountain Biodiversity: A Global Assessment. London: Parthenon Publishing.

Griffith B, Scott JM, Adamcik R, Ashe D, Czech B, Fischman R, Gonzalez P, Lawler J, McGuire AD, Pidgorna A (2009) Climate change adaptation for the US National Wildlife Refuge System. Environ Manage 44:1043–1052.

Hampe A, Petit RJ (2005) Conserving biodiversity under climate change: The rear edge matters. Ecol Lett 8:461–467.

Hannah L, Midgley GF, Millar D (2002) Climate change-integrated conservation strategies. Glob Ecol Biogeogr doi: 10.1111/j.1466–8238.2010.00558.X 11:485–495.

Hickling R, Roy DB, Hill JK, Thomas CD (2005) A northward shift of range margins in British Odonata. Glob Chang Biol 11:502–506.

Hill JK, Thomas CD, Blakeley DS (1999) Evolution of flight morphology in a butterfly that has recently expanded its geographic range. Oecologia 121:165–170.

Hughes CL, Hill JK, Dytham C (2003) Evolutionary trade-offs between reproduction and dispersal in populations at expanding range boundaries. Proc R Soc Lond B Biol Sci 270:S147–S150.

Inouye DW, Barr B, Armitage KB, Inouye BD (2000) Climate change is affecting altitudinal migrants and hibernating species. Proc Natl Acad Sci USA 97:1630–1633.

Intergovernmental Panel on Climate Change (IPCC) (2001a) Climate Change 2001: The Scientific Basis. Cambridge: Cambridge University Press.

Intergovernmental Panel on Climate Change (IPCC) (2001b) Climate Change 2001: Impacts, Adaptation, and Vulnerability. Cambridge: Cambridge University Press.

Intergovernmental Panel on Climate Change (IPCC) (2007a) Climate Change 2007: The Physical Science Basis. Cambridge: Cambridge University Press.

Intergovernmental Panel on Climate Change (IPCC) (2007b) Climate Change 2007: Impacts, Adaptation, and Vulnerability. Cambridge: Cambridge University Press.

Intergovernmental Panel on Climate Change (IPCC) (2007c) Climate Change 2007: Mitigation of Climate Change. Cambridge: Cambridge University Press.

Jones C, Lowe J, Liddicoat S, Betts R (2009) Committed terrestrial ecosystem changes due to climate change. Nat Geosci 2:484–487.

Kasischke ES, Turetsky MR (2006) Recent changes in the fire regime across the North American boreal region—Spatial and temporal patterns of burning across Canada and Alaska. Geophy Res Lett 33:L09703. doi:10.1029/2006GL025677.

Kirkbride MP (1995) Relationships between temperature and ablation on the Tasman Glacier, Mount Cook National Park, New Zealand. N Z J Geol Geophys 38:17–27.

Knowles N, Dettinger MD, Cayan DR (2006) Trends in snowfall versus rainfall in the western United States. J Clim 19:4545–4559.

Kullman L, Öberg L (2009) Post-little ice age tree line rise and climate warming in the Swedish Scandes: A landscape ecological perspective. J Ecol 97:415–429.

Kurz WA, Dymond CC, Stinson G, Rampley GJ, Neilson ET, Carroll AL, Ebata T, Safranyik L (2008) Mountain pine beetle and forest carbon feedback to climate change. Nature 452:987–990.

Le Quéré C, Raupach MR, Canadell JG, Marland G, Bopp L, Ciais P, Conway TJ, Doney SC, Feely RA, Foster P, et al.

(2009) Trends in the sources and sinks of carbon dioxide. Nat Geosci 2:831–836.

Lenoir J, Gégout JC, Marquet PA, de Ruffray P, Brisse H (2008) A significant upward shift in plant species optimum elevation during the 20th century. Science 320:1768–1771.

Lüthi D, Le Floch M, Bereiter B, Blunier T, Barnola JM, Siegenthaler U, Raynaud D, Jouzel J, Fischer H, Kawamura K, Stocker TF (2008) High-resolution carbon dioxide concentration record 650,000–800,000 years before present. Nature 453:379–382.

Mann ME, Zhang Z, Hughes MK, Bradley RS, Miller SK, Rutherford S, Ni F (2008) Proxy-based reconstructions of hemispheric and global surface temperature variations over the past two millennia. Proc Natl Acad Sci USA 105:13252–13257.

Mann ME, Woodruff JD, Donnelly JP, Zhang Z (2009) Atlantic hurricanes and climate over the past 1,500 years. Nature 460:880–883.

Menzel A, Dose V (2005) Analysis of long-term time-series of beginning of flowering by Bayesian function estimation. Meteorologische Zeitschrift 14:429–434.

Menzel A, Sparks T, Estrella N, Roy DB (2006) Geographic and temporal variability in phenology. Glob Ecol Biogeogr 15:498–504.

Midgley GF, Hannah L, Millar D, Rutherford MC, Powrie LW (2002) Assessing the vulnerability of species richness to anthropogenic climate change in a biodiversity hotspot. Glob Ecol Biogeogr 11:445–451.

Millennium Ecosystem Assessment (2006) Ecosystems and Human Well-Being: Current State and Trends. Washington, DC: Island Press.

Moritz C, Patton JL, Conroy CJ, Parra JL, White GC, Beissinger SR (2008) Impact of a century of climate change on small-mammal communities in Yosemite National Park, USA. Science 322:261–264.

National Academy of Sciences of the USA (NAS) (2008) Ecological Impacts of Climate Change. Washington, DC: National Academies Press.

Nemani RR, Keeling CD, Hashimoto H, Jolly WM, Piper SC, Tucker CJ, Myneni RB, Running SW (2003) Climate-driven increases in global terrestrial net primary production from 1982 to 1999. Science 300:1560–1563.

Nitschke CR, Innes JL (2008) Integrating climate change into forest management in South-Central British Columbia: An assessment of landscape vulnerability and development of a climate-smart framework. For Ecol Manage 256:313–327.

Norby RJ, DeLucia EH, Gielen B, Calfapietra C, Giardina CP, King JS, Ledford J, McCarthy HR, Moore DJP, Ceulemans R, et al. (2005) Forest response to elevated CO_2 is conserved across a broad range of productivity. Proc Natl Acad Sci USA 102:18052–18056.

Overpeck JT, Whitlock C, Huntley B (2003) Terrestrial biosphere dynamics in the climate system: Past and future. In K Alverson, R Bradley, T Pedersen (Eds), Paleoclimate, Global Change, and the Future. Berlin, Germany: Springer.

Pacala S, Socolow R (2004) Stabilization wedges: Solving the climate problem for the next 50 years with current technologies. Science 305:968–972.

Parmesan C, Yohe G (2003) A globally coherent fingerprint of climate change impacts across natural systems. Nature 421:37–42.

Parmesan C, Ryrholm N, Stefanescu C, Hill JK, Thomas CD, Descimon H, Huntley B, Kaila L, Kullberg J, Tammaru T, Tennent WJ, Thomas JA, Warren M (1999) Poleward shifts in geographical ranges of butterfly species associated with regional warming. Nature 399:579–583.

Peñuelas J, Boada M (2003) A global change-induced biome shift in the Montseny mountains (NE Spain). Glob Chang Biol 9:131–140.

Peterson AT, Ortega-Huerta MA, Bartley J, Sánchez-Cordero V, Soberón J, Buddemeier RH, Stockwell DRB (2002) Future projections for Mexican faunas under global climate change scenarios. Nature 416:626–629.

Phillips OL, Martínez RV, Arroyo L, Baker TR, Killeen T, Lewis SL, Malhi Y, Mendoza AM, Neill D, Vargas PN, et al. (2002) Increasing dominance of large lianas in Amazonian forests. Nature 418:770–774.

Pitelka LF, Gardner RH, Ash J, Berry S, Gitay H, Noble IR, Saunders A, Bradshaw RHW, Brubaker L, Clark JS, Davis MB (1997) Plant migration and climate change. Am Sci 85:464–473.

Pounds JA, Bustamante MR, Coloma LA, Consuegra JA, Fogden MPL, Foster PN, La Marca E, Masters KL, Merino-Viteri A, Puschendorf R, Ron SR, Sanchez-Azofeifa GA, Still CJ, Young BE (2006) Widespread amphibian extinctions from epidemic disease driven by global warming. Nature 439:161–167.

Raffa KF, Aukema BH, Bentz BJ, Carroll AL, Hicke JA, Turner MG, Romme WH (2008) Cross-scale drivers of natural disturbances prone to anthropogenic amplification: The dynamics of bark beetle eruptions. BioScience 58:501–517.

Randerson JT, Liu H, Flanner MG, Chambers SD, Jin Y, Hess PG, Pfister G, Mack MC, Treseder KK, Welp LR, Chapin FS, Harden JW, Goulden ML, Lyons E, Neff JC, Schuur EAG, Zender CS (2006) The impact of boreal forest fire on climate warming. Science 314:1130–1132.

Raupach MR, Marland G, Ciais P, Le Quéré C, Canadell JG, Klepper G, Field CB (2007) Global and regional drivers of accelerating CO_2 emissions. Proc Natl Acad Sci USA 104:10288–10293.

Richardson DM, Hellmann JJ, McLachlan JS, Sax DF, Schwartz MW, Gonzalez P, Brennan EJ, Camacho A, Root TL, Sala OE, Schneider SH, Ashe DM, Clark JR, Early R, Etterson JR, Fielder ED, Gill JL, Minteer BA, Polasky S, Safford HD, Thompson AR, Vellend M (2009) Multidimensional evaluation of managed relocation. Proc Natl Acad Sci USA 106:9721–9724.

Ron SR, Duellman WE, Coloma LA, Bustamante MR (2003) Population decline of the Jambato toad *Atelopus ignescens* (Anura: Bufonidae) in the Andes of Ecuador. J Herpetol 37:116–126.

Root TL, Price JT, Hall KR, Schneider SH, Rosenzweig C, Pounds JA (2003) Fingerprints of global warming on animals and plants. Nature 421:57–60.

Rosenzweig C, Karoly D, Vicarelli M, Neofotis P, Wu Q, Casassa G, Menzel A, Root TL, Estrella N, Seguin B, Tryjanowski P, Liu C, Rawlins S, Imeson A (2008) Attributing physical and biological impacts to anthropogenic climate change. Nature 453:353–357.

Schultz MG, Heil A, Hoelzemann JJ, Spessa A, Thonicke K, Goldammer JG, Held AC, Pereira JMC, van het Bolscher M (2008) Global wildland fire emissions from

1960 to 2000. Global Biogeochem Cycles 22:GB2002. doi: 10.1029/2007GB003031.

Scott JM, Griffith B, Adamcik RS, Ashe DM, Czech B, Fischman RL, Gonzalez P, Lawler JJ, McGuire AD, Pidgorna A (2008) National Wildlife Refuges. In U.S. Climate Change Science Program. Preliminary Review of Adaptation Options for Climate-Sensitive Ecosystems and Resources. Washington, DC: U.S. Environmental Protection Agency.

Simmons RE, Barnard P, Dean WRJ, Midgley GF, Thuiller W, Hughes G (2004) Climate change and birds: Perspectives and prospects from southern Africa. Ostrich 75:295–308.

Sitch S, Huntingford C, Gedney N, Levy PE, Lomas M, Piao SL, Betts R, Ciais P, Cox P, Friedlingstein P, Jones CD, Prentice IC, Woodward FI (2008) Evaluation of the terrestrial carbon cycle, future plant geography and climate-carbon cycle feedbacks using five Dynamic Global Vegetation Models (DGVMs). Glob Chang Biol 14:2015–2039.

Stervander M, Lindstrom K, Jonzen N, Andersson A (2005) Timing of spring migration in birds: Long-term trends, North Atlantic Oscillation, and the significance of different migration routes. J Avian Biol 36:210–221.

Stirling I, Lunn NJ, Iacozza J (1999) Long-term trends in the population ecology of polar bears in western Hudson Bay in relation to climatic change. Arctic 52:294–306.

Thomas DSG, Leason HC (2005) Dunefield activity response to climate variability in the southwest Kalahari. Geomorphology 64:117–132.

Thomas CD, Cameron A, Green RE, Bakkenes M, Beaumont LJ, Collingham YC, Erasmus BF, De Siqueira MF, Grainger A, Hannah L, Hughes L, Huntley B, Van Jaarsveld AS, Midgley GF, Miles L, Ortega-Huerta MA, Peterson AT, Phillips OL, Williams SE (2004) Extinction risk from climate change. Nature 427:145–148.

Thompson R, Clark RM (2008) Is spring starting earlier? The Holocene 18:95–104.

Thuiller W, Midgley GF, Hughes GO, Bomhard B, Drew G, Rutherford MC, Woodward FI (2006) Endemic species and ecosystem sensitivity to climate change in Namibia. Glob Chang Biol 12:759–776.

Tidemann CR, Vardon MJ, Loughland RA, Brocklehurst PJ (1999) Dry season camps of flying-foxes (*Pteropus spp.*) in Kakadu World Heritage Area, north Australia. J Zool 247:155–163.

Tryjanowski P, Sparks TH, Profus P (2005) Uphill shifts in the distribution of the white stork *Ciconia ciconia* in southern Poland: The importance of nest quality. Divers Distrib 11:219–223.

Tucker CJ (1979) Red and photographic infrared linear combinations for monitoring vegetation. Remote Sens Environ 8:127–150.

United States Climate Change Science Program (US CCSP) (2008) Preliminary Review of Adaptation Options for Climate-Sensitive Ecosystems and Resources. Washington, DC: U.S. Environmental Protection Agency.

United States Global Change Research Program (USGCRP) (2009) Global Change Impacts in the United States. Cambridge: Cambridge University Press.

van Vuuren DP, Meinshausen M, Plattner GK, Joos F, Strassmann KM, Smith SJ, Wigley TML, Raper SCB, Riahi K, de la Chesnaye F, den Elzen MGJ, Fujino J, Jiang K, Nakicenovic N, Paltsev S, Reilly JM (2008) Temperature increase of 21st century mitigation scenarios. Proc Natl Acad Sci USA 105:15258–15262.

Westerling A, Hidalgo HG, Cayan DR, Swetnam TW (2006) Warming and earlier Spring increase western U.S. forest wildfire activity. Science 313:940–943.

Wigley TML (2005) The climate change commitment. Science 307:1766–1769.

Williams JW, Jackson ST, Kutzbach JE (2007) Projected distributions of novel and disappearing climates by 2100 AD. Proc Natl Acad Sci USA 104:5738–5742.

Williams DW, Liebhold AM (2002) Climate change and the outbreak ranges of two North American bark beetles. Agric For Entomol 4:87–99.

Williams SE, Bolitho EE, Fox S (2003) Climate change in Australian tropical rainforests: An impending environmental catastrophe. Proc R Soc Lond B Biol Sci 270: 1887–1892.

Wilson PR, Ainley DG, Nur N, Jacobs SS, Barton KJ, Ballard G, Comiso JC (2001) Adélie Penguin population change in the Pacific sector of Antarctica: Relation to sea-ice extent and the Antarctic circumpolar current. Mar Ecol Prog Ser 213: 301–309.

Wilson RJ, Gutierrez D, Gutierrez J, Martinez D, Agudo R, Monserrat VJ (2005) Changes to the elevational limits and extent of species ranges associated with climate change. Ecol Lett 8:1138–1146.

Wullschleger SD, Norby RJ (2001) Sap velocity and canopy transpiration for a 12-year-old sweetgum stand exposed to free-air CO_2 enrichment. New Phytol 150: 489–498.

Zeng N, Neelin JD, Lau K-M, Tucker CJ (1999) Enhancement of interdecadal climate variability in the Sahel by vegetation interaction. Science 286:1537–1540.

Fire in the Earth System

S.P. Harrison, J.R. Marlon, and P.J. Bartlein

Abstract Fire is an important component of the Earth System that is tightly coupled with climate, vegetation, biogeochemical cycles, and human activities. Observations of how fire regimes change on seasonal to millennial timescales are providing an improved understanding of the hierarchy of controls on fire regimes. Climate is the principal control on fire regimes, although human activities have had an increasing influence on the distribution and incidence of fire in recent centuries. Understanding of the controls and variability of fire also underpins the development of models, both conceptual and numerical, that allow us to predict how future climate and land-use changes might influence fire regimes. Although fires in fire-adapted ecosystems can be important for biodiversity and ecosystem function, positive effects are being increasingly outweighed by losses of ecosystem services. As humans encroach further into the natural habitat of fire, social and economic costs are also escalating. The prospect of near-term rapid and large climate changes, and the escalating costs of large wildfires, necessitates a radical re-thinking and the development of approaches to fire management that promote the more harmonious co-existence of fire and people.

Keywords Wildfire · Fire regimes · Fire patterns

S.P. Harrison (✉)
Department of Biological Sciences, Macquarie University, North Ryde, NSW 2109 Australia
e-mail: sharriso@science.mq.edu

Introduction

Fire is a natural, recurring episodic event in almost all types of vegetation, although most prominent in savannas, Mediterranean woodlands, and boreal forests. Climate has a strong influence on all aspects of the fire regime, from the seasonal timing of lightning ignitions, through temperature and humidity control of fuel drying, to wind-driven fire spread. Climate also influences the nature and availability of fuel, through its influence on the productivity and type of vegetation. As the major form of vegetation disturbance, wildfires are important in regulating ecosystem dynamics, diversity and carbon cycling. Trace gas and particle emissions associated with wildfires have a major impact on atmospheric composition and chemistry and through this on climate itself. The increase in large fires seen in recent years in many parts of the world has caused major concern – not least because, while these large wildfires have probably become more frequent in response to anthropogenic climate change, the devastation caused has been exacerbated by land-use and socioeconomic changes. Indeed, the natural fire regime is being increasingly influenced by human activities. Humans set and suppress fires and use them to manage agricultural and natural ecosystems. Humans also have indirect effects on natural fire regimes – the modification and fragmentation of the natural vegetation cover through agricultural expansion and urbanization during the twentieth century, for example, has tended to reduce the incidence of wildfires. However, when wildfires do occur in proximity to settlements, the consequences from a human perspective are much greater.

J. Dodson (ed.), *Changing Climates, Earth Systems and Society*, International Year of Planet Earth,
DOI 10.1007/978-90-481-8716-4_3, © Springer Science+Business Media B.V. 2010

Wildfires are fires that are uncontrolled and unplanned, regardless of whether they are ignited by lightning or humans. Wildfires are thus differentiated from controlled burns and agricultural fires, although agricultural fires that escape may become wildfires. The complexity of fire behaviour is often simplified by characterizing wildfires in terms of regional *fire regimes* (see, e.g., Lavorel et al., 2007). The term fire regime is loosely used to describe multiple characteristics of the regional fire record, including whether the fires are a result of human or natural ignitions, the timing or seasonality of fire, the type of fire (surface or crown), the typical size of the fire (which is related to fire type), the intensity of the fire and hence its severity in terms of the amount of biomass burnt, and the characteristic frequency of fires, or return time (see, e.g., Gill, 1977; Bond and Keeley, 2005). Changes in the external controls on fire may provoke simultaneous changes in several of these characteristics, but they can also change independently. Furthermore, given that climate varies continuously on decadal to millennial timescales, characterizing regional regimes solely in terms of frequency or return time is not helpful. Thus, while it is useful to continue to use the term fire regime to describe various aspects of fire behaviour, it is important to realize the dynamic and ever-changing nature of fire regimes.

From an Earth System perspective, fire is an important global process that is tightly coupled with climate, vegetation, biogeochemical cycles, and human activities. Improved understanding of the role of fire requires an approach that recognizes this interconnectedness, and as Bowman et al. (2009) have recently argued, there is an urgent need to develop a coordinated and holistic approach to fire science in order to manage fire better. In this chapter, we draw on empirical observations and modeling work to characterize pyrogeography, emphasizing the links between climate and fire (e.g., via examination of the variation of fire in time and space), fire and climate (e.g., via effects on radiative forcing), fire and vegetation (e.g., via feedbacks to the climate system that in turn drive further vegetation changes), and fire and human activities (e.g., via explicit as well as hidden economic costs). This holistic approach provides a long-term and broad-scale context for current fire regime changes and draws attention to the role of fire as a catalyst as well as a consequence of global environmental change.

Observations of Wildfire Regimes

There are many different sources of information on wildfire. However, these sources differ widely in their purpose, type, scale, quality, temporal and spatial resolution, and processing methods, making comparisons and integration of information about fire across a range of temporal and spatial scales challenging. Nevertheless, there are global-scale data sets that document how the spatial patterns of fire change on a variety of different timescales ranging from seasonal, through interannual, decadal or centennial, and to the multi-millennial or longer-term variability shown in geologic records (Fig. 1).

Contemporary Fire Patterns

Satellite remote sensing systems are now the primary means for studying the contemporary and recent temporal (Fig. 1a) and spatial (Fig. 2) variability of fires because only remote sensing can provide detailed and consistent information from regional (e.g., Sukhinin et al., 2004; JRC-EU, 2005; van der Werf et al., 2008a) to global scales (Carmona-Moreno et al., 2005; Giglio et al., 2006; Mota et al., 2006; Randerson et al., 2007; Frolking et al., 2009). Such data are especially critical for documenting fire in remote areas, such as in the boreal forest, where ground-based data are exiguous. Remote sensing systems identify the areas burnt by fires and also can detect active fires (i.e., the energy radiated by fire: Ellicott et al., 2009). Emissions of trace gases and particulates are not measured directly but are commonly estimated as the product of burnt area, fuel load, and combustion completeness for a particular time interval and spatial domain (van der Werf et al., 2006). Burnt areas from past fires are determined by analyzing the unique signatures left by charred vegetation and patterns of regrowth after fires (Giglio et al., 2006).

Most of the fire-related global products available document variations in burnt area (e.g., Carmona-Moreno et al., 2005; van der Werf et al., 2006; Riaño et al., 2007; Roy et al., 2008; Tansey et al., 2008) or in the presence and timing of active fires (Giglio et al., 2006). There are still substantial differences between the available products, in part a result of the different sensors and approaches used and in part because

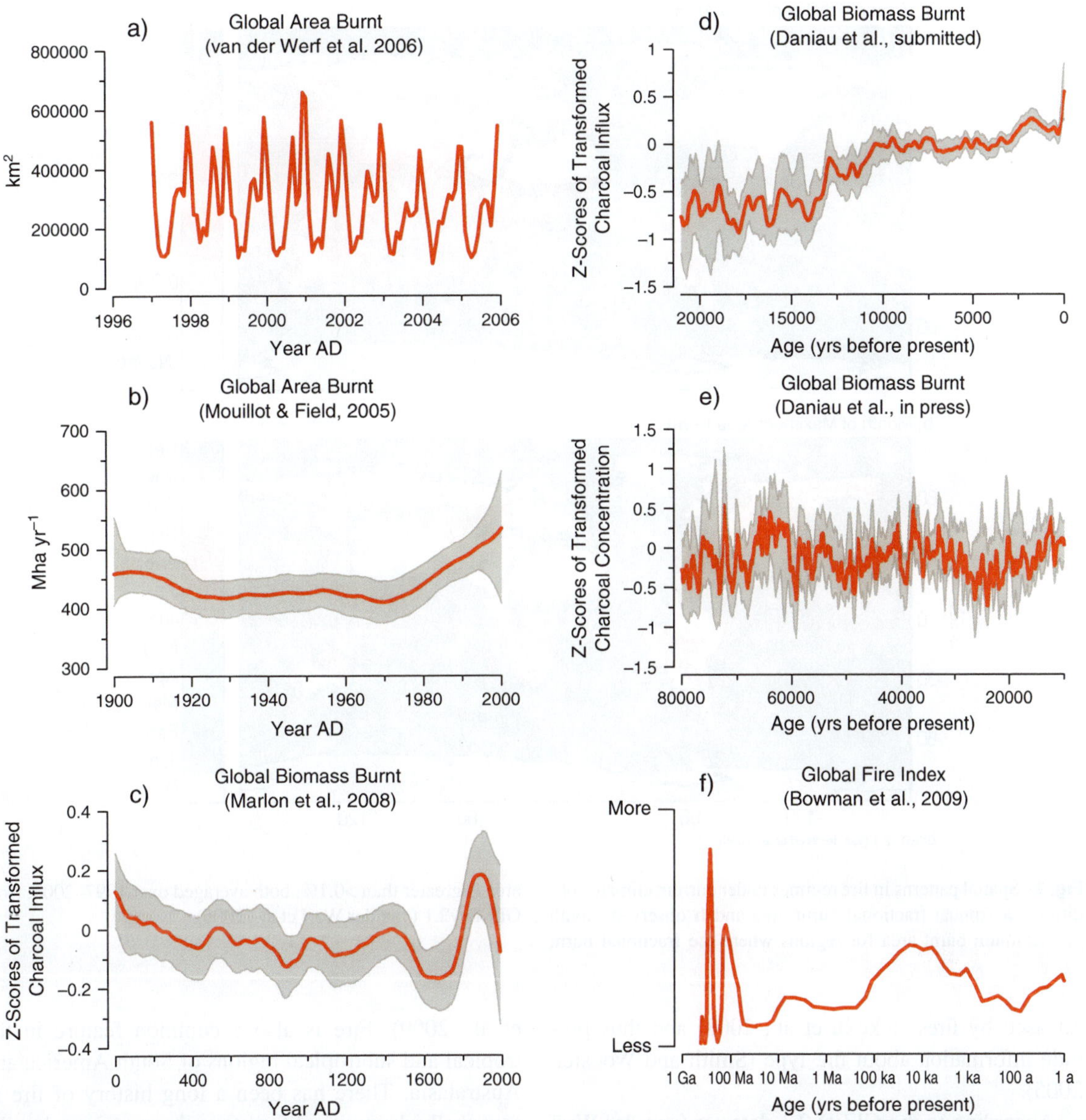

Fig. 1 Variability in global fire on different timescales. **a** Global area burnt from 1997 to 2006 derived from satellite-based remote sensing (GFED v2.1, van der Werf et al., 2006). **b** Global area burnt over the twentieth century, estimated by combining data from tree-ring, historical and remotely sensed sources (Mouillot and Field, 2005). **c** Estimates of global biomass burnt over the past two millennia, based on compositing ca. 400 sedimentary charcoal records worldwide (Marlon et al., 2008). **d** Estimates of global biomass burnt since the Last Glacial Maximum, ca. 21,000 years ago, based on compositing ca. 700 sedimentary charcoal records worldwide (Daniau et al., submitted). **e** Estimates of global biomass burnt during the last glacial (ca. 80–11,000 years ago), based on compositing 30 sedimentary charcoal records worldwide (Daniau et al., in press). **f** A qualitative index of global fire over the past 1 billion years based on discontinuous sedimentary charcoal records (Berner, 1999)

each product covers a different interval of time. van der Werf et al. (2006) have reconstructed global variations in fire from 1997 to 2004 (Fig. 1a) by combining fire data from several remote-sensing instruments including the Moderate Resolution Imaging Spectroradiometer (MODIS), Along Track Scanning Radiometer (ATSR), the Visible Infrared Scanner (VIRS) and the Advanced Very High Resolution Radiometer (AVHRR). Newer satellite products provide additional information about the radiative energy

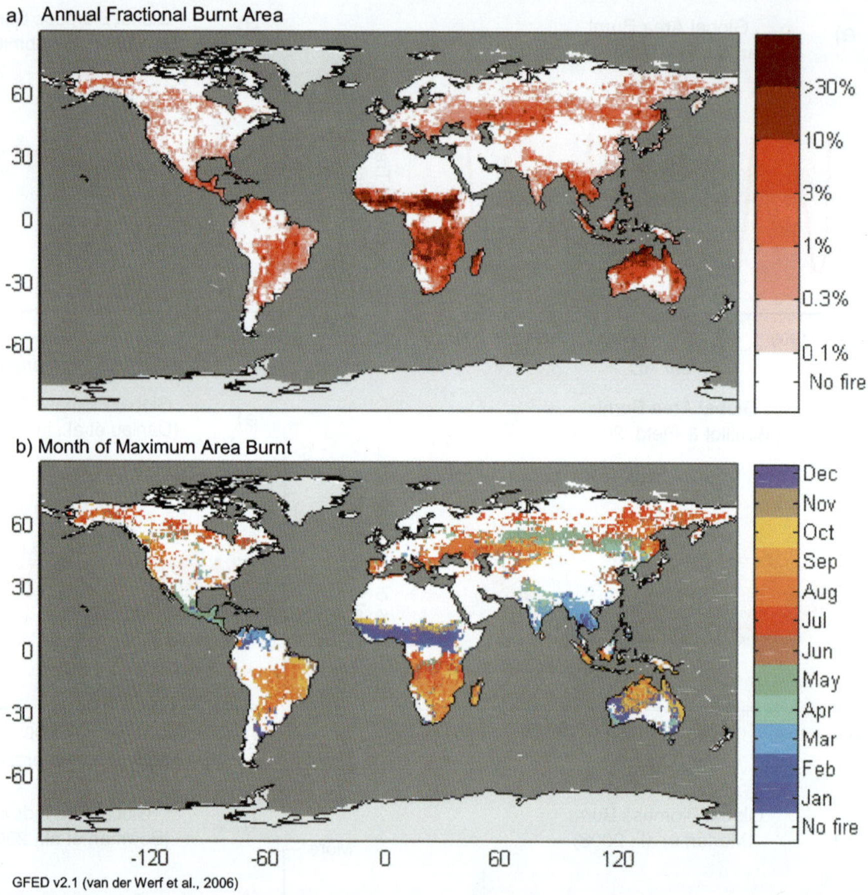

Fig. 2 Spatial patterns in fire regimes under current climate conditions. **a** Annual fractional burnt area and **b** observed month of maximum burnt area for regions where the fractional burnt area is greater than >0.1%, both averaged over 1997–2006, from GFED v2.1 (van der Werf et al., 2006)

released by fires (Ickoku et al., 2008) and thus provide information about fire type (Smith and Wooster, 2005).

According to the GFED 2.1 data set (van der Werf et al., 2006), 3–4 Mm^2 of land burns every year, and fires are relatively frequent on about a third of the land surface (Chuvieco et al., 2008). The largest fractional areas burnt occur in tropical and subtropical regions of South America, Africa, Southeast Asia, Indonesia, and Australia (Fig. 2a). Africa accounts for 40% of the global area burnt. Most African fires are concentrated in the Sahelian and Sudanian regions of sub-Saharan Africa and in southern Africa (Silva et al., 2003). There is much heterogeneity within these broad regions, with semi-arid regions burning less due to fuel limitations, and wetter areas burning more as a result of greater fuel availability (van der Werf et al., 2008b; Archibald et al., 2009). Fire is also a common feature in the tropical and subtropical regions of South America and Australasia. There has been a long history of fire in seasonally dry tropical forests in these regions, but the present level of burning in Amazonia and Indonesia was attained only in the last few decades and reflects human use of fire to clear forests for crops and pastures (so-called deforestation fire). Fires in temperate and boreal forests are less frequent than those in the tropics but, when they occur, consume large quantities of biomass (Kasischke and Bruhwiler, 2002; Campbell et al., 2007).

The most striking pattern in the timing of the fire season is the transition between a tropical region with winter (dry) season dominance and an extratropical region with summer (warm) season dominance (Fig. 2b). This pattern is seen in both hemispheres.

Superimposed on this pattern is the tendency for drier, fuel-limited regions in the subtropics to experience peak burning in the autumn, as fuel dries after the peak growing season. Thus, maximum burning tends to occur from December to February in the Sahel, India, and Central America, during October and November on the Saharan margin, and during July through September in southern Europe and North America. In the southern hemisphere, peak burning in the tropics and subtropics of southern Africa, Australia and South America occurs during July through September, while December through February marks the peak fire season in the southern mid-latitude regions.

The global record of fire over the last decade (Fig. 1a) is dominated by the northern hemisphere seasonal cycle. Nevertheless, the GFED record shows interannual differences of the order of 10% (ca. 0.3 Mm^2) around the long-term mean value of area burnt (van der Werf et al., 2006). There is no discernible trend in the total area burnt over this relatively short interval (1997–2006), but an AVHRR-based reconstruction of relative changes in burnt area covering the period 1982–2000 (Riaño et al., 2007) shows a trend toward increasing fire over this longer interval. This latter data set also shows substantial interannual variability, including a marked transient reduction in global burnt area after the El Chichon (1982) and Pinatubo (1991) volcanic eruptions.

Historical Records of Fire

Although many countries now collect statistics on fires, these records rarely extend back more than a few decades (http://www.fire.uni-freiburg.de/) although there are some records, e.g., national parks or managed forests that extend back to the first decades of the twentieth century (see, e.g., Fulé et al., 2003). Information on earlier fires is largely anecdotal, inferred from historical documents, photographs, ethnographic records or other archives (see, e.g., Habeck, 1994). Much of the available information is biased toward fires that resulted in destruction of infrastructure, even if these fires started naturally. More consistent longer-term fire records can be obtained from fire scars in tree-ring records (Swetnam, 1993; Kipfmueller and Baker, 2000; Kitzberger et al., 2001) and forest stand-age data in areas with high-severity, stand-replacing fires (Kipfmueller and Baker, 2000; Hallett et al., 2003). Both sources can provide records covering several centuries or even, in rare instances, millennia but only provide information at local to subcontinental scales.

Mouillot and Field (2005) combined data from historical, dendrochronological, and satellite-derived fire records with information about vegetation, population, and land-use changes to reconstruct trends in annual area burnt during the twentieth century (Fig. 1b). Interpolation and smoothing techniques were used to fill gaps and remove artifacts introduced by the different data sources. As a result the estimates are highly uncertain but, despite these problems, a smoothed summary trend of the data shows features consistent with independent evidence of variations in twentieth century burning, including a decline in fire activity at the beginning of the century and an increase in area burnt during recent decades. The early twentieth century decline in burning is consistent with a long-term global decrease in biomass burnt that has been independently inferred from charcoal data (Fig. 1c) and dendrochronological studies (Veblen et al., 1999; Grissino-Mayer et al., 2004; Drobyshev et al., 2008), and the late twentieth century increase in area burnt is consistent with local- to regional-scale observations (Cochrane, 2003; Girardin, 2007; Fisher et al., 2009) and with satellite and model-based reconstructions (Pechony and Shindell, 2009).

Paleofires

Burning vegetation produces charcoal, black carbon, and carbon spherules that accumulate naturally in lake sediments, peat, soils, and marine sediments (Tolonen, 1986; Ohlson and Tryterud, 2000; Conedera et al., 2009; Power et al., 2010). The rate of sediment accumulation, which depends on the type of site, its geomorphic setting, and the control on erosion rates exerted by topographic, land cover, and climatic factors, determines the temporal resolution of these records. In some cases, annually laminated sediments have been used to explore year-to-year or multi-year variation in biomass burning (Clark, 1990; Atahan et al., 2004; Power et al., 2006) but in general paleo-records provide continuous reconstructions of changing fire regimes on decadal or centennial timescales (e.g., Walsh et al., 2008). Many hundreds of individual charcoal records have been generated, from all regions of the world. Although these records cannot provide quantitative estimates of biomass burnt, they can be interpreted in terms of relative changes in biomass burning at subcontinental to global scales (see e.g., Haberle and Ledru, 2001; Carcaillet et al., 2002; Marlon et al., 2008; Power et al., 2008; Daniau et al., in press; Daniau et al., submitted).

The global sedimentary charcoal records for the past two millennia (Fig. 1c) show a long-term decline in biomass burning between 1 AD and ~1750 AD, with centennial-scale variability associated with known climate fluctuations including the Medieval Warm Period and the Little Ice Age (Marlon et al., 2008). The largest changes in global biomass burning occur during the past ~250 years, when warming after the Little Ice Age coincided with major changes in population and land use. The records show an initial increase in biomass burning, but after ca. 1850 AD there is a pronounced decrease. Marlon et al. (2008) have argued that because this downturn pre-dates the introduction of active fire-fighting, it is most plausibly explained as an inadvertent consequence of human activities, resulting from landscape fragmentation and reduction of fuel in intensively managed agricultural landscapes.

Charcoal records also provide information for long-term trends during the last climatic cycle (Power et al., 2008; Daniau et al., in press; Daniau et al., submitted). Biomass burning was low during the coldest intervals during the past climatic cycle, Marine Isotope Stage 4 (73.5–59.4 ka) and Marine Isotope Stage 2 (27.8–14.7 ka), increased during the warmer, interstadial interval of Marine Isotope Stage 3 (59.4–27.8 ka), and has generally increased toward modern levels since the Last Glacial Maximum ca. 21,000 years ago (Fig. 1d). There are sufficient sites covering the last deglaciation to show that fire regimes in the northern and southern hemispheres during the deglaciation (Daniau et al., in press), a relationship first identified in temperature records from the polar ice caps (EPICA Community Members, 2006) and consistent with the association of these temperatures with changes in the strength of the Meridional Overturning Circulation. Both the past 21,000 years and earlier periods of the last glacial (Fig. 1e) are characterized by millennial-scale variability in fire regimes, tracking the Dansgaard-Oeschger (D-O) climate cycles seen in the Greenland ice core record and the Heinrich Stadials recorded in the North Atlantic (Marlon et al. 2009; Daniau et al., in press). There is an increase in fire, seen both at a global scale and at a continental scale in North America, during the rapid D-O warming events (Fig. 1e). Fire initially decreases during D-O or Heinrich cooling intervals, but then recovers to pre-cooling levels. Analyses of multiple climate cycles suggest that the response of fire to the recorded climate changes is extremely fast (Fig. 3).

There are fewer highly resolved records of biomass burning over time scales longer than the last climatic cycle. The longest continuous record of biomass burning is a million-year-long black carbon record from marine sediments in the eastern equatorial Atlantic (Bird and Cali, 1998; Bird and Cali, 2002). This record represents fires in sub-Saharan West Africa and shows large episodic increases in fire activity. Periods of high burning occur during some interglacials; no fire is recorded during glacial periods.

Older geological records of fire (Fig. 1f, Bowman et al., 2009) show that burning occurred on Earth at least as long ago as the Silurian (ca. 420 million years ago) and Carboniferous (ca. 400 million years ago) (Glasspool et al., 2004). The Permian (ca. 250–300 million years ago) was also characterized by high levels of fire. On such long time scales, variations in atmospheric oxygen are thought to be the primary control on whether and how much fire was possible (Scott, 2000; Scott and Glasspool, 2006). Experimental studies suggest fire cannot be sustained when the oxygen content of the atmosphere is below about 15% (Jones and Chaloner, 1991; Belcher and McElwain, 2008). At oxygen levels above ca. 35%, wildfires would consume most terrestrial vegetation (Scott, 2000; Lenton, 2001). Modeling experiments suggest that the abundant charcoal found in Carboniferous and Permian sediments may reflect widespread wildfires that occurred once atmospheric oxygen increased above ~21% (Berner, 1999). The Paleocene-Eocene thermal maximum (ca. 65 Ma) was also characterized by major increases in biomass burning (Collinson et al., 2007; Marynowski and Simoneit, 2009). Peaks in charcoal in marine sediments ca. 7–8 million years ago (Fig. 1f) are contemporaneous with the expansion of open grasslands and savanna vegetation globally (Keeley and Rundel, 2005).

There is a high degree of variability in fire-regime patterns at different temporal scales, and this is more

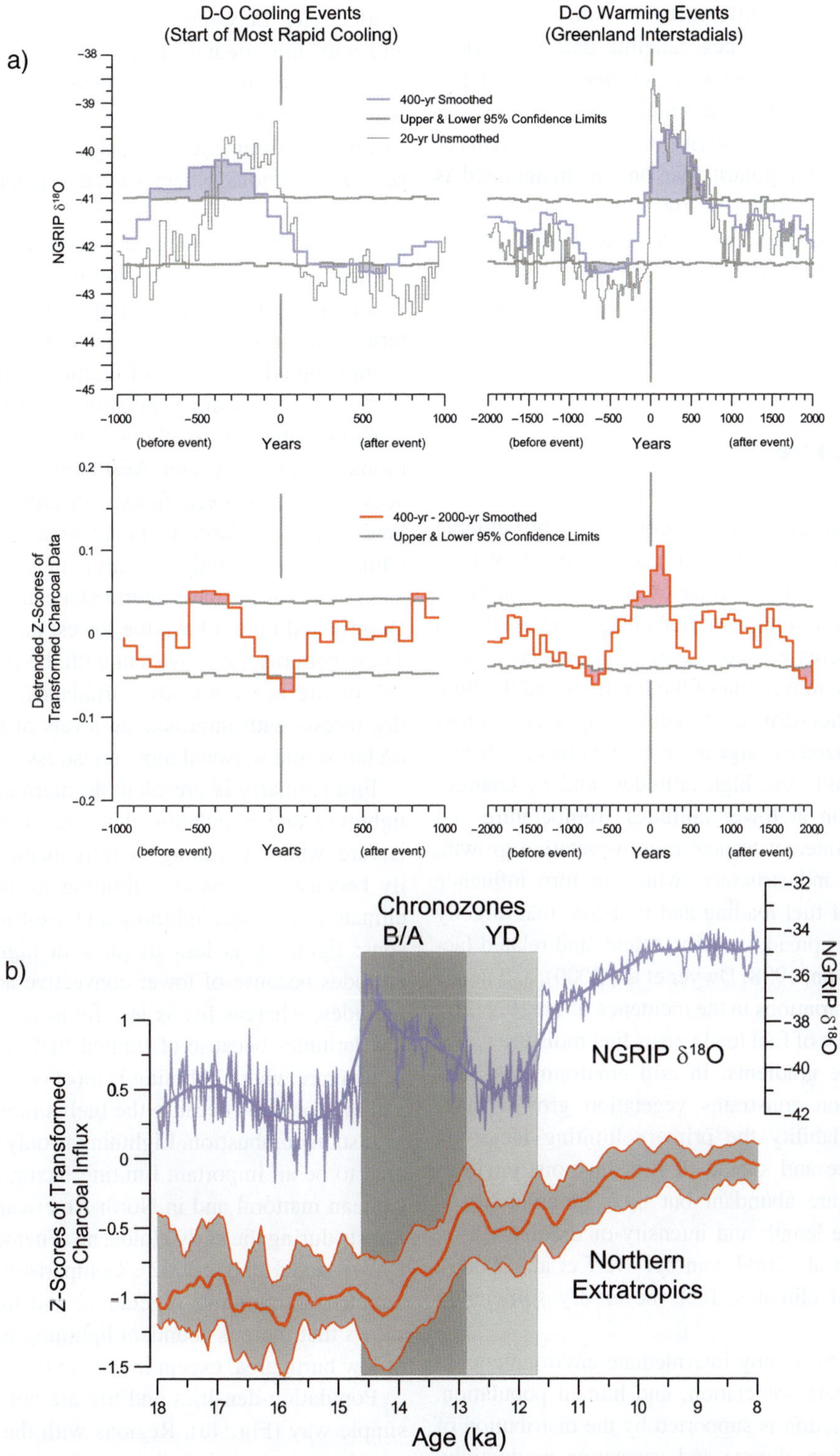

Fig. 3 The response of fire regimes to rapid climate changes. **a** Superposed epoch analysis of ice-core and biomass-burning records over the interval 80–10 ka, showing the response of fire to rapid cooling and warming intervals during the Dansgaard-Oeschger cycles (Daniau et al., in press). **b** Comparison of the temporal evolution of composite charcoal records from the northern extratropics through the Younger Dryas chronozone (Daniau et al., submitted), which, as shown by the ice core record, was characterized by rapid cooling and subsequent rapid warming on the order of 10°C in Greenland in about a decade

than a reflection of differences in data sources and methodological approaches. Satellite data show clear seasonal cycles in fire, but at longer timescales changes in fire are more episodic than periodic. Although regional fire regimes are often characterized in terms of return times, such regularity can only be maintained as long as the major drivers of fire activity, including climate conditions, vegetation cover and ignition sources remain unchanged. Both historical and paleo-records suggest that stability is the exception rather than the norm.

Controls of Fire

Globally (Fig. 2a), fire is controlled by climate (Carmona-Moreno et al., 2005), which on short time periods produces the familiar cycle of seasonal burning (Fig. 1a). Seasonal variations in climate are driven by gradual increases and decreases in insolation that result from the movement of the Earth around the Sun. These insolation-driven climate changes are generally characterized by large temperature changes during the year at mid- and high latitudes, and by changes in precipitation at lower latitudes. Temperature and precipitation interact to determine vegetation growth, composition, and structure, which in turn influence the patterns of fuel loading and moisture that directly determine fire spread, severity, extent, and related factors (Stephenson, 1998; Dwyer et al., 2000), and hence the seasonal variations in the incidence of fire (Fig. 2b). The importance of fuel load versus fuel moisture varies across climate gradients. In arid environments, lack of precipitation constrains vegetation growth making fuel availability the primary limiting factor on the occurrence and spread of fire. In moist environments, fuels are abundant but their flammability is limited by the length and intensity of the dry season (Westerling et al., 2003; van der Werf et al., 2008b). In the wettest climates, fuels never dry sufficiently to burn.

Fire tends to occupy intermediate environments in terms of climate, vegetation, and human population. This generalization is supported by the distribution of area burnt along climate and vegetation-productivity gradients, such as available moisture versus net primary productivity (Fig. 4a). The largest annual area burnt tends to be in regions with net primary productivity between about 400 and 1,000 gC/m^2 and with intermediate levels of moisture availability (between 0.3 and 0.8 on an index of actual to potential evaporation, AE/PE). Fire is largely absent from extreme environments, such as low tundra, deserts, and semi-arid regions where sparse vegetation results in discontinuous fuel loads that prevent the spread of fire. Fire is also rare in temperate oceanic margins and in tropical rainforests, where the fuel is too wet to burn. It is remarkable that the extremely heterogeneous pattern of fire in geographical space (Fig. 2a) reduces to a single broad maximum of fractional area burnt when plotted in climate and vegetation space (Fig. 4a).

Comparing the distribution of fire against those of biomes along NPP and AE/PE gradients shows that seasonally dry tropical forests, savanna and dry woodlands have the largest annual area burnt (Fig. 4b), while grassland and dry shrublands, and temperate and boreal forests have intermediate annual area burnt. Deserts and tundra have the lowest annual area burnt. These patterns are consistent with the preferred "habitat" of fire in woodlands, shrublands, and seasonally dry forests with intermediate levels of biomass accumulation and seasonal moisture stress.

Fire similarly is prevalent at intermediate levels of lightning and population densities (Fig. 4c, d). Fire is rare where lightning is infrequent, not necessarily because ignition is a limiting factor but because climate limits both lightning and fire for different reasons: lightning is less frequent in high than in low latitudes because of lower convective activity in high latitudes, whereas fire is less frequent at high than at low latitudes because of limited fuel availability. Fire is also rare where lightning is most common – in tropical rainforests – because the fuel is rarely dry enough to sustain combustion. Lightning is only rarely considered to be an important limiting factor, such as in the Chilean mattoral and in North American coastal rainforests during the early Holocene (Fuentes et al., 1994; Brown and Hebda, 2002). Comparison of the distributions of lightning, vegetation, and fire (Fig. 4a, c) shows that there is abundant lightning in most regions of low burnt area, except in the tundra.

Population densities and fire are not related in any simple way (Fig. 4d). Regions with the highest burnt area have intermediate levels of population. Overall, the distribution of population resembles that of vegetation more fire, with low population densities in tundra, boreal forest, and tropical forest biomes. In

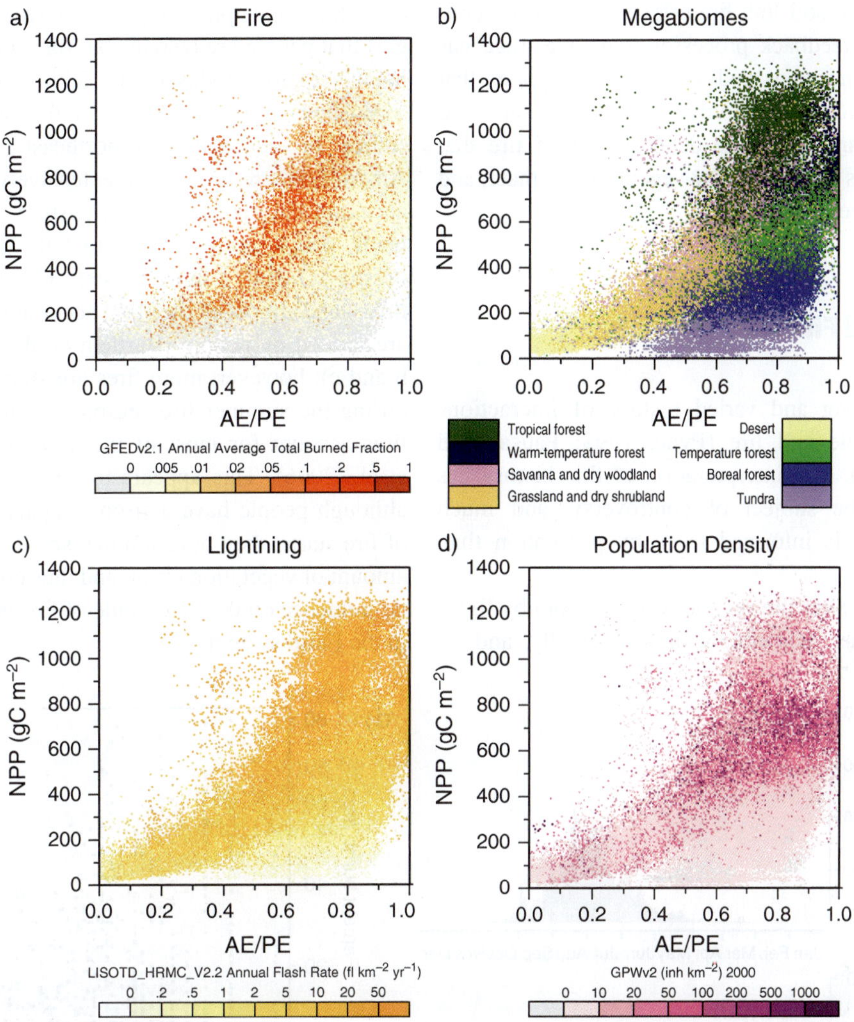

Fig. 4 Theoretical relationships between climate, lightning, vegetation, population density, and fire at a global scale based on satellite remotely sensed burnt-fraction data (GFEDv2.1) (van der Werf et al., 2006). Model-based estimates of net primary productivity (NPP) are from Cramer et al. (1999); values of actual to equilibrium evapotranspiration (AE/PE) were calculated using the CRU CL 2.0 data set (New et al., 2002). The distribution of biomes is plotted in climate space for comparison. Lightning data are from the LIS/OTD HRMC V2.2 dataset (http://gcmd.nasa.gov/records/GCMD_lohrmc.html) and population density from the HYDE 3.0 data set (Klein Goldewijk and van Drecht, 2006)

many areas, fire is uncommon at both very low and very high population densities (Archibald et al., 2009).

A conceptual model of the major controls on fire must include climate, vegetation, and human activity (e.g., Lafon and Grissino-Mayer, 2007; Balshi et al., 2007; Krawchuk et al., 2008). Climate controls fire activity through changes in fire weather, which include factors such as temperature, precipitation, wind, and the length of dry spells between storms, all of which show complex geographic variation. Fire weather also determines the occurrence of natural ignitions through lightning. Vegetation regulates fire because its composition and structure determine fuel characteristics (amount and flammability). Net primary productivity (NPP), for example, is a strong predictor of fire occurrence (Krawchuk and Moritz, 2009). However, at broad spatial scales and on multi-annual timescales, vegetation type and productivity are themselves determined by climate. Topography is a secondary control on fire that strongly influences fire behavior at finer spatial and temporal scales (Heinselman, 1973; Gavin et al., 2006; Parisien and Moritz, 2009). The atmosphere and soil link fire effects back to fire controls through feedback processes (e.g., via trace gas and aerosol emissions and via nutrient supplies that affect vegetation productivity). Fuels, ignitions, and fire weather are proximate determinants of fire that are themselves determined by vegetation, climate, and human activities.

People and Fire

There is a long and varied history of interactions between people and fire (Pyne, 1995; Pausas and Keeley, 2009). The interpretation of this history has often been the subject of controversy, and much of the debate is informed more by perception than observation.

People can alter fire regimes directly both by lighting fires, either deliberately or accidentally, and by excluding or suppressing fires. Available data suggest that people are responsible for igniting the largest *number* of fires today (FAO, 2007), except in boreal forests where lightning is still the primary ignition source. In southeast Asia, northeast Asia, and South America, where fire is used extensively for land clearance and the maintenance of pasture and agricultural lands, people are estimated to light over 80% of all fires (FAO, 2007). In the western US, about half of all fires – and almost all fires that occur during the winter – are ignited by people (Bartlein et al., 2008) (Fig. 5a, b and c); however, more fires are started by lightning during the summer fire season, and lightning-ignited fires account for most of the *area* burnt (Kasischke et al., 2006). This appears to be a common pattern: although people have a strong impact on the number of fire starts, they have a much smaller impact on the amount of vegetation burnt, and thus comparatively little influence on the magnitude of fire-induced changes in the Earth system.

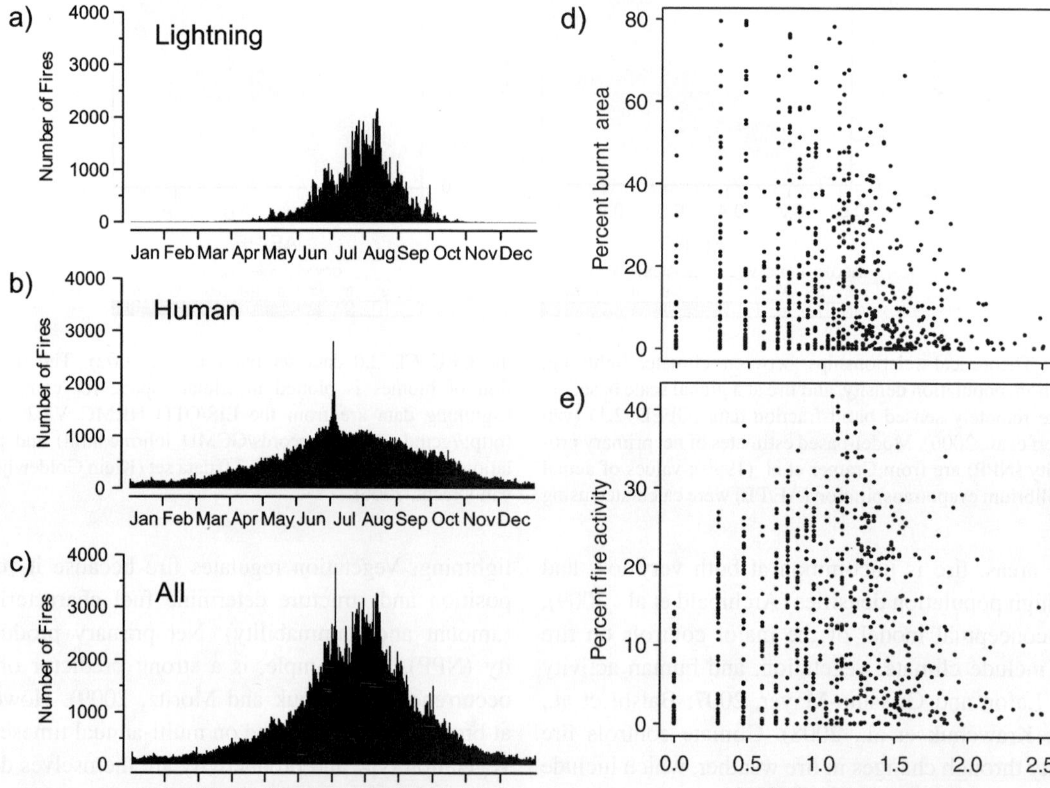

Fig. 5 The influence of people on key aspects of fire regimes. Number and timing of fires caused by **a** lightning, and **b** humans compared to **c** total fires in the western USA (redrawn from Bartlein et al., 2008). Percent **d** burnt area and **e** fire activity (number of fires) as a function of population density in South Africa (redrawn from Archibald et al., 2009)

The high economic and social costs of fires have led to active policies of fire suppression in many countries (FAO, 2007). The impacts of fire suppression, however, are not always as intended. The exclusion or suppression of fire has been claimed to contribute to dangerous fuel build-up (Donovan and Brown, 2007; Mitchell et al., 2009), changes in forest stand structure (including a paradoxical reduction in carbon storage: Fellows and Goulden, 2008), undesirable changes in species composition, and pest and disease outbreaks (Castello et al., 1995). Fires that do occur in areas where fire has been suppressed by fire fighting can burn larger and hotter, and thus become increasingly destructive rather than restorative (Moritz and Stephens, 2008).

People also have an indirect but important impact on fire regimes, through altering land use (e.g., Heyerdahl et al., 2001; Marlon et al., 2008; Carcaillet et al., 2009; Krawchuk and Moritz, 2009). Similar changes in land use do not necessarily have the same effect in different ecosystems. Evidence from South America indicates reductions in fire occurrence associated with increased grazing in prairies and savannas, but increases in fire are associated with increased grazing in rain forests (Bella et al., 2006). In fire-sensitive regions, such as in seasonally dry tropical forests, the effects of human "deforestation" fires are well known. Large fires are novel to these ecosystems (Cochrane et al., 1999; Bush et al., 2007a; Bush et al., 2007b; Field et al., 2009), and there is concern about the effects of such fires on ecosystem services and biodiversity (Cochrane and Barber, 2009).

The pervasive use of fire by people have led some modellers to estimate past fires and emissions by scaling changes in fire to changes in population growth (e.g., Dentener et al., 2006). Fire data from South Africa, however, indicate that population density is not linearly related to burnt area or number of fires (Archibald et al., 2009). Instead, burnt area was found to be inversely correlated with changes in population density; burnt area declines very slightly at low levels and then more rapidly declines as densities increase further (Fig. 5d). Numbers of fires increase with population density at low levels, but then decreases as population density continues to rise (Fig. 5e). Temporal trends in fire history at the global scale show a similar correlation between increasing population levels and increasing biomass burnt up to a certain level (i.e., when global population reached about 1.6 billion at 1900 AD) followed by a rapid reduction in fire activity as global population grew from 1.6 to over 6 billion in 2000 AD, significantly reducing the "available habitat" for fire through land-use changes (Marlon et al., 2008).

Fire and Vegetation Dynamics

Wildfire is an agent of vegetation disturbance, initiating succession and promoting short-term changes in biogeochemical cycling (Liu et al., 2005). Fire also shapes population dynamics and interactions (Mutch, 1970; Biganzoli et al., 2009). In the short-term, smoke haze from large-scale fires reduce light availability sufficiently to noticeably decrease photosynthesis (Schafer et al., 2002) and thus potentially jeopardize the yield of annual species and crops. Paleoecological data shows that changes in fire regime associated with rapid climate changes have lead to abrupt reorganizations of vegetation communities (Shuman et al., 2002; Williams et al., 2008). Abrupt changes in fire regimes associated with the relatively recent human colonization of regions such as New Zealand appear to have promoted large-scale and permanent changes in vegetation (McGlone and Wilmshurst, 1999; McWethy et al., 2009). Similarly, the use of fire for deforestation in tropical forests or the introduction of invasive grasses in the more recent past have led to the conversion of forests and woodlands to more open, flammable communities (D'Antonio and Vitousek, 1992; Cochrane et al., 1999). In tropical rainforests, deforestation fires initiate a positive feedback cycle, whereby fire creates canopy openings, leading to reduced fuel moisture and low relative humidity and hence increasing the risk of fire (Cochrane et al., 1999; Nepstad et al., 2004). In contrast, fires in arid regions can destroy vegetation cover and fuel continuity and prevent a recurrence of fire until the system recovers. Increases in human-caused ignitions can increase fire frequency beyond the range of natural variability thereby altering species composition and decreasing soil fertility (Tilman and Lehman, 2001; Syphard et al., 2009). Deliberate protection from fire promotes the growth of closed forests. In some regions, these changes in vegetation lead to changes in the type, intensity and frequency of fires. For example, regrowth of forests after fire exclusion leads to a switch from frequent, low-intensity surface to less frequent but high-intensity crown fires (Allen et al., 2002). Local circumstances that exclude fire may favour the persistence of fire-sensitive vegetation types, such as tropical rainforest, such that the occurrence or absence of fire may be responsible for alternate stable ecosystems within the same climate zone (Grimm, 1984; Wilson and King, 1995; Bowman, 2000).

Many species possess survival or reproductive traits that allow them to persist in fire-prone areas (Bond and van Wilgen, 1996; Schwilk and Ackerly, 2001; Pausas et al., 2004; Paula et al., 2009). The development of thick bark, for example, allows trees to survive low-intensity fires, while epicormic resprouting after damage favours rapid regrowth after fire. The association of plant species with distinct reproductive and survival characteristics under different fire regimes suggests that fire is a powerful biological filter on plant distribution, and has led to speculation that fire has had a pronounced evolutionary effect on the development of biotas. It has been suggested that the expansion of savannas around 7–8 million years ago, for example, was a result of increases in wildfire that led to a drier climate through biophysical and biogeochemical feedbacks to climate, and thus promoted expansion of drought-tolerant vegetation (Beerling and Osborne, 2006). Similar arguments have been made for the aridification of Australian vegetation following aboriginal colonization ca. 45–50 thousand years ago (Miller et al., 1999), although in this case there is evidence for similar changes in fire regime in regions not colonized until much later (Dodson et al., 2005; Stevenson and Hope, 2005) and the observed aridification can be explained by external forcing alone (Hope et al., 2004; Pitman and Hesse, 2007). Traits that promote fire or recovery following fire are not unambiguously the result of natural selection by fire (Schwilk and Kerr, 2002), and both the expansion of savannas and drought-tolerant vegetation at various stages in Earth's history is susceptible to a more direct climatic explanation.

Potential Feedbacks to Climate

Changing climate and human activities lead to change in fire regimes, but wildfires could in principle themselves contribute to changes in climate – via emissions of long-lived greenhouse gases (CO_2, CH_4, N_2O), other trace gases including biogenic volatile organic compounds (BVOCs), and aerosol precursors, and physical land-surface changes resulting from changes in vegetation structure induced by fire (Galanter et al., 2000; Winkler et al., 2008; DeFries et al., 2008). Disentangling the relative importance of different putative feedbacks among vegetation, fire, and climate is challenging (Bonan, 2008) and represents an area of Earth System Science that remains poorly quantified.

Carbon Cycle Feedbacks

Fires consume vast quantities of vegetation annually and in the process release carbon dioxide (CO_2), methane (CH_4), carbon monoxide (CO), and nitrous oxide (N_2O) (Cofer et al., 1998; Rinsland et al., 2006; Bian et al., 2007), which are primary, long-lived greenhouse gases (Fig. 6). Terrestrial biomass represents about a quarter to a third of the terrestrial carbon reservoir (Field and Raupach, 2004; Houghton et al., 2009). Biomass burning for deforestation is estimated to have contributed ca. 19% of the atmospheric CO_2 increase since pre-industrial times (Randerson et al., 1997; Bowman et al., 2009). Annually, fires release about 2.5 Pg of carbon per year (Pg C/year) to the atmosphere (Lavoué et al., 2000; van der Werf et al., 2006; Schultz et al., 2008). This amount is larger during severe drought years (van der Werf et al., 2004; van der Werf et al., 2006). During the 1997–1998 El Niño, burning in Indonesia alone contributed about 40% of all carbon emitted globally that year, which showed the largest annual increase in CO_2 since precise records began in 1957 (Page et al., 2002). The trend toward increasing fires in Indonesia and in the tropical Americas is mainly attributed to deforestation fires (Morton et al., 2008; Field et al., 2009), but climate-induced increases in droughts are expected to exacerbate the problem (van der Werf et al., 2008a; Cochrane and Barber, 2009).

Fire emissions of CO_2 add to the global atmospheric CO_2 budget when more carbon is added than is sequestered by vegetation regrowth (Fig. 6). As a result, fires in savanna and grassland ecosystems, for example, where post-fire vegetation recovery is rapid, have no net effect on the atmospheric CO_2 budget in the absence of a fire-regime shift. When post-fire vegetation recovery is slower as in temperate or boreal forests, CO_2 emissions may have long-lasting effects

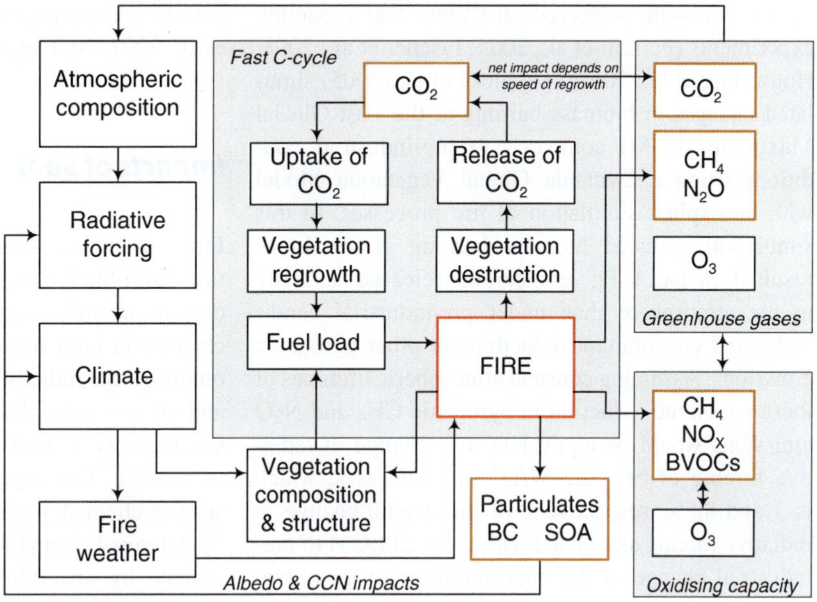

Fig. 6 Schematic showing carbon-cycle, atmospheric chemistry and biogeophysical feedbacks from fire to the climate system

(Amiro et al., 2001; Litvak et al., 2003; Balshi et al., 2009). Fire is becoming an increasing concern in boreal and tundra regions in particular because of the high sensitivity of these ecosystems to global warming. Increased surface heating from higher fire frequencies in tundra ecosystems associated with rapid warming there could contribute to the release of large quantities of carbon from melting permafrost (Zoltai, 1993; Mazhitova, 2000) increased fire activity in boreal forests has the potential to significantly increase carbon emissions there (Kasischke et al., 1995; Kasischke and Bruhwiler, 2002; Flannigan et al., 2005).

Atmospheric Chemistry Feedbacks

The main atmospheric impacts of fire (Fig. 6) are related to emissions of particles, nitrogen oxides (NO_x), methane (CH_4), and other volatile hydrocarbons, either directly or through secondary effects including the formation of ozone (O_3) and aerosols (Andreae and Merlet, 2001). Methane is 25 times more powerful than CO_2 as a greenhouse gas, but has a shorter atmospheric lifetime. Biomass burning is only one source of CH_4; emissions from wetlands and agricultural activities are larger. Bowman et al. (2009) estimate that biomass burning from deforestation fires currently contributes about 4% to the current total radiative forcing of CH_4 (Fig. 7). However, there is substantial uncertainty regarding the contribution of biomass burning to variations in CH_4, owing to the variety of methane sources. Furthermore, the interannual variability of global CH_4 concentrations is not well understood.

Attempts have been made to disentangle the relative importance of contributing factors to past changes in atmospheric composition as shown in ice core

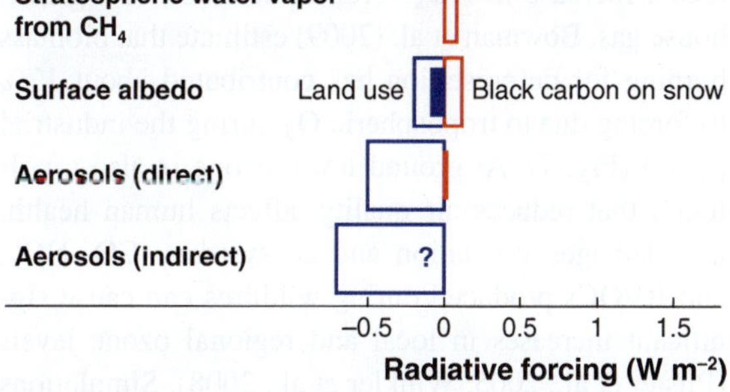

Fig. 7 Estimated contribution of changes in fire regime due to deforestation to radiative forcing over the industrial era, redrawn from Bowman et al. (2009)

records through paleoreconstructions and modeling experiments (Ferretti et al., 2005; Fischer et al., 2008; Houweling et al., 2008). Thonicke et al. (2005) simulated changes in biomass burning at the Last Glacial Maximum (LGM) compared to pre-industrial conditions using a Dynamic Global Vegetation Model with an explicit simulation of fire processes. In this simulation, reduced biomass burning at the LGM resulted in ca. 1 Pg less carbon released annually to the atmosphere than under pre-industrial conditions, and concomitant reductions in other pyrogenic emissions. Assuming constant atmospheric lifetimes of these gases, the reduction in pyrogenic CH_4 and N_2O emissions would be equivalent to a change in radiative forcing of ca. –0.2 W/m^2 in both cases, which is 7 and 8.5%, respectively, of the overall change in radiative forcing associated with the total LGM to pre-industrial change in the concentrations of these two gases. The possible impact of reduced pyrogenic emissions of CO_2 at the LGM on the glacial-interglacial change in atmospheric CO_2 concentration is negligible, compared to the large uptake of CO_2 by the land biosphere during the deglaciation and the even larger release of CO_2 by the ocean. However, changes in pyrogenic emissions could have had further minor impacts on radiative forcing through influencing ozone production and atmospheric oxidizing capacity, with implications, e.g., the lifetime of CH_4 (Thonicke et al., 2005).

Nitrogen oxides (NO_x) other than N_2O and volatile organic carbons (BVOCs) from biomass burning contribute to the formation of ozone (O_3) in the troposphere. Global biomass burning emissions represent ca. 25% of the global NO_x emissions (Jaegle et al., 2005). Hastings et al. (2009) used ice core data to demonstrate that human activities have caused a rapid recent increase in NO_x. Tropospheric O_3 is a greenhouse gas. Bowman et al. (2009) estimate that biomass burning for deforestation has contributed about 17% to forcing due to tropospheric O_3 during the industrial period (Fig. 7). At ground level, ozone is also a pollutant that reduces air quality, affects human health, and damages vegetation and ecosystems. CO, NO_x, and BVOCs produced during wildfires can cause significant increases in local and regional ozone levels (Pfister et al., 2005; Winkler et al., 2008). Simulations have demonstrated substantial sensitivity of radiative forcing to variation in overall pyrogenic emissions and relative changes in warming (black carbon, BC) and cooling (secondary organic aerosol, SOA) aerosol (Ito et al., 2007; Naik et al., 2007).

Impacts of Soot

Fires affect land-surface albedo by altering vegetation cover, darkening land through soot (black carbon) deposition, and increasing snow exposure. Small-scale changes in land-surface albedo have very little effect on the global radiative budget (Randerson et al., 2006) and in any case the impacts of these changes persist for only a short time until the ecosystems begin to recover. The impact of soot production on atmospheric chemistry may be a more important influence. Soot aerosols from biomass burning can affect regional climate by absorbing radiation and heating the air (Fig. 6); this in turn can alter regional atmospheric stability and vertical motions, and thus influence large-scale circulation and the hydrologic cycle (Menon et al., 2002). Open biomass burning associated with deforestation and crop residue removal produces ~40% of global black carbon emissions (Bond et al., 2004). Indirect effects on radiative forcing occur primarily because aerosols affect cloud properties (Sherwood, 2002; Guyon et al., 2005). Unlike trace gases, aerosols are not well mixed: aerosol distribution is heterogeneous spatially and through the atmospheric column vertically. As a result, the effects of aerosols on climate are difficult to model and remain one of the largest uncertainties in the simulation of global climate. Bowman et al. (2009) estimate a slight negative forcing from direct aerosol effects from biomass burning but consider the indirect aerosol effects too uncertain to estimate (Fig. 7).

Emissions of soot from wildfires decrease the albedo of snow, ice, and land and thus can increase surface temperatures (Hansen and Nazarenko, 2004). Increasing concentrations of black carbon, in combination with decreasing concentrations of sulfate aerosols, have substantially contributed to rapid Arctic warming during the past three decades (Shindell and Faluvegi, 2009). Additionally, black carbon within soot that is deposited over snow and ice significantly increases solar absorption and melting (Ramanathan and Carmichael, 2008), which may be one of the important contributors to Arctic sea ice retreat (Flanner et al., 2007). The impact of even small shifts in albedo on snow has large potential consequences for radiative forcing due to positive feedbacks. Koch and Hansen (2005) used a global climate model to show that soot in the Arctic today comes from both industrial combustion and biomass burning. Model experiments indicate that soot has substantially contributed to rapid Arctic warming during the past three decades (Shindell and Faluvegi, 2009), and may be responsible for half of the 1.9°C increase in Arctic temperatures between 1890 and 2007 because of its effects on snow and ice albedo at high latitudes (Ramanathan and Carmichael, 2008).

Costs

Fires in Indonesia during the 1997–1998 droughts brought the potential costs of uncontrolled wildfires to global attention (Tacconi, 2003). The fires were largely set to clear land for oil palm and timber plantations and affected about 5 million hectares. Smoke and haze from the burning peat and forest engulfed millions of people in Southeast Asia, reducing visibility, causing respiratory illness, and increasing infant mortality (Jayachandran, 2005). The economic costs were estimated at between US$4.4–9 billion – an amount that exceeds the funds needed annually to provide basic sanitation, water, and sewerage services to all of Indonesia's 120 million rural poor (EEPSEA and WWF, 1998). The fires have been widely regarded as the largest environmental disaster of the century (Glover, 2001).

The costs of these fires have been studied in detail (Tomich et al., 2004; Lohman et al., 2007). Costs related directly to suppression and control of the fires were less than 1% of the total costs (Fig. 8c). Direct losses incurred during and immediately after the fire event, including the costs of lost timber, agriculture, and tourist revenue amounted to ca. 28% of the total cost. The less tangible indirect costs, which include direct forest benefits (i.e., food, local medicines, raw materials and recreation), indirect forest benefits (e.g., storm protection, water supply and regulation, erosion control, soil formation, nutrient cycling, and waste treatment), and costs associated with biodiversity losses, and the release of carbon to the atmosphere, were estimated as comprising the largest (ca. 48%) economic impact (Fig. 8c) while short-term health costs were estimated as ca. 23% of the total cost. Although this accounting appears comprehensive, there are large uncertainties associated with the estimates of indirect costs. The indirect costs associated with carbon release and biodiversity loss, for example, reflect income lost to Indonesia for conserving its forest; but these are hypothetical, being estimated as the price that agencies and organizations are willing to pay to conserve tropical forests (EEPSEA and WWF, 1998). On the other hand, additional costs associated with evacuations, long-term health damages or loss of life, or effects from the haze that may have reduced crop productivity, photosynthesis, pollination, and other biological processes (see e.g., Schafer et al., 2002), were not taken into account. The estimated biodiversity costs do not take into account the intrinsic value of species that have been extirpated or whose extinction has been accelerated, the potential but currently unexploited value of ecotourism or medicines, or losses of cultural diversity of indigenous forest-based culture (EEPSEA and WWF, 1998).

There are few reliable data on the cost of wildfires for many of the regions where fire is prevalent. Even in developed countries, data on fire-related costs are often uncertain and incomplete. According to the FAO global fire assessment (FAO, 2007), the direct costs of fire are in the hundreds of millions of dollars for many countries, and in the billions for larger countries and regions. India reported costs of US$107 million in one year, for example. Russia reported costs of US$4.2 billion in 1998, and timber losses were estimated at US$0.5–1 billion yearly in northeast Asia. Mexico estimated losses of US$337 million in wood, US$6.6 million in firewood and US$39 million in reforestation costs.

Wildfires are recognized as environmental disasters by the international community; this fact provides an alternative tool for the assessment of costs and impacts (International Disaster Database: http://www.emdat.be/). Fifteen countries declared wildfire disasters during the decade 2000–2009 (Fig. 8b). The predominance of countries with Mediterranean-type climates in this list reflects the strong control of climate on wildfires and points to the vulnerability of these environments to fire in the face of global warming. The analyses show large discrepancies in the overall cost of wildfires between countries. The estimated economic losses from wildfires in the USA dwarf the losses experienced by all of the other countries (Fig. 8a). However, when expressed

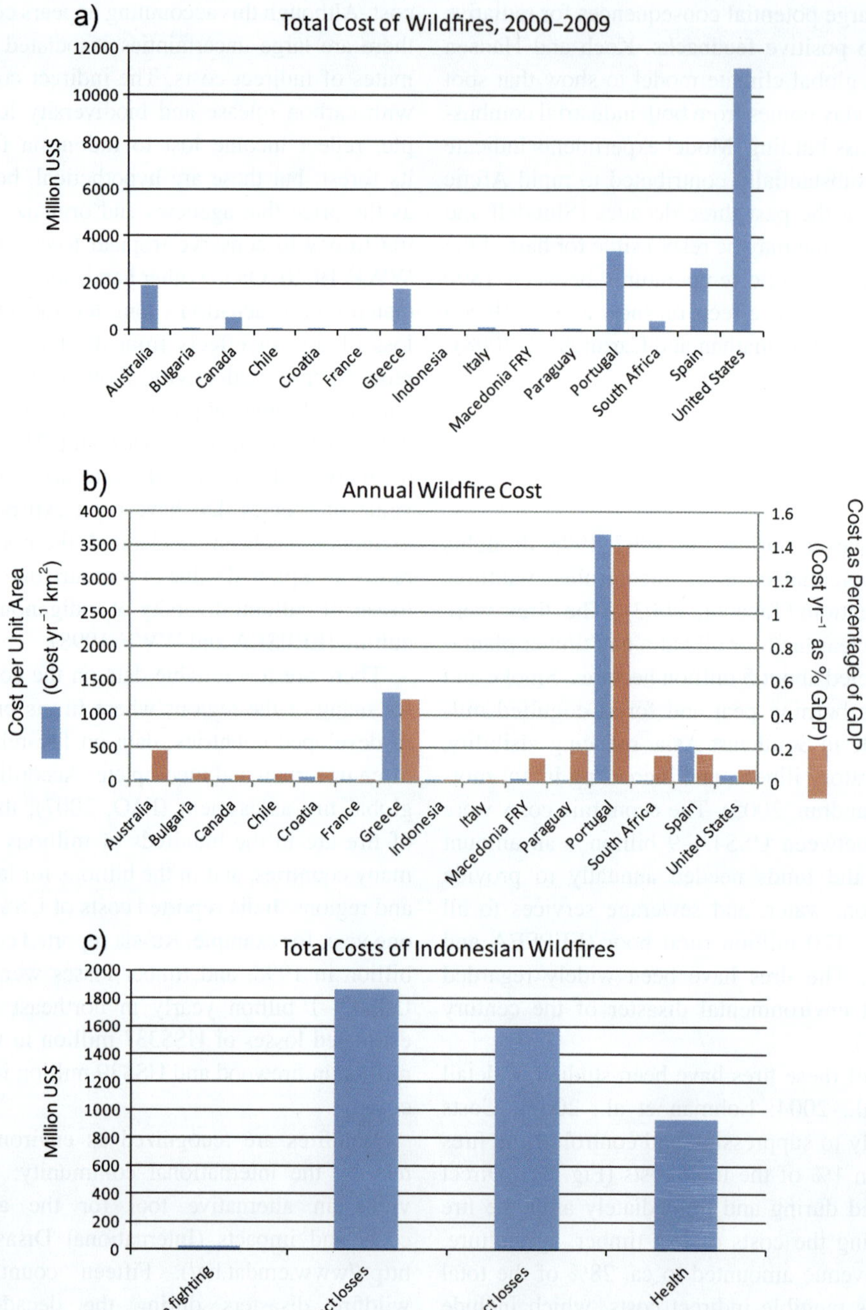

Fig. 8 The costs of wildfires. **a** Estimated costs of wildfire disasters in the decade 2000–2009 by country in millions of US$, where a wildfire disaster is defined as an event where more than ten people were killed or more than 100 injured, where a state of emergency was declared or international aid was solicited. Data from the International Disaster Database (http://www.emdat.be/). These costs are re-expressed **b** as a yearly cost per area where area is total land area of the country and as a percentage of gross domestic product (GDP) in 2008 (The World Bank: World Development Indicators database: http://web.worldbank.org/WBSITE/EXTERNAL/DATASTATISTICS/0,,menuPK:232599~pagePK:64133170~piPK:64133498~theSitePK:239419,00.html). **c** Estimated costs of the fires in Indonesia during the 1997/1998 ENSO season, showing fire-fighting costs, direct costs from loss of timber, agriculture or tourist revenue, indirect costs including ecosystem services provided by forests, indirect benefits from forest, carbon costs, and loss of biodiversity, and short-term health costs. Redrawn from data in Lohman et al. (2007)

either in terms of areal impact or as a proportion of gross national product (GDP), the economic impact of fire in the USA is much smaller than its impact in, e.g., Australia or several countries in southern Europe (Fig. 8b). The wildfire disasters in Portugal, in particular, cost the equivalent of 1.4% of GDP. This form of accounting does not include indirect costs, and the actual economic impact of wildfires may be much larger than these numbers suggest.

The fire-fighting costs for the Indonesian fires were US$ 34 million, or less than 1% of the total estimated costs. By comparison, the fire-fighting costs in the USA for the Yellowstone National Park fires of 1988 were US$ 120 million. After the "Black Saturday" fires in February 2009, the Australian government budgeted nearly US$ 1 billion for costs associated with fire-fighting, clean up, and rebuilding. Fire-fighting is a much larger and better-funded endeavor in developed countries such as the US and Australia, typically involving large hand-crews, tankers, bulldozers, helicopters, and airplanes to protect lives, homes, and structures even though a change in weather conditions is always, in reality, the main agency that stops large wildfires.

Fire suppression activities are relatively easy to quantify because they occur during and immediately after the fire event. Other direct costs, such as loss of property or timber are also relatively easy to assess (Dale, 2009). However, the longer-term, indirect consequences of fires have far-reaching effects that can dwarf the costs related to fire suppression. Long-term damages to watershed values and other ecosystem services, such as the regulation of water and soil quality and quantity, and of carbon sequestration, for example, may represent the most significant costs from large fire events (Lynch, 2004). The California Fire Plan delineates a comprehensive list of both cultural and environmental assets potentially at risk from wildfire in order to identify areas where the potential cost of a major wildfire event is largest, and thereby help prioritize pre-fire management activities. The asset analysis includes items such as air and water quality, hydropower generation, flood control, water storage capacity, historic buildings, and scenic areas. Areas are ranked based on the potential physical effects of fire as well as the human valuation of those effects.

A recent analysis estimated the total cost of fire in Australia at over US$7.7 billion per annum, or about 1.15% of the country's GDP (Ashe et al., 2009). Similar analyses show that total fire-related costs in the UK, USA, Canada, and Denmark all fall in the range of 0.9–2% of GDP. In the USA, fire-related costs are ca. 2% of GDP and rapidly increasing not only because of an increase in the number of big fires but because of rapid development of homes in the wildland-urban interface (USDA, 2006; Rasker, 2009). Ashe et al. (2009) have shown that Australia is investing about US$6.6 billion (or 85% of the total cost of fire) to manage a loss of about US$1.2 billion (or 15% of the total cost of fire). Such analyses raise critical questions about current approaches to fire management, in Australia and elsewhere.

Future Fire Regimes

Increases in the number of large fires and in the area burnt have been observed during recent years in many regions, including Canada (Stocks et al., 2002; Gillett et al., 2004), the United States (Westerling et al., 2006), southern Europe (Piñol et al., 1998), Siberia (Kajii et al., 2002), eastern Eurasia (Balzter et al., 2007; Groisman et al., 2007), and Australia (Cary, 2002). It seems increasingly likely that these increases are a result of anthropogenic climate changes (e.g., see Running, 2006), and this has led to a growing concern for predicting how potential changes in climate over the twenty-first century might affect natural fire regimes.

Most of the assessments of the impact of anthropogenic climate changes on regional wildfire regimes have been based on observed relationships between climate and some aspect of the fire regime at a regional or sub-continental scale (e.g., Flannigan et al., 2001; Cardoso et al., 2003; Balshi et al., 2009; Girardin and Mudelsee, 2008). Krawchuk et al. (2009) have used various statistical relationships to project future changes in fire regimes globally. Although they show major increases in the probability of fire in for example the boreal zone, this increase is offset by reductions in the incidence of fire elsewhere, and overall there is no discernible change in fire at the global scale. A key issue with all of these future predictions is the degree to which statistical relationships characteristic of modern fire-climate interactions hold under potentially different climates and CO_2 concentrations. Flannigan et al. (2001) addressed this by showing that application of the modern statistical relationship between fire weather and fire incidence to simulated climates for 6,000 years ago produced patterns of changes in fire regimes across Canada consistent with charcoal-based reconstructions of changes in biomass burning during the mid-Holocene. In general, however, there have been few attempts to test whether statistical relationships observed under modern conditions are likely to hold when climate changes beyond the modern observed range.

Process-based modeling provides a way of overcoming the limitations of statistical modeling. There are now many fire models, some adapted for specific ecosystems (e.g., Ito and Penner, 2005; Crevoisier et al., 2007) while others are global in scope (Lenihan et al., 1998; Kucharik et al., 2000; Thonicke et al., 2001; Venevsky et al., 2002). Some models have been designed to be coupled within climate models (e.g., Arora and Boer, 2005; Krinner et al., 2005) in order to investigate feedbacks. There have been only a few attempts to apply such models to estimate future changes in fire regimes either at a regional (Bachelet et al., 2005; Lenihan et al., 2008) or at a global scale (Scholze et al., 2006). Nevertheless, some robust features are beginning to emerge. In general, there is an overall increase in fire in response to warming over the twenty-first century (Fig. 9). Some regions, however, experience reduced fires: in tropical regions, this decrease reflects the fact that warming is accompanied by an increase in precipitation while in semi-arid regions the decrease occurs because regional warming results in considerable reduction of fuel loads and decreases in fuel connectivity. The changes in fire regimes, whether these are regional increases or decreases in fire, become more marked through the century (compare Fig. 9a, b, and c) and with the magnitude of the global increase in temperature (compare, e.g., Fig. 9c with f).

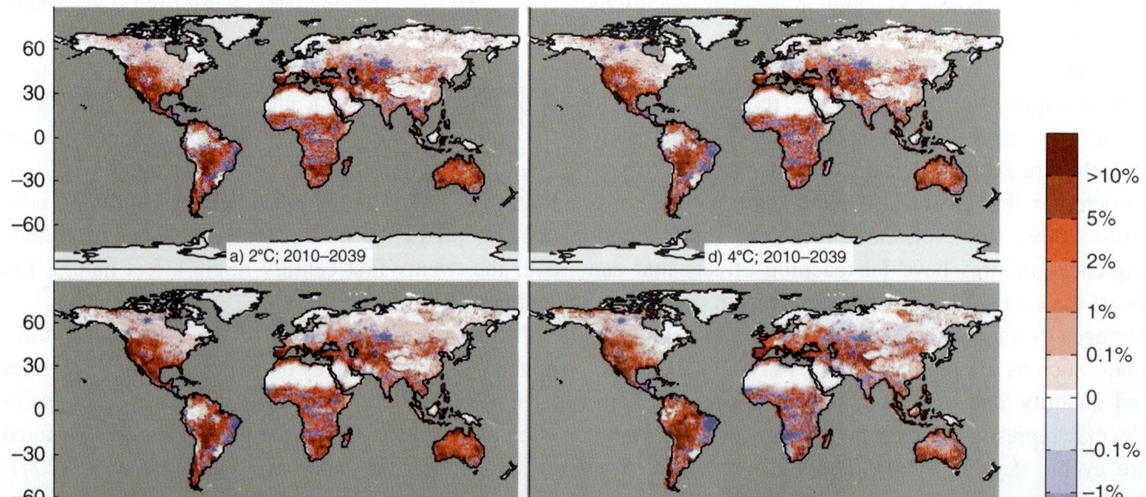

Fig. 9 Changes in global fire regimes during the twenty-first century as simulated by the coupled fire-vegetation model LPX (Prentice et al., in preparation) driven by output from the HADCM3 coupled climate model pattern-scaled to produce either a 2°C or a 4°C warming by 2050 (Harrison et al., in preparation)

The recognition that there will be regional differences in the direction of future changes in fire regime is common to both statistical and process-based modeling projections, as is the general idea that the changes in fire regime will become more severe during the twenty-first century. However, detailed comparisons of the predictions show few other similarities. Krawchuk et al. (2009) show increased fire across much of the boreal zone, whereas Scholze et al. (2006) show a decrease in fire in large parts of the boreal zone and major increases in fire in the temperate forests and in semi-arid regions. The simulations presented here (Fig. 9) show more extreme changes in fire regimes than either of these two previous studies. There is an urgent need to evaluate and improve approaches to predicting changes in future wildfires in response to anthropogenic climate changes, both because of the large costs and because of the potential feedbacks from fire to climate (Flannigan et al., 2009). It is also important because future climate changes could amplify direct human impacts on fire regimes in that human-set fires are more likely to escape control and to persist for longer during drought conditions (van der Werf et al., 2008b).

Prognostic modeling of fire is relatively new and there are, as yet, no model-based investigations of the potential magnitude of pyrogenic feedbacks through atmospheric chemistry to climate. It is difficult to estimate the overall impact of potential future changes in biomass burning on radiative forcing without targeted model experiments because the changes in biomass burning are regionally specific and the influence of changes in pyrogenic emissions on atmospheric chemistry are sensitive to location (Naik et al., 2007). Although Bowman et al. (2009) have suggested that deforestation fires during the industrial era were responsible for ca. 19% of the total increase in radiative forcing since pre-industrial times, Arneth et al. (in revision) have estimated that the additional net radiative forcing over the twenty-first century would be small, of the order of 0.02 W/m^2.

Mitigation and Adaptation: Can We Manage Future Fire?

The observed increases in large wildfires in recent decades, together with predictions of changes in regional fire regimes during the twenty-first century, raise urgent questions as to how fires should be managed. Policies geared toward the total suppression of wildfire are not tenable as a strategy for dealing with projected changes in fire regimes during the twenty-first century. Even in the most developed countries, fire-fighting capacity has been insufficient to deal with wildfires in recent years (Lohman et al., 2007; Bowman et al., 2009) and the costs of protection and fire-fighting continue to escalate (FAO, 2007). There is also increasing awareness of the deleterious effects of fire suppression, and particularly of the fact that fire suppression may lead to bigger and more devastating fires through the build-up of dangerous levels of fuel (Dombeck et al., 2004; Lynch, 2004). However, even when property is not at risk and there are obvious benefits to be gained from letting fires burn naturally, existing incentives, policy restrictions, and personnel limitations can overwhelmingly predispose institutions to continue to suppress fires (USDA, 2006).

The impact of fire suppression can be seen in the contrasting histories of recent fire in Arizona (Skroch and Swetnam, 2008). As a result of a change in management policy, lightning-ignited fires have been allowed to burn in the Rincon Mountains of Saguaro National Park since 1983. The forests now contain more mature trees and less understory fuels. However, fire-suppression policy still remains the policy in the more populated Santa Catalina Mountains nearby. As a result, build-up of fuel helped promote high-intensity crown fires which swept through the Catalinas in the summer of 2008, destroying hundreds of homes and costing millions of dollars to extinguish.

Fire suppression, and indeed controlled burning to some extent, may have a deleterious impact on biodiversity (Gill et al., 1999; Brown and Smith, 2000; Granström, 2001; Tilman and Lehman, 2001; Hutto, 2008). Many plant species are fire tolerant, require fire for regeneration, or promote fire as a competitive strategy (Bond and Midgley, 1995; Whelan, 1995; Schwilk and Ackerly, 2001). Savanna ecosystems, which occupy ca. 20% of the land surface, support a large proportion of the world's human population and most of its rangeland, livestock, and wild herbivore biomass (Scholes and Archer, 1997; Grace, 2006). Most savannas only exist as a result of fire (Bond, 2008; Lehmann et al., 2008). Thus, preserving natural fire regimes may be a necessary component of biodiversity management in the future (Shlisky et al., 2007; Hutto, 2008).

Reducing the risk of large wildfires through controlled burning (Pyne et al., 1996; Fernandez and Botelho, 2003) is increasingly seen as a management option in many parts of the world. Initiatives such as Firescape in southeast Arizona and the Four Forests Initiative on the Mogollon Rim, for example, are coordinating fire and fuel reduction activities at the landscape scale (Skroch and Swetnam, 2008). For such initiatives to be effective, however, the public needs to accept more frequent (albeit less destructive) fires despite the inherent risks associated with implement controlled burns and the inconvenience of smoke.

Controlled burning has also been proposed as a climate-change mitigation strategy (Myneni et al., 2001). Wildfires release substantial amounts of carbon. Bushfires in the savanna areas of northern Australia, for example, result in the loss of ca. 340,000 km^2 of vegetation annually and produce between 2 and 4% of Australia's accountable greenhouse gas emissions (Russell-Smith et al., 2007; Cook and Meyer, 2009). Similarly, the amount of CO_2 emitted by fires in the USA is equivalent to 4–6% of anthropogenic emissions at the continental scale (Wiedinmyer and Neff, 2007). It is not surprising, then, that the idea that mitigation of wildfire impacts through prescribed burning could potentially lead to major abatement in pyrogenic emissions has been discussed in the context of forests in Australia (Williams et al., 2004; Bushfire Cooperative Research Centre, 2006; Beringer et al., 2007), Europe (Narayan et al., 2007), and North America (Hurteau et al., 2008). The underlying idea is that prescribed burning decreases both the intensity and extent of subsequent wildfires by reducing fuel loads. Decreasing the intensity of fires, particularly in savannas, means that biomass is lost through decomposition rather than combustion, resulting in reduced emissions of CH_4 and N_2O. In savannas, prescribed burning in the early dry season, even though it may not affect the total area burnt, has nevertheless been shown to reduce emissions by reducing average fire intensity (Williams et al., 2002). On the other hand, low intensity (incomplete) burns may actually release more methane than intense complete burns (Crutzen and Andreae, 1990; Hao and Ward, 1993). Prescribed fire has been used successfully in northern Australia to achieve greenhouse-gas emission abatement (Russell-Smith et al., 2007). However, in general the magnitude of any reduction in emissions through fire management and prescribed burning is dependent on the nature of the vegetation and its dynamics, carbon stocks and flows, the efficacy of prescribed burning and the trade-offs between the different fire regimes. The gains in terms of climate-change abatement are probably small unless prescribed burning is useful to achieve other land-management goals. Indeed, many countries have already decided not to use fire management as a climate-change mitigation strategy because current agreements do not account for or protect against future changes in natural disturbances whether by fire or fire-promoted insect attack (Kurz et al., 2008). Additional agreements would be required to encourage the inclusion of forest management practices as a mitigation strategy.

Perhaps the most urgent aspect of future fire management concerns the reduction of economic damage. Part of the escalating cost of fire management has resulted from expansion of housing and urban infrastructure in inappropriate areas (i.e., adjacent to potentially lethal fire regimes) such as the wildland-urban interface in fire-prone ecosystems (van Wagtendonk, 2007). In many cases, people moving into such areas are unaware of the wildfire risk and are unprepared for it. The adoption of strategies to promote fire preparedness, such as creating defendable space around homes and other structures by clearing brush and keeping tree branches trimmed above the ground, using fire-proof building materials, and developing evacuation plans (Dombeck et al., 2004), only provide a limited solution to the fire management problem. The two primary options in addressing development in the wildland-urban interface are to deforest areas to remove the threat of fire, or to avoid development in such locations. In either case, a change in fire policy and management is becoming increasingly urgent as all indicators suggest that wildfires will continue to burn larger, hotter, and longer for the near future in such fire-prone ecosystems. At local scales, human values, about fire are at the root of many of our most serious fire-related problems. Issues such as the social, economic, and environmental impacts of fire and fire suppression, public safety, smoke, hazard mitigation, and fire communication and education must be examined within the mesh of political, social, and economic contexts where they exist (Gill, 2005). However, instituting policies that promote increased fire use in locations with risk to private property, such as cost/benefit analyses that aim to balance community and ecosystem needs and redesigning urban growth plans (Moritz and Stephens, 2008) are often considered politically infeasible (USDA, 2006).

The growing number, severity, and cost of fires in recent decades across all latitudes and countries have made it clear that governments and institutions are unable to adequately manage fire (FAO, 2007; Lohman et al., 2007). Networks of organizations have taken on increasing responsibilities for disaster policy design and administration, but their capacity is mixed, and most lack the ability to assess risk and evaluate their own activities (Charles and Michael, 2009). Models of successful fire-management strategies exist in some areas (Arno and Fiedler, 2004), and some of these models may be applicable in broader contexts. However, investment is needed in identifying others, and in documenting existing efforts so that we can learn from them.

Conclusions

Although highly variable in space and time, in general fire is tightly linked to climate and climatic variations. Global changes of the twenty-first century make it inevitable that the impact of large fires will continue and increase in magnitude. Changing patterns of wildfire activity present significant environmental, economic, and political challenges to communities and nations. At the core of these challenges is a need to increase public awareness of and political support for restoring fire to ecosystems that need it, to protect environments from fire only where fire is not needed or wanted, and to form new relationships with fire in an uncertain future. Adapting to climate change impacts on wildfire will require greater emphasis on flexible, adaptable approaches to fire management. New strategies may be required that acknowledge the limitations of fire-fighting capabilities and consequently focus on avoidance rather than attack. A greater emphasis on fire preparedness will also involve a shift in responsibility for creating defensible space to those who benefit most from fire protection services.

References

Allen CD, Savage M, Falk DA, Suckling KF, Swetnam TW, Schulke T, Stacey PB, Morgan P, Hoffman M, Klingel JT (2002) Ecological restoration of southwestern ponderosa pine ecosystems: A broad perspective. Ecol Appl 12:1418–1433.

Amiro BD, Todd JB, Wotton BM, Logan KA, Flannigan MD, Stocks BJ, Mason JA, Martell DL, Hirsch KG (2001) Direct carbon emissions from Canadian forest fires 1959–1999. Can J For Res 31:512–525.

Andreae MO, Merlet P (2001) Emission of trace gases and aerosols from biomass burning. Glob Biogeochem Cycles 15:955–966.

Archibald S, Roy DP, van Wilgen B, Scholes RJ (2009) What limits fire? An examination of drivers of burnt area in Southern Africa. Glob Chang Biol 15:613–630.

Arneth A, Harrison SP, Zaehle S, Tsigaridis K, Menon S, Bartlein PJ, Feichter H, Korhola A, Kulmala M, O'Donnell D, Schurgers S, Sorvari S, Vesala T (in revision) Terrestrial biogeochemical feedbacks in the climate system. Nat Geosci.

Arno SF, Fiedler CE (2004) Mimicking Nature's Fire: Restoring Fire-Prone Forests in the West (242 pp). Washington, DC: Island Press.

Arora VK, Boer GJ (2005) Fire as an interactive component of dynamic vegetation models. J Geophys Res 110:1–20.

Ashe B, McAneney KJ, Pitman AJ (2009) Total cost of fire in Australia. J Risk Res 12:121–136.

Atahan P, Dodson JR, Itzstein-Davey F (2004) A fine-resolution Pliocene pollen and charcoal record from Yallalie, southwestern Australia. J Biogeogr 31:199–205.

Bachelet D, Lenihan J, Neilson R, Drapek R, Kittel T (2005) Simulating the response of natural ecosystems and their fire regimes to climatic variability in Alaska. Can J For Res 35:2244–2257.

Balshi MS, McGuire AD, Duffy P, Flannigan M, Kicklighter DW, Melillo J (2009) Vulnerability of carbon storage in North American boreal forests to wildfires during the 21st century. Glob Chang Biol 15:1491–1510.

Balshi MS, McGuire AD, Zhuang Q, Melillo J, Kicklighter DW, Kasischke E, Wirth C, Flannigan M, Harden J, Clein JS, Burnside TJ, McAllister J, Kurz WA, Apps M, Shvidenko A (2007) The role of historical fire disturbance in the carbon dynamics of the pan-boreal region: A process-based analysis. J Geophys Res 112:G02029.

Balzter H, Gerard F, Weedon G, Grey W, Combal B, Bartholome E, Bartalev S, Los S (2007) Coupling of vegetation growing season anomalies with hemispheric and regional scale climate patterns in Central and East Siberia. J Clim 20:3713–3729.

Bartlein P, Hostetler SW, Shafer SL, Holman JO, Solomon AM (2008) Temporal and spatial structure in a daily wildfire-start data set from the western United States (1986–96). Int J Wildland Fire 17:8–17.

Beerling DJ, Osborne CP (2006) The origin of the savanna biome. Glob Chang Biol 12:2023–2031.

Belcher CM, McElwain JC (2008) Limits for combustion in low O_2 redefine paleoatmospheric predictions for the Mesozoic. Science 321:1197–1200.

Bella CM, Jobbágy EG, Paruelo JM, Pinnock S (2006) Continental fire density patterns in South America. Glob Ecol Biogeogr 15:192–199.

Beringer J, Hutley L, Tapper N, Cernusak L (2007) Savanna fires and their impact on net ecosystem productivity in North Australia. Glob Chang Biol 13:990–1004.

Berner RA (1999) Atmospheric oxygen over Phanerozoic time. Proc Natl Acad Sci 96:10955–10957.

Bian H, Chin M, Kawa SR, Duncan B, Arellano A, Kasibhatla P (2007) Sensitivity of global CO simulations to uncertainties in biomass burning sources. J Geophys Res Atmos 112. doi: 23310.21029/22006JD008376.

Biganzoli F, Wiegand T, Batista WB (2009) Fire-mediated interactions between shrubs in a South American temperate savannah. Oikos 118:1383–1395.

Bird MI, Cali JA (1998) A million-year record of fire in sub-Saharan Africa. Nature 394:767–769.

Bird MI, Cali JA (2002) A revised high-resolution oxygen-isotope chronology for ODP-668B: Implications for Quaternary biomass burning in Africa. Glob Planet Change 33:73–76.

Bonan GB (2008) Forests and climate change: Forcings, feedbacks, and the climate benefits of forests. Science 320:1444–1449.

Bond WJ, Midgley JJ (1995) Kill thy neighbour — an individualistic argument for the evolution of flammability. Oikos 73:79–85.

Bond WJ, van Wilgen BW (1996) Fire and Plants (263 pp). London: Chapman & Hall.

Bond T, Streets D, Yarber K, Nelson S, Wo J-H, Klimont Z (2004) A technology-based global inventory of black and organic carbon emissions from combustion. J Geophys Res 109. doi: 10.1029/2003JD003697.

Bond WJ, Keeley JE (2005) Fire as a global 'herbivore': The ecology and evolution of flammable ecosystems. Trends Ecol Evol 20:387–394.

Bond WJ (2008) What limits trees in C_4 grasslands and savannas? Annu Rev Ecol Evol Syst 39:641–659.

Bowman DMJS (2000) Australian Rainforests: Islands of Green in a Land of Fire (344 pp). Cambridge: Cambridge University Press.

Bowman DMJS, Balch JK, Artaxo P, Bond WJ, Carlson JM, Cochrane MA, D'Antonio CM, DeFries RS, Doyle JC, Harrison SP (2009) Fire in the Earth system. Science 324:481–484.

Brown JK, Smith JK (2000) Wildland Fire in Ecosystems: Effects of Fire on Flora (p. 257). Gen. Tech. Rep. RMRS-GTR-42-vol. 2. Ogden, UT: U.S. Department of Agriculture, Forest Service, Rocky Mountain Research Station.

Brown KJ, Hebda RJ (2002) Origin, development, and dynamics of coastal temperate conifer rainforests of southern Vancouver Island, Canada. Can J For Res 32:353–372.

Bush MB, Listopad MCS, Silman MR (2007a) A regional study of Holocene climate change and human occupation in Peruvian Amazonia. J Biogeogr 34:1342–1356.

Bush MB, Silman MR, de Toledo MB, Listopad C, Gosling WD, Williams C, de Oliveira PE, Krisel C (2007b) Holocene fire and occupation in Amazonia: Records from two lake districts. Philos Trans R Soc B 362:209–218.

Bushfire Cooperative Research Centre (2006) Climate change and its impacts on the management of bushfire. Fire Note 4,http://www.bushfirecrc.com/publications/downloads/Fire note_climate190906.pdf Firenote_climate190906.pdf.

Campbell J, Donato D, Azuma D, Law B (2007) Pyrogenic carbon emission from a large wildfire in Oregon, United States. J Geophys Res 112. doi: 10.1029/2007JG000451.

Carcaillet C, Ali AA, Blarquez O, Genries A, Mourier B, Bremond L (2009) Spatial variability of fire history in subalpine forests: From natural to cultural regimes. Ecoscience 16:1–12.

Carcaillet C, Almquist H, Asnong H, Bradshaw RHW, Carrión JS, Gaillard MJ, Gajewski K, Haas JN, Haberle SG, Hadorn P, Müller SD, Richard PJH, Richoz I, Rösch M, Sánchez Goñi MF, Von Stedingk H, Stevenson AC, Talon B, Tardy C, Tinner W, Tryterud E, Wick L, Willis KJ (2002) Holocene biomass burning and global dynamics of the carbon cycle. Chemosphere 49:845–863.

Cardoso MF, Hurtt GC, Moore B, Nobre CA, Prins EM (2003) Projecting future fire activity in Amazonia. Glob Chang Biol 9:656 – 669.

Carmona-Moreno C, Belward A, Malingreau J-P, Hartley A, Garcia-Alegre M, Antonovskiy M, Buchshtaber V, Pivovarov V (2005) Characterizing interannual variations in global fire calendar using data from Earth observing satellites. Glob Chang Biol 11:1537.

Cary G (2002) Importance of a changing climate for fire regimes in Australia. In R Bradstock, J Williams, AM Gill (Eds), Flammable Australia: The Fire Regimes and Biodiversity of a Continent (pp. 26–46). Cambridge: Cambridge University Press.

Castello JD, Leopold DJ, Smallidge PJ (1995) Pathogens, patterns, and processes in forest ecosystems. BioScience 45:16–24.

Charles RW, Michael M (2009) Wildland fire mitigation networks in the western United States. Disasters 33:721–746.

Chuvieco E, Giglio L, Justice C (2008) Global characterization of fire activity: Toward defining fire regimes from Earth observation data. Glob Chang Biol 14:1488–1502.

Clark JS (1990) Fire and climate change during the last 750 yr in northwestern Minnesota. Ecol Monogr 60:135–159.

Cochrane M (2003) Fire science for rainforests. Nature 421:913–919.

Cochrane MA, Alencar A, Schulze MD, Souza CM Jr, Nepstad DC, Lefebvre P, Davidson EA (1999) Positive feedbacks in the fire dynamic of closed canopy tropical forests. Science 284:1832–1835.

Cochrane MA, Barber CP (2009) Climate change, human land use and future fires in the Amazon. Glob Chang Biol 15:601–612.

Cofer W, Winstead E, Stocks B, Goldammer J, Cahoon DR (1998) Crown fire emissions of CO_2, CO, H_2, CH_4 and TNNMHC from a dense jack pine boreal forest fire. Geophys Res Lett 25:3919–3922.

Collinson ME, Steart DC, Scott AC, Glasspool IJ, Hooker JJ (2007) Episodic fire, runoff and deposition at the Palaeocene-Eocene boundary. J Geol Soc 164:87–97.

Conedera M, Tinner W, Neff C, Meurer M, Dickens AF, Krebs P (2009) Reconstructing past fire regimes: Methods, applications, and relevance to fire management and conservation. Quat Sci Rev 28:555–576.

Cook GD, Meyer CP (2009) Culture, ecology, and economy of fire management in North Australian savannas: Rekindling the Wurrk tradition. In J Russell-Smith, P Whitehead, P Cooke (Eds), Fire, Fuels and Greenhouse Gases (p. 416). Melbourne, VIC: CSIRO Publishing.

Cramer W, Kicklighter DW, Bondeau A, Moore III B, Churkina G, Nemry B, Ruimy A, Schloss A and the participants of the Potsdam NPP Model Intercomparison (1999) Comparing global models of terrestrial net primary productivity (NPP): Overview and key results. Glob Chang Biol 5(Suppl. 1):1–15.

Crevoisier C, Shevliakova E, Gloor M, Wirth C, Pacala S (2007) Drivers of fire in the boreal forests: Data constrained design of a prognostic model of burned area for use in dynamic global vegetation models. Geophys Res Lett 112. doi: 10.1029/2006JD008372.

Crutzen PJ, Andreae MO (1990) Biomass burning in the tropics: Impact on atmospheric chemistry and biogeochemical cycles. Science 250:1669–1678.

D'Antonio CM, Vitousek PM (1992) Biological invasions by exotic grasses, the grass/fire cycle, and global change. Annu Rev Ecol Syst 23:63–87.

Dale L (2009) The True Cost of Wildfire in the Western U.S. (p. 16). Lakewood, CO: Western Forestry Leadership Coalition.

Daniau A-L, Harrison SP, Bartlein PJ (in press) Fire regimes during the last glacial. Quat Sci Rev.

Daniau A-L, Tinner W, Bartlein PJ, Harrison SP, Prentice IC, Brewer S, Friedlingstein P, Harrison-Prentice TI, Inoue J, Marlon JR, et al. (submitted) Hemispherically-asynchronous changes in fire consistent with deglacial climate changes. Nat Geosci.

DeFries RS, Morton DC, Werf GRVd, Giglio L, Randerson JT, Collatz GJ, Houghton RA, Kasibhatla PS, Shimabukuro Y (2008) Fire-related carbon emissions from land use transitions in Southern Amazonia. Geophys Res Lett 35. doi: 10.1029/2008GL035689.

Dentener F, Kinne S, Bond T, Boucher O, Cofala J, Generoso S, Ginoux P, Gong S, Hoelzemann JJ, Ito A, Marelli L, Penner JE, Putaud JP, Textor C, Schulz M, van der Werf GR, Wilson J (2006) Emissions of primary aerosol and precursor gases in the years 2000 and 1750 prescribed data-sets for AeroCom. Atmospheric Chemistry and Physics 6:4321–4344.

Dodson JR, Robinson M, Tardy CT (2005) Two fine-resolution Pliocene charcoal records and their bearing on pre-human fire frequency in south-western Australia. Austral Ecol 30:592–599.

Dombeck MP, Williams JE, Wood CA (2004) Wildfire policy and public lands: Integrating scientific understanding with social concerns across landscapes. Conserv Biol 18: 883–889.

Donovan GH, Brown TC (2007) Be careful what you wish for: The legacy of Smokey Bear. Front Ecol Environ 5:73–79.

Drobyshev I, Goebel PC, Hix DM, Corace RG, Semko-Duncan ME (2008) Pre- and post-European settlement fire history of red pine dominated forest ecosystems of Seney National Wildlife Refuge, Upper Michigan. Can J For Res 38:2497–2514.

Dwyer E, Gregoire J-M, Pereira JMC (2000) Climate and vegetation as driving factors in global fire activity. In JB Innes, MB Beniston, MM Verstraete (Eds), Biomass Burning and Its Inter-Relationships with the Climate System (pp. 171–191). Advances in Global Change Research. New York, NY: Springer.

EEPSEA and WWF (1998) The Indonesian fires and haze of 1997: The economic toll. EEPSEA Research Report rr1998051. Singapore: Economic and Environment Program for Southeast Asia (EEPSEA)/World Wide Fund for Nature (WWF).

Ellicott E, Vermote E, Giglio L, Roberts G (2009) Estimating biomass consumed from fire using MODIS FRE. Geophys Res Lett 36. doi: 10.1029/2009GL038581.

EPICA Community Members (2006) One-to-one coupling of glacial climate variability in Greenland and Antarctica. Nature 444:195–198.

FAO (2007) Fire Management: Global Assessment 2006, FAO Forestry Paper 0258–6150 (p. 151). Food and Agriculture Organization of the United Nations.

Fellows AW, Goulden ML (2008) Has fire suppression increased the amount of carbon stored in western U.S. forests? Geophys Res Lett 35. doi: 10.1029/2008GL033965.

Fernandez P, Botelho H (2003) A review of prescribed burning effectiveness in fire hazard reduction. Int J Wildland Fire 12:117–128.

Ferretti DF, Miller JB, White JWC, Etheridge DM, Lassey KR, Lowe DC, Meure CMM, Dreier MF, Trudinger CM, van Ommen TD, Langenfelds RL (2005) Unexpected changes to the global methane budget over the past 2000 years. Science 309:1714–1717.

Field CB, Raupach MR (2004) The Global Carbon Cycle: Integrating Humans, Climate, and the Natural World (526 pp). Washington, DC: Island Press.

Field RD, Van der Werf G, Shen SSP (2009) Human amplification of drought-induced biomass burning in Indonesia since 1960. Nat Geosci 2:185–188.

Fischer H, Behrens M, Bock M, Richter U, Schmitt J, Loulergue L, Chappellaz J, Spahni R, Blunier T, Leuenberger M, Stocker TF (2008) Changing boreal methane sources and constant biomass burning during the last termination. Nature 452. doi: 10.1038/nature06825.

Fisher JL, Loneragan WA, Dixon K, Delaney J, Veneklaas EJ (2009) Altered vegetation structure and composition linked to fire frequency and plant invasion in a biodiverse woodland. Biol Conserv 142:2270–2281.

Flanner MG, Zender CS, Randerson JT, Rasch PJ (2007) Present-day forcing and response from black carbon in snow. J Geophys Res 112. doi: 10.1029/2006JD008003.

Flannigan M, Campbell I, Wotton M, Carcaillet C, Richard PJH, Bergeron Y (2001) Future fire in Canada's boreal forest: Paleoecology results and general circulation model – regional climate model simulations. Can J For Res 31:854–864.

Flannigan MD, Logan KA, Amiro BD, Skinner WR, Stocks BJ (2005) Future area burned in Canada. Clim Change 72:1–16.

Flannigan MD, Krawchuk MA, de Groot WJ, Wotton BM, Gowman LM (2009) Implications of changing climate for global wildland fire. Int J Wildland Fire 18:483–507.

Frolking S, Palace MW, Clark DB, Chambers JQ, Shugart HH, Hurtt GC (2009) Forest disturbance and recovery: A general review in the context of space-borne remote sensing of impacts on aboveground biomass and canopy structure. J Geophys Res 114:1–27.

Fuentes ER, Segura AM, Holmgren H (1994) Are the responses of matorral shrubs different from those in an ecosystem with a reputed fire history? In JM Moreno, WC Oechel (Eds), The Role of Fire in Mediterranean-Type Ecosystems (pp. 16–25). New York, NY: Springer.

Fulé PZ, Crouse JE, Heinlein TA, Moore MM, Covington WW, Verkamp G (2003) Mixed-severity fire regime in a high-elevation forest: Grand Canyon, Arizona. Landsc Ecol 18:465–486.

Galanter M, Levy II H, Carmichael G (2000) Impacts of biomass burning on tropospheric CO, NO_x, and O_3. J Geophys Res Atmos 105:6633–6653.
Gavin DG, Hu FS, Lertzman K, Corbett P (2006) Weak climatic control of stand-scale fire history during the late Holocene. Ecology 87:1722–1732.
Giglio L, Randerson JT, van der Werf GR, Collatz GJ, Kasibhatla P (2006) Global estimation of burned area using MODIS active fire observations. Atmos Chem Phys Discuss 5:11091–11141.
Gill AM (1977) Management of fire-prone vegetation for plant species conservation in Australia. Search 8:20–26.
Gill AM, Woinarski JCZ, York A (1999). Australia's Biodiversity – Responses to Fire – Plants, Birds and Invertebrates. Biodiversity Technical Paper, Commonwealth of Australia.
Gill MA (2005) Landscape fires as social disasters: An overview of 'the bushfire problem'. Environ Hazards 6:65–80.
Gillett NP, Weaver AJ, Zwiers FW, Flannigan MD (2004) Detecting the effect of climate change on Canadian forest fires. Geophys Res Lett 31. doi: 10.1029/2004GL020876.
Girardin MP (2007) Interannual to decadal changes in area burned in Canada from 1781 to 1982 and the relationship to Northern Hemisphere land temperatures. Glob Ecol Biogeogr 16:557–566.
Girardin MP, Mudelsee M (2008) Past and future changes in Canadian boreal wildfire activity. Ecol Appl 18:391–406.
Glasspool IJ, Edwards D, Axe L (2004) Charcoal in the Silurian as evidence for the earliest wildfire. Geology 32:381–383.
Glover D (2001) The Indonesian fires and haze of 1997: The economic toll. In P Eaton, M Radojevic (Eds), Forest Fires and Regional Haze in Southeast Asia (pp. 227–236). New York, NY: Nova Science Publishers.
Grace JB (2006) Structural Equation Modeling and Natural Systems (365 pp). Cambridge: Cambridge University Press.
Granström A (2001) Fire management for biodiversity in the European boreal forest. Scand J For Res 16(Suppl. 3):62–69.
Grimm EC (1984) Fire and other factors controlling the Big Woods vegetation of Minnesota in the mid-nineteenth century. Ecological Monographs 54:291–311.
Grissino-Mayer HD, Romme WH, Floyd ML, Hanna DD (2004) Climatic and human influences on fire regimes of the southern San Juan Mountains, Colorado, USA. Ecology 85:1708–1724.
Groisman PY, Sherstyukov BG, Razuvaev VN, Knight RW, Enloe JG, Stroumentova NS, Whitfield PH, Forland E, Hannsen-Bauer I, Tuomenvirta H, Aleksandersson H, Mescherskaya AV, Karl TR (2007) Potential forest fire danger over Northern Eurasia: Changes during the 20th century. Glob Planet Change 56:371–386.
Guyon P, Frank G, Welling M, Chand D, Artaxo P, Rizzo L, Nishioka G, Kolle O, Fritsch H, Dias MAFS, Gatti LV, Cordova M, Andreae MO (2005) Airborne measurements of trace gas and aerosol particle emissions from biomass burning in Amazonia. Atmos Chem Phys Discuss 5: 2791–2831.
Habeck J (1994) Using general land office records to assess forest succession in Ponderosa Pine/Douglas-fir forests in western Montana. Northwest Sci 68:69–78.
Haberle SG, Ledru MP (2001) Correlations among charcoal records of fires from the past 16,000 years in Indonesia, Papua New Guinea, and Central and South America. Quat Res 55:97–104.
Hallett DJ, Mathewes RW, Walker RC (2003) A 1000-year record of forest fire, drought and lake-level change in southeastern British Columbia, Canada. Holocene 13: 751–761.
Hansen J, Nazarenko L (2004) Soot climate forcing via snow and ice albedos. Proc Natl Acad Sci USA 101:423–428.
Hao W, Ward D (1993) Methane production from global biomass burning. J Geophys Res 98:20657–20661.
Hastings MG, Jarvis JC, Steig EJ (2009) Anthropogenic impacts on nitrogen isotopes of ice-core nitrate. Science 324. DOI: 10.1126/science.1170510.
Heinselman ML (1973) Fire in the virgin forests of the Boundary Waters Canoe Area, Minnesota. Quat Res 3:329–382.
Heyerdahl EK, Brubaker LB, Agee JK (2001) Spatial controls of historical fire regimes: A multiscale example from the Interior West, USA. Ecol Soc 82:660–678.
Hope G, Kershaw AP, van der Kaars S, Xiangjun S, Liew PM, Heusser LE, Takahara H, McGlone M, Miyoshi N, Moss PT (2004) History of vegetation and habitat change in the Austral-Asian region. Quat Int 118–119:103–126.
Houghton RA, Hall F, Goetz SJ (2009) Importance of biomass in the global carbon cycle. J Geophys Res Biogeosci 114. doi: 10.1029/2009JG000935.
Houweling S, van der Werf G, Klein Goldewijk K, Röckmann T, Aben I (2008) Early anthropogenic emissions and the variation of CH_4 and $^{13}CH_4$ over the last millennium. Glob Biogeochem Cycles 22. doi: 10.1029/2007GB002961.
Hurteau MD, Koch GW, Hungate BA (2008) Carbon protection and fire risk reduction: Toward a full accounting of forest carbon offsets. Front Ecol Environ 6:493–498.
Hutto RL (2008) The ecological importance of severe wildfires: Some like it hot. Ecol Appl 18:1827–1834.
Ickoku C, Giglio L, Wooster MJ, Remner LA (2008) Global characterization of biomass-burning patterns using satellite measurements of fire radiative energy. Remote Sensing Environ 112:2950–2962.
Ito A, Penner JE (2005) Historical emissions of carbonaceous aerosols from biomass and fossil fuel burning for the period 1870–2000. Glob Biogeochem Cycles 19. doi: 10.1029/2004GB002374.
Ito A, Ito A, Akimoto H (2007) Seasonal and interannual variations in CO and BC emissions from open biomass burning in Southern Africa during 1998–2005. Glob Biogeochem Cycles 21:1–13.
Jaegle L, Steinberger L, Martin RV, Chance K (2005) Global partitioning of NO_x sources using satellite observations: Relative roles of fossil fuel combustion, biomass burning and soil emissions. Faraday Discuss 130:407–423.
Jayachandran S (2005). Air quality and infant mortality during Indonesia's massive wildfires in 1997. Bureau for Research in Economic Analysis of Development, Working Paper No. 95.
Jones TP, Chaloner WG (1991) Fossil charcoal, its recognition and paleoatmospheric significance. Palaeogeogr Palaeoclimatol Palaeoecol 97:39–50.
JRC-EU (2005) SAFARI 2000 global burned area map, 1-km, Southern Africa, 2000. Data online from Oak Ridge National Laboratory Distributed Active Archive Center, Oak Ridge, Tennessee (http://daac.ornl.gov/S2K/guides/

spot_veg_burned.html). Joint Research Centre, European Commission (www.jrc.ec.europa.eu/). Ispra, Varese, Italy.

Kajii Y, Kato S, Streets DG, Tsai NY, Shvidenko A, Nilsson S, McCallum I, Minko NP, Abushenko N, Altyntsev D, Khodzer TV (2002) Boreal forest fires in Siberia in 1998: Estimation of area burned and emissions of pollutants by advanced very high resolution radiometer satellite data. J Geophys Res 107:ACH 4-1–ACH 4-8.

Kasischke ES, Christensen NL, Stocks BJ (1995) Fire, global warming, and the carbon balance of boreal forests. Ecol Appl 5:437–451.

Kasischke ES, Bruhwiler LP (2002) Emissions of carbon dioxide, carbon monoxide, and methane from boreal forest fires in 1998. J Geophys Res 108:FFR 2-1–FFR 2-14.

Kasischke ES, Rupp TS, Verbyla DL (2006) Fire trends in the Alaskan boreal forest. In FS Chapin (Ed), Alaska's Changing Boreal Forest (p. 368). New York, NY: Oxford University Press.

Keeley JE, Rundel PW (2005) Fire and the Miocene expansion of C-4 grasslands. Ecol Lett 8:683–690.

Kipfmueller K, Baker WL (2000) A fire history of a subalpine forest in south-eastern Wyoming, USA. J Biogeogr 27:71–85.

Kitzberger T, Swetnam TW, Veblen TT (2001) Inter-hemispheric synchrony of forest fires and the El Niño-Southern Oscillation. Glob Ecol Biogeogr 10:315–326.

Klein Goldewijk K, van Drecht G (2006) HYDE3: Current and historical population and land cover. In AF Bouwman, T Kram, K Klein Goldewijk (Eds), Integrated Modelling of Global Environmental Change. An Overview of IMAGE 2.4. Bilthoven, The Netherlands: Netherlands Environmental Assessment Agency.

Koch D, Hansen J (2005) Distant origins of Arctic black carbon: A goddard institute for space studies modelE experiment. J Geophys Res 110. doi: 10.1029/2004JD005296.

Krawchuk M, Moritz M, Parisien M-A, Van Dorn J, Hayhoe K (2009) Global pyrogeography: The current and future distribution of wildfire. PLoS ONE 4:e5102.

Krawchuk MA, Cumming SG, Flannigan MD, Wein RW (2008) Biotic and abiotic regulation of lightning fire initiation in the mixed wood boreal forest. Ecology 87:458–468.

Krawchuk MA, Moritz MA (2009) Fire regimes of China: Inference from statistical comparison with the United States. Glob Ecol Biogeogr 18:626–639.

Krinner G, Viovy N, Noblet-Ducoudré Nd, Ogé J, Polcher J, Friedlingstein P, Ciais P, Sitch S, Prentice IC (2005) A dynamic global vegetation model for studies of the coupled atmosphere-biosphere system. Glob Biogeochem Cycles 19. doi: 10.1029/2003GB002199.

Kucharik CJ, Foley JA, Delire C, Fisher VA, Coe MT, Lenters JD, Young-Molling C, Ramankutty N (2000) Testing the performance of a dynamic global ecosystem model: Water balance, carbon balance, and vegetation structure. Glob Biogeochem Cycles 14:795–825.

Kurz WA, Stinson G, Rampley GJ, Dymond CC, Neilson ET (2008) Risk of natural disturbances makes future contribution of Canada's forests to the global carbon cycle highly uncertain. Proc Natl Acad Sci USA 105: 1551–1555.

Lafon CW, Grissino-Mayer HD (2007) Spatial patterns of fire occurrence in the central Appalachian Mountains and implications for wildland fire management. Phys Geogr 28: 1–20.

Lavorel S, Flannigan M, Lambin E, Scholes M (2007) Vulnerability of land systems to fire: Interactions among humans, climate, the atmosphere, and ecosystems. Mitig Adapt Strat Glob Change 12:33–53.

Lavoué D, Liousse C, Cachier H, Stocks BJ, Goldammer JG (2000) Modeling of carbonaceous particles emitted by boreal and temperate wildfires at northern latitudes. J Geophys Res 105:26871–26890.

Lehmann CER, Prior LD, Williams RJ, Bowman DMJS (2008) Spatio-temporal trends in tree cover of a tropical mesic savanna are driven by landscape disturbance. J Appl Ecol 45:1304–1311.

Lenihan J, Bachelet D, Neilson R, Drapek R (2008) Response of vegetation distribution, ecosystem productivity, and fire to climate change scenarios for California. Clim Change 87:215–230.

Lenihan JM, Daly C, Bachelet D, Neilson RP (1998) Simulating broad-scale fire severity in a dynamic global vegetation model. Northwest Sci 72:91–103.

Lenton TM (2001) The role of land plants, phosphorus weathering and fire in the rise and regulation of atmospheric oxygen. Global Chang Biol 7:613–629.

Litvak ME, Miller S, Wofsy SC, Goulden M (2003) Effect of stand age on whole ecosystem CO_2 exchange. J Geophys Res 108. doi: 10.1029/2001JD000854.

Liu H, Randerson JT, Lindfors J, Chapin III FS (2005) Changes in the surface energy budget after fire in boreal ecosystems of interior Alaska: An annual perspective. J Geophys Res 110. doi: 10.1029/2004JD005158.

Lohman DJ, Bickford D, Sodhi NS (2007) The burning issue. Science 316:376.

Lynch DL (2004) What do forest fires really cost? J For 102:42–49.

Marlon J, Bartlein P, Carcaillet C, Gavin DG, Harrison SP, Higuera PE, Joos F, Power MJ, Prentice CI (2008) Climate and human influences on global biomass burning over the past two millennia. Nat Geosci 1:697–701.

Marlon J, Bartlein P, Walsh MK, Harrison SP, Brown KJ, Edwards ME, Higuera PE, Power MJ, Anderson RS, Briles CE, et al. (2009) Wildfire responses to abrupt climate change in North America. Proc Natl Acad Sci 106:2519–2524.

Marynowski L, Simoneit BRT (2009) Widespread Upper Triassic to Lower Jurassic wildfire records from Poland: Evidence from charcoal and pyrolytic polycyclic aromatic hydrocarbons. Palaios 24:784–798.

Mazhitova GG (2000) Pyrogenic dynamics of permafrost-affected soils in the Kolyma upland. Eurasian Soil Sci 33:542–551.

McGlone MS, Wilmshurst JM (1999) Holocene record of climate, vegetation change and peat bog development, east Otago, New Zealand.

McWethy D, Whitlock C, Wilmshurst JM, McGlone MS, Li X (2009) Rapid deforestation of South Island, New Zealand by early Polynesian fires. Holocene 19:883–897.

Menon S, Hansen J, Nazarenko L, Luo Y (2002) Climate effects of black carbon aerosols in China and India. Science 297:2250–2253.

Miller GH, Magee JW, Johnson BJ, Fogel ML, Spooner NA, McCulloch MT, Ayliffe LK (1999) Pleistocene

extinction of Genyornis newtoni: Human impact on Australian megafauna. Science 283:205–208.
Mitchell SR, Harmon ME, O'Connell KEB (2009) Forest fuel reduction alters fire severity and long-term carbon storage in three Pacific Northwest ecosystems. Ecol Appl 19:643–655.
Moritz MA, Stephens SL (2008) Fire and sustainability: Considerations for California's altered future climate. Clim Change 87:S265–S271.
Morton DC, DeFries RS, Randerson JT, Giglio L, Schroeder W, Werf GRvd (2008) Agricultural intensification increases deforestation fire activity in Amazonia. Glob Chang Biol 14:2262–2275.
Mota BW, Pereira JMC, Oom D, Vasconcelos MJP, Schultz M (2006) Screening the ESA ATSR-2 World Fire Atlas (1997–2002). Atmos Chem Phys 6:1409–1424.
Mouillot F, Field CB (2005) Fire history and the global carbon budget: A 1° × 1° fire history reconstruction for the 20th century. Glob Chang Biol 11:398–420.
Mutch RW (1970) Wildland fires and ecosystems – A hypothesis. Ecology 51:1046–1051.
Myneni RB, Dong J, Tucker CJ, Kaufmann RK, Kauppi PE, Liski J, Zhou L, Alexeyev V, Hughes MK (2001) A large carbon sink in the woody biomass of northern forests. Proc Natl Acad Sci USA 98:14784–14789.
Naik V, Mauzerall DL, Horowitz LW, Schwarzkopf MD, Ramaswamy V, Oppenheimer M (2007) On the sensitivity of radiative forcing from biomass burning aerosols and ozone to emission location. Geophys Res Lett 34:1–5.
Narayan C, Fernandes P, van Brusselen J, Schuck A (2007) Potential for CO_2 emissions mitigation in Europe through prescribed burning in the context of the Kyoto Protocol. For Ecol Manage 251:164–173.
Nepstad D, Lefebvre P, Da Silva UL, Tomasella J, Schlesinger P, Solórzano L, Moutinho P, Ray D, Benito JG (2004) Amazon drought and its implications for forest flammability and tree growth: A basin-wide analysis. Glob Chang Biol 704–717.
New M, Lister D, Hulme M, Makin I (2002) A high-resolution data set of surface climate over global land areas. Clim Res 21:1–25.
Ohlson M, Tryterud E (2000) Interpretation of the charcoal record in forest soils: Forest fires and their production and deposition of macroscopic charcoal. Holocene 10: 519–525.
Page SE, Siegert F, Rieley JO, Boehm H-DV, Jaya A, Limin S (2002) The amount of carbon released from peat and forest fires in Indonesia during 1997. Nature 420:61–65.
Parisien M-A, Moritz MA (2009) Environmental controls on the distribution of wildfire at multiple spatial scales. Ecol Monogr 79:127–154.
Paula, S, Arianoutsou, M, Kazanis, D, Tavsanoglu, C, Lloret, F, Buhk, C, Ojeda, F, Luna, B, Moreno, JM, Rodrigo, A, Espelta, JM, Palacio, S, Fernández-Santos, B, Fernandes, PM (2009) Fire-related traits for plant species of the Mediterranean basin. Ecology 90:1420–1420.
Pausas JG, Bradstock RA, Keith DA, Keeley JE (2004) Plant functional traits in relation to fire in crown-fire ecosystems. Ecology 85:1085–1100.
Pausas JG, Keeley JE (2009) A burning story: The role of fire in the history of life. BioScience 59:593–601.
Pechony O, Shindell DT (2009) Fire parameterization on a global scale. J Geophys Res 114:1–10.
Pfister G, Hess PG, Emmons LK, Lamarque J-F, Wiedinmyer C, Edwards DP, Pétron G, Gille JC, Sachese GW (2005) Quantifying CO emissions from the 2004 Alaskan wildfires using MOPITT CO data. Geophys Res Lett 32. doi: 10.1029/2005GL022995.
Piñol J, Terradas J, Lloret F (1998) Climate warming, wildfire hazard, and wildfire occurrence in coastal eastern Spain. Clim Change 38:345–357.
Pitman AJ, Hesse P (2007) The significance of large scale land cover change on the Australian palaeomonsoon. Quat Sci Rev 26:189–200.
Power MJ, Whitlock C, Bartlein PJ, Stevens LR (2006) Fire and vegetation history during the last 3800 years in northwestern Montana. Geomorphology 75:420–436.
Power MJ, Marlon J, Ortiz N, Bartlein PJ, Harrison SP, Mayle FE, Ballouche A, Bradshaw RHW, Carcaillet C, Cordova C, et al. (2008) Changes in fire regimes since the Last Glacial Maximum: An assessment based on a global synthesis and analysis of charcoal data. Clim Dyn 30:887–907.
Power MJ, Marlon JR, Bartlein PJ, Harrison SP (2010) Fire history and the Global Charcoal Database: A new tool for hypothesis testing and data exploration. Palaeogeogr Palaeoclimatol Palaeoecol 291:52–59.
Pyne SJ (1995) World Fire: The Culture of Fire on Earth (384 pp). Seattle, WA: University of Washington Press.
Pyne SJ, Andrews PL, Laven RD (1996) Introduction to Wildland Fire (769 pp). New York, NY: Wiley.
Ramanathan V, Carmichael G (2008) Global and regional climate changes due to black carbon. Nat Geosci 1:221–227.
Randerson JR, Thompson MV, Conway TJ, Fung I, Field CB (1997) The contribution of terrestrial sources and sinks to trends in the seasonal cycle of atmospheric carbon dioxide. Glob Biogeochem Cycles 11:535–560.
Randerson JT, Liu H, Flanner MG, Chambers SD, Jin Y, Hess PG, Pfister G, Mack MC, Treseder KK, Welp LR, Chapin FS, Harden JW, Goulden ML, Lyons E, Neff JC, Schuur EAG, Zender CS (2006) The impact of boreal forest fire on climate warming. Science 314:1130–1132.
Randerson JT, Van der Werf GR, Giglio L, Collatz GJ, Kasibhatla PS (2007). Global Fire Emissions Database, Version 2 (GFEDv2.1). Available on-line [http://daac.ornl.gov/] from Oak Ridge National Laboratory Distributed Active Archive Center, Oak Ridge, Tennessee, USA. doi: 10.3334/ORNLDAAC/849.
Rasker R (2009) Cost of fighting fires on public lands (p. 4). Bozeman, MT: Headwaters Economics.
Riaño D, Ruiz J, Isidoro D, Ustin S (2007) Global spatial patterns and temporal trends of burned area between 1981 and 2000 using NOAA-NASA Pathfinder. Glob Chang Biol 13:40–50.
Rinsland CP, Luo M, Logan JA, Beer R, Worden HM, Worden JR, Bowman K, Kulawik SS, Rider D, Osterman G, Gunson M, Goldman A, Shephard M, Clough SA, Rodgers C, Lampel M, Chiou L (2006) Measurements of carbon monoxide (CO) distributions by the tropospheric emission spectrometer instrument onboard the Aura spacecraft: Overview of analysis approach and examples of initial results. Geophys Res Lett 33. doi: 10.1029/2006GL027000.

Roy DP, Boschetti L, Justice CO, Ju J (2008) The Collection 5 MODIS Burned Area Product – Global Evaluation by Comparison with the MODIS Active Fire Product. Remote Sensing Environ 112:3690–3707.

Running SW (2006) Is global warming causing more, larger wildfires? Science 313:927–928.

Russell-Smith J, Yates C, Whitehead P, Smith R, Craig R, Allan G, Thackway R, Frakes I, Cridland S, Meyer M (2007) Bushfires 'down under': Patterns and implications of contemporary Australian landscape burning. Int J Wildland Fire 16:361–377.

Schafer JS, Eck TF, Holben BN, Artaxo P, Yamasoe MA, Procopio AS (2002) Observed reductions of total solar irradiance by biomass-burning aerosols in the Brazilian Amazon and Zambian Savanna. Geophys Res Lett 29. doi: 10.1029/2001GL014309.

Scholes RJ, Archer SR (1997) Tree-grass interactions in savannas. Annu Rev Ecol Syst 28:517–544.

Scholze M, Knorr W, Arnell NW, Prentice IC (2006) A climate-change risk analysis for world ecosystems. Proc Natl Acad Sci USA 103:13116–13120.

Schultz MG, Heil A, Hoelzemann JJ, Spessa A, Thonicke K, Goldammer JG, Held AC, Pereira JMC, van het Bolscher M (2008) Global wildland fire emissions from 1960 to 2000. Glob Biogeochem Cycles 22:1–17.

Schwilk DW, Ackerly DD (2001) Flammability and serotiny as strategies: Correlated evolution in pines. Oikos 94: 326–336.

Schwilk DW, Kerr B (2002) Genetic niche hiking: An alternative explanation for the evolution of flammability. Oikos 99:431–442.

Scott AC (2000) The Pre-Quaternary history of fire. Palaeogeogr Palaeoclimatol Palaeoecol 164:281–329.

Scott AC, Glasspool IJ (2006) The diversification of Paleozoic fire systems and fluctuations in atmospheric oxygen concentration. Proc Natl Acad Sci USA 103: 10861–10865.

Sherwood S (2002) A microphysical connection among biomass burning, cumulus clouds, and stratospheric moisture. Science 295:1272–1275.

Shindell D, Faluvegi G (2009) Climate response to regional radiative forcing during the twentieth century. Nat Geosci 2:294–300.

Shlisky A, Waugh J, Gonzalez P, Gonzalez M, Manta M, Santoso H, Alvarado E, Nuruddin AA, Rodríguez-Trejo DA, Swaty R, Schmidt D, Kaufmann M, Myers R, Alencar A, Kearns F, Johnson D, Smith J, Zollner D (2007). Fire, ecosystems and people: Threats and strategies for global biodiversity conservation. In The Nature Conservance Global Fire Initiative (p. 17). The Nature Conservancy Global Fire Initiative Technical Report 2007-2, Boulder, CO.

Shuman B, Webb III T, Bartlein PJ, Williams JW (2002) The anatomy of a climatic oscillation: Vegetation change in eastern North America during the Younger Dryas chronozone. Quat Sci Rev 21:1777–1791.

Silva JMN, Pereira JMC, Cabral AI, Sá ACL, Vasconcelos MJP, Mota B, Grégoire J-M (2003) An estimate of the area burned in southern Africa during the 2000 dry season using SPOT-VEGETATION satellite data. J Geophys Res 108. doi: 10.1029/2002JD002320.

Skroch M, Swetnam TW (2008) Public Should Accept More Frequent, but Less Destructive, Forest Fires. Tucson, AZ: Arizona Daily Star.

Smith AMS, Wooster MJ (2005) Remote classification of head and backfire types from MODIS fire radiative power observations. Int J Wildland Fire 14:249–254.

Stephenson N (1998) Actual evapotranspiration and deficit: Biologically meaningful correlates of vegetation distribution across spatial scales. J Biogeogr 25:855–870.

Stevenson J, Hope G (2005) A comparison of late Quaternary forest changes in New Caledonia and northeastern Australia. Quat Res 64:372–383.

Stocks BJM, Mason JA, Todd JB, Bosch EM, Wotton BM, Amiro BD, Flannigan MD, Hirsch KG, Logan KA, Martell DL, Skinner WR (2002) Large forest fires in Canada, 1959–1997. J Geophys Res Atmos. doi: 10.1029/2001JD000484, 2003

Sukhinin AI, French NHF, Kasischke ES, Hewson JH, Soja AJ, Csiszar IA, Hyer EJ, Loboda T, Conrad SG, Romasko VI, Pavlichenko EA, Miskiv SI, Slinkina OA (2004) AVHRR-based mapping of fires in Russia: New products for fire management and carbon cycle studies. Remote Sensing Environ 93:546– 564.

Swetnam TW (1993) Fire history and climate change in giant sequoia groves. Science 262:885–889.

Syphard AD, Radeloff VC, Hawbaker TJ, Stewart SI (2009) Conservation threats due to human-caused increases in fire frequency in Mediterranean-climate ecosystems. Conserv Biol 23:758–769.

Tacconi L (2003). Fires in Indonesia: Causes, Costs and Policy Implications. Occasional Paper No. 38. Jakarta, Indonesia: Center for International Forestry Research.

Tansey K, Grégoire J-M, Defourny P, Leigh R, Pekel J-F, van Bogaert E, Bartholomé E (2008) A new, global, multi-annual (2000–2007) burnt area product at 1 km resolution. Geophys Res Lett 35. doi: 10.1029/2007GL031567.

Thonicke K, Venevsky S, Sitch S, Cramer W (2001) The role of fire disturbance for global vegetation dynamics: Coupling fire into a dynamic global vegetation model. Glob Ecol Biogeogr 10:661–677.

Thonicke K, Prentice IC, Hewitt C (2005) Modeling glacial-interglacial changes in global fire regimes and trace gas emissions. Glob Biogeochem Cycles 19. doi: 10.1029/2004GB002278.

Tilman D, Lehman C (2001) Human-caused environmental change: Impacts on plant diversity and evolution. Proc Natl Acad Sci 98:5433–5440.

Tolonen K (1986) Charred particle analysis. In BE Berglund (Ed), Handbook of Holocene Palaeoecology and Palaeohydrology (pp. 485–490). New York, NY: Wiley.

Tomich TP, Thomas DE, van Noordwijk M (2004) Environmental services and land use change in Southeast Asia: From recognition to regulation or reward? Agric Ecosyst Environ 104:229–244.

USDA (2006) Audit Report: Forest Service Large Fire Suppression Costs (p. 48). United States Department of Agriculture.

van der Werf GR, Randerson JT, Collatz GJ, Giglio L, Kasibhatla PS, Avelino A, Olsen SC, Kasischke ES (2004) Continental-scale partitioning of fire emissions during the 1997–2001 El Nino/La Nina period. Science 303:73–76.

van der Werf GR, Randerson JT, Giglio L, Collatz GJ, Kasibhatla PS (2006) Interannual variability in global biomass burning emission from 1997 to 2004. Atmos Chem Phys 6:3423–3441.

van der Werf GR, Dempewolf J, Trigg SN, Randerson JT, Kasibhatla PS, Giglio L, Murdiyarso D, Peters W, Morton DC, Collatz GJ, Dolman AJ, DeFries RS (2008a) Climate regulation of fire emissions and deforestation in equatorial Asia. Proc Natl Acad Sci USA 105:20350–20355.

van der Werf GR, Randerson JT, Giglio L, Gobron N, Dolman AJ (2008b) Climate controls on the variability of fires in the tropics and subtropics. Glob Biogeochem Cycles 22. doi: 10.1029/2007GB003122, 2008

van Wagtendonk JW (2007) The history and evolution of wildland fire use. Fire Ecol Spec Issue 3:3–17.

Veblen TT, Kitzberger T, Villalba R, Donnegan J (1999) Fire history in northern Patagonia: The roles of humans and climatic variation. Ecol Monogr 69:47–67.

Venevsky S, Thonicke K, Sitch S, Cramer W (2002) Simulating fire regimes in human-dominated ecosystems: Iberian Peninsula case study. Glob Chang Biol 8:984–998.

Walsh MK, Whitlock C, Bartlein PJ (2008) A 14,300-year-long record of fire-vegetation-climate linkages at Battle Ground Lake, southwestern Washington. Quaternary Research 70:251–264.

Westerling AL, Gershunov A, Brown TJ, Cayan DR, Dettinger MD (2003) Climate and wildfire in the western United States. Bull Am Meteor Soc 84:595–604.

Westerling AL, Hidalgo HG, Cayan DR, Swetnam TW (2006) Warming and earlier spring increase western US forest wildfire activity. Science 313:940–943.

Whelan RJ (1995) The Ecology of Fire (346 pp). Cambridge, NY: Cambridge University Press.

Wiedinmyer C, Neff J (2007) Estimates of CO_2 from fires in the United States: Implications for carbon management. Carbon Balance Manag 2. doi: 10.1186/1750–0680–1182–1110.

Williams R, Griffiths A, Allan G (2002) Fire regimes and biodiversity in the wet-dry tropical savanna landscapes of northern Australia. In R Bradstock, J Williams, A Gill (Eds), Flammable Australia: Fire Regimes and Biodiversity of a Continent (pp. 281–304). Cambridge: Cambridge University Press.

Williams R, Hutley L, Cook G, Russell-Smith J, Edwards A, Chen X (2004) Assessing the carbon sequestration potential of mesic savannas in the Northern Territory, Australia: Approaches, uncertainties and potential impacts of fire. Funct Plant Biol 31:415–422.

Williams J, Shuman B, Bartlein P (2008) Rapid responses of the prairie-forest ecotone to early Holocene aridity in mid-continental North America. Glob Planet Change 66:195–207.

Wilson JB, King WM (1995) Human-mediated vegetation switches as processes in landscape ecology. Landsc Ecol 10:191–196.

Winkler H, Formenti P, Esterhuyse D, Swap RJ, Helas G, Annegarn H, Andreae M (2008) Evidence for large-scale transport of biomass burning aerosols from sunphotometry at a remote South African site. Atmos Environ 42: 5569–5578.

Zoltai S (1993) Cyclic development of permafrost in the peatlands of northwestern Alberta, Canada. Arct Alp Res 25:240–246.

Vanishing Polar Ice Sheets

Nancy A.N. Bertler and Peter J. Barrett

Abstract Global temperatures have increased by 0.8°C since instrumental records began, and the last decade has been the warmest. This warming has been closely linked to increasing greenhouse gas concentrations due to human activities. Arctic warming is leading to thinning and disappearance of sea ice on the Arctic Ocean and the increased melting of the Greenland ice sheet. In the last decade the Antarctic ice sheet has also begun to respond. Shrinking ice sheets have significant implications for sea level rise. Knowledge of their behaviour in the geological past, along with climate and ice sheet modelling, provides guidance on what we can expect in the future. Over the past 50 million years (m.y.), the world has evolved from a largely ice-free Greenhouse Earth with atmospheric CO_2 concentrations 3–6 times pre-industrial levels, and sea level over 60 m higher, into an Icehouse Earth with lower CO_2 concentrations. Concentrations were low enough by 34 m.y. ago to reduce temperatures and allow the formation of the first big Antarctic ice sheets, and further cooling and reduction in greenhouse gases followed until the continental Northern Hemisphere ice sheets formed about 2.5 m.y. ago. Trapped gases in ice cores allow us to estimate past atmospheric CO_2 levels quite accurately for the last 800,000 years. Geochemical estimates of earlier CO_2 levels are much less accurate but nevertheless consistently indicate these have not exceeded 400 ppm in the last 24 m.y. Atmospheric CO_2 concentration has increased from 280 ppm in the pre-industrial times to 390 ppm in early 2010. Earth's atmosphere last had such concentrations ~3 million years ago when geochemical measurements indicate global surface temperature was ~3°C warmer than today, and global sea-level was ~25±5 m higher. This implies the loss of the Greenland ice sheet, the West Antarctic ice sheet and a part of the much larger East Antarctic ice sheet. Both field evidence and modelling studies imply rates of ice loss are slow, of the order of a meter or two of sea level equivalent per century, but there are significant uncertainties because key boundary conditions for ice sheet collapse are not well understood. Feed-back effects on atmospheric warming and ice loss could well accelerate the process. The Arctic is today warming as fast as, or faster than, any other area on earth, with the imminent loss of summer sea-ice and increasing loss of ice from Greenland (adding to rising global sea level). Antarctic warming has also begun to rise, despite its suppression from the loss of the ozone hole in the last three decades. Nevertheless the warming of shallow to intermediate Southern Ocean waters and the rise of Antarctic surface temperatures has led to the collapse of ice shelves around the Antarctic Peninsula since 1990 and in the last decade the acceleration of ice loss from West Antarctica, as well as parts of East Antarctica. The thermal inertia of oceans and ice sheets means that they take decades to centuries to respond but once ice loss has begun it could continue for millennia. The ice sheets have now begun to shrink. Both modelling and palaeodata indicate a significant risk of setting in train events leading to their inevitable loss unless the release of CO_2 from fossil fuels is curtailed and CO_2 sinks are found to return atmospheric CO_2 levels to well below 400 ppm.

N.A.N. Bertler (✉)
Antarctic Research Centre, Victoria University of Wellington, Wellington 6012, New Zealand; GNS Science, National Isotope Centre, Lower Hutt 5040, New Zealand
e-mail: Nancy.Bertler@vuw.ac.nz

J. Dodson (ed.), *Changing Climates, Earth Systems and Society*, International Year of Planet Earth,
DOI 10.1007/978-90-481-8716-4_4,

Keywords Antarctica · Arctic · Sea-level · Ice sheet stability · Marine sediment cores · Ice cores · Sea-ice

Polar Ice Sheets – Drivers and Recorders of the Global Climate System

The two polar regions host the last great ice sheets of the planet, the Antarctic Ice Sheets in the South and Greenland Ice Sheet in the North. The dimension, characteristics and behaviour of these ice sheets are remarkably different, owing largely to the asymmetry of the hemispheres (Fig. 1). While the North Pole lies in the middle of the Arctic Ocean, surrounded by the Eurasian and North American continents, the Antarctic continent occupies the South Pole and is surrounded by the Southern Ocean.

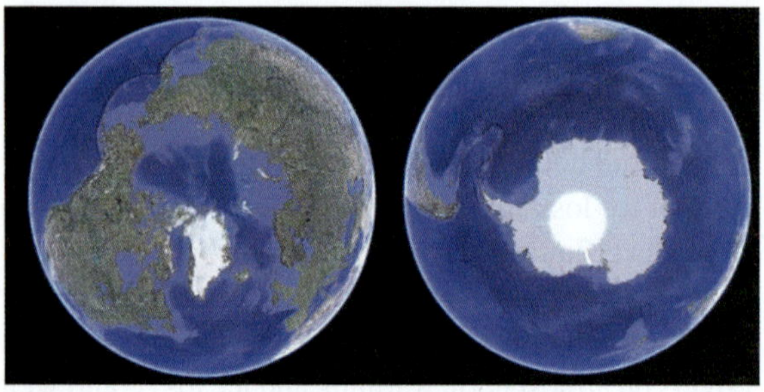

Fig. 1 The great ice sheets: Greenland (*left image*) in the Northern Hemisphere and Antarctica (*right image*) in the Southern Hemisphere. Images obtained from Google Earth 2009 (Data SIO, NOAA, U.S. Navy, NGA, GEBCO)

The Polar-Tropical Power Struggle

Due to the tilt of the Earth's axis (currently 23.4° relative to our orbital plane around the sun), we experience seasons, with the most extreme seasonal change occurring in latitudes above 66° North and South. These areas experience 24 h of light during polar summers and 24 h of darkness during polar winters. As a consequence during the Southern Hemisphere summer, Antarctica receives more solar radiation than any other region on the planet. However, during the depths of the austral winter, the sun never rises and Antarctica experiences its polar night, with the coldest temperatures ever measured at Vostok Station (–89.2°C). While Antarctica endures winter, Greenland enjoys summer, and vice versa. The change over from the full polar day to the full polar night occurs at the South Pole from mid-September to end of October and at the North Pole from early March to mid-April with rapidly shortening days and wonderful protracted sunsets. The reverse is true for polar sunrises.

The tropical regions in contrast are more or less perpendicular to the incoming solar radiation and hence receive the global annual maximum radiation with little seasonal variability. Consequently, the tropics receive an almost constant amount of solar radiation (~342 W/m^2). Averaged over the year, this is about five times the amount received at the poles. In addition, the white surface of the ice sheets and sea-ice reflect up to 90% of the solar radiation back into space, which further cools the polar regions. This effect is enhanced by the dry atmosphere and clear skies over the poles allows even more energy to escape back into space (King and Turner, 1997).

Thus the tropical regions warm and the polar regions cool the Earth, with the temperature gradient between tropics and poles driving the major atmospheric and ocean circulation pattern. The gradient is currently greater in the Southern Hemisphere because of the huge pole-centered Antarctic ice sheet. However, this has also varied over geological time as ice sheets have come and gone in both polar regions. A larger gradient, i.e. a greater temperature difference between polar and tropical regions results in more vigorous atmospheric circulation (King and Turner, 1997). An increased gradient also enhances sea-ice formation in the polar regions and evaporation in the tropical oceans, which together with the wind forcing of the ACC and associated upwelling, help drive the thermohaline ocean circulation, transporting vast amounts of heat, salt, nutrients and gases around the globe (Broecker, 1997).

Antarctica

Antarctica is the highest, driest, windiest and coldest continent on Earth, with an average elevation of approximately 2,200 m. The continent hosts two distinct ice sheets: the predominantly land-based East Antarctic Ice Sheet (EAIS) and the smaller largely marine-based West Antarctic Ice Sheet (WAIS), which is grounded to depths of almost 2,000 m below sea-level (Fig. 2). The two ice sheets are separated by the

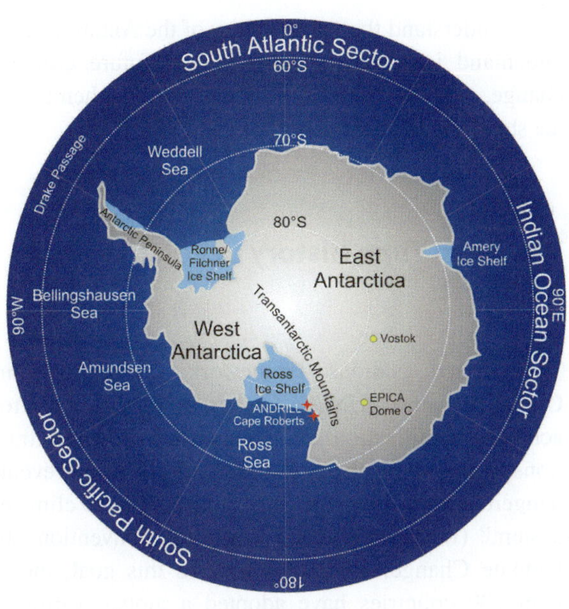

Fig. 2 Map of Antarctica, showing major ice shelves (*light blue shading*), Vostok and EPICA Dome C Ice Core records (*yellow circles*), and ANDRILL and Cape Roberts marine sediment cores (*red stars*)

Transantarctic Mountains, which extend over 3,500 km from the Ross Sea to the Weddell Sea. The WAIS occupies a significantly smaller area (1,970,000 km^2) than the EAIS (10,350,000 km^2). Together they hold about 70% of the world's fresh water, the equivalent of about 57 m global sea-level increase, with about 5 m stored within the WAIS and about 52 m within the EAIS (Lythe et al., 2001). A third distinct area is the Antarctic Peninsula with small ice caps and glaciers holding the equivalent of just 0.5 m of sea level rise (Lythe et al., 2001). This most northern part of the Antarctic has recorded over the past decades the strongest temperature increase on the globe (0.5°C/decade) (Turner et al., 2009).

There are three different types of ice in Antarctica: glacial ice that forms the large ice sheets, ice shelves, which are formed by glacial ice sliding off the continent into the ocean, where it floats northward but remains attached to the continent, and sea-ice, which is frozen sea water. Snow accumulates slowly in the Antarctic interior due to the cold dry air over the continent, the annual snow accumulation averaging between 3 and 5 cm of water equivalent per year. As snow accumulates the overlying weight of new snow compacts older snow to firn and with burial under tens of metres of firn eventually to ice. In this way annual layers of snow end up forming ice sheets over 4,000 m thick.

While the EAIS as a feature is thought to have covered East Antarctica for many millions of years, the ice itself is much younger. The reason for this is that the ice is not stagnant but moves from the central ridgelines to the ocean like slow-moving rivers of ice. If the ice is lost faster to the ocean than replenished through snowfall in the interior, the ice sheet as a whole is losing mass and shrinks. If snow accumulates faster than ice slides into the ocean, the ice sheet is gaining mass and grows. As Antarctica as a whole is holding the equivalent of 57 m of global sea-level, even a 1% change in its mass balance equates to over half a metre change in global sea-level.

The oldest continuous ice record was drilled by the European drilling consortium EPICA at Dome C in 2004. The 3,270 m long ice core record spans almost 1,000,000 years (EPICA Community Members, 2004). However, it has been suggested that the EAIS might contain even older ice in other places, perhaps up to 1.4 m.y. (million years) old (IPICS, 2008). Even older ice has been found in the form of remnants from ancient glaciers that once flowed through the McMurdo Dry Valleys. At one location the ice lies beneath rock debris with an overlying volcanic ash dated at 8 m.y. (Sugden et al., 2002).

Antarctic ice sheets are colder than the Greenland Ice Sheet, as the southern continent is more or less centered over the South Pole and covers a much greater area than Greenland. In addition, the circumpolar currents of the Southern Ocean and circumpolar westerly winds further isolate Antarctica through their zonal pattern (parallel to lines of latitude) (Turner et al., 2009). In contrast, Greenland receives warmer air masses and ocean currents that are forced northward by the American continent from warmer regions at lower latitudes (King and Turner, 1997).

Greenland

Greenland is the world's largest island, and hosts an ice sheet about the same size (1,700,000 km^2, Bamber et al., 2009) as the WAIS. The Greenland ice sheet holds the equivalent of approximately 7 m global sea-level rise. In contrast to the marine based WAIS, the Greenland ice sheet rests on bedrock that lies predominantly above sea-level. The average height of the ice sheet is about 2,100 m, with a maximum thickness of 3,411 m. Owing to warmer temperatures, Greenland receives substantially higher snow precipitation than Antarctica, which affords higher resolution ice core records (Mayewski and White, 2002; Steffensen et al., 2008) (Fig. 3). However, this causes ice to flow faster from the Greenland interior to the ocean, and hence the ice is much younger at the base (maximum age of around 150,000–200,000 years) (Buchardt and Dahl-Jensen, 2008) than the ice at the base of the Antarctic Ice Sheet (EPICA Community Members, 2004). Greenland has only small ice shelves. While almost all ice loss in Antarctica occurs through calving of icebergs from ice shelves into the ocean, in Greenland ice loss occurs through calving and surface ablation (surface melt and evaporation) in equal amounts (Pritchard et al., 2009). Moreover in contrast to Antarctica, Greenland has an indigenous population, which are thought to have reached the island about 4,500 years ago. Greenland's population has recently exceeded 50,000 inhabitants, with most living in coastal settlements (ACIA, 2005).

Fig. 3 Map of Greenland, showing locations of major towns (*red circles*) and ice core records (*yellow circles*)

To understand the vulnerability of the Antarctic and Greenland ice sheets to current and future climate change, we have to travel back to the time when these ice sheets first came into existence.

The Deep Time History of Polar Ice Sheets

The United Nations Framework Convention on Climate Change (UNFCCC) has as its objective to achieve "stabilization of greenhouse gas concentrations in the atmosphere at a level that would prevent dangerous anthropogenic interference with the climate system" (United Nations Framework Convention on Climate Change, 1992). To achieve this goal, more than 100 countries have adopted a global warming limit of 2°C or below (relative to pre-industrial levels) as a guiding principle for mitigation efforts to reduce climate change risks, impacts and damages, according to Meinshausen et al. (2009). More recently, the 2°C target was also noted in the UNFCCC Conference of the Parties meeting in Copenhagen in December 2009 under the "Copenhagen Accord." However, even if emissions in 2050 are reduced to half of 1990 levels, Meinshausen and colleagues (2009) estimate a 12–45% probability of exceeding 2°C for a range of climate sensitivities. This equates to a greenhouse gas concentration roughly equivalent to 450 parts per million (CO_{2e} ppm), However, the IPCC "business as usual" scenario (A1F1 – no mitigation efforts) projects an increase of CO_{2e} concentrations centred on 1,000 ppm, more than 3.5 times pre-industrial levels (280 ppm) by the end of the century (IPCC, 2007c). To study a world with such high CO_2 concentrations and its effect on temperature and sea-level, we have to look back more than 30 million years to a world without ice sheets. This section looks at that world, and how the Antarctic and Greenland ice sheets first formed, along with their history in more recent times.

Continents in Motion

The distribution of continents and oceans has changed dramatically over long geological periods. From 500 to 200 m.y. ago, the Earth's landmasses were combined in the supercontinent Pangaea. About 200 m.y. ago, seafloor spreading tore Pangaea into two pieces: Laurasia, which drifted northward to form the modern continents of North America, Greenland, Europe and Asia (Eurasia), and southward drifting Gondwana, which comprised the modern fragments of Antarctica, Africa, South America, India, Australia and New Zealand (Fig. 4). About 180 m.y. ago, seafloor spreading began to break Gondwana apart, beginning with the drift of Africa and India northward from Antarctica, the keystone continent, followed later by the other pieces. As this was happening Antarctica moved southward, reaching the polar region around 120 m.y. ago. It has remained there to this day.

As the other fragments drifted away from Antarctica, they opened the South Atlantic, Indian and Southern Oceans (Lawver and Gahagan, 2003) (Fig. 5). These geographic changes had a significant effect on climate because of the changing shape and size of the ocean basins and also through the growth of mountain ranges where continents collided. The movements of the continents have been tracked and dated using magnetic lineations on the ocean floor and ages of new ocean crust, so that past geography for any particular time is known with some confidence (Fig. 5). This allows us to use climate models developed for understanding weather and climate of today's Earth and also to understand climate behaviour on Earth in the distant past. However, these models also need to be checked and constrained by data from past climates, particularly past temperatures and CO_2 levels. The ways in which these are gauged are explained in section "Gauging Past Temperatures and CO_2 Levels."

Gauging Past Temperatures and CO_2 Levels

The ice sheets that are currently a permanent feature of our polar regions developed a little over 30 m.y. ago, as a consequence of a major change in Earth's climate. Our review of this history leans heavily on the framework that has been developed over the last two centuries as geologists have measured, correlated and dated layers of sedimentary strata and their features around the world. In the last few decades, however,

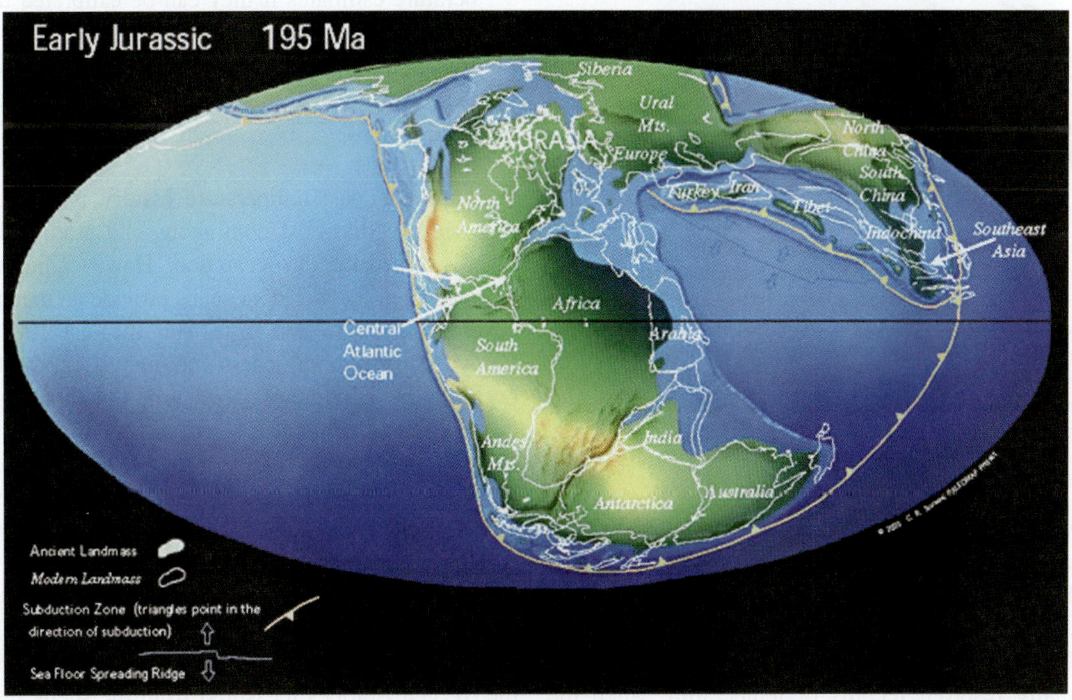

Fig. 4 Location and constellation of the Gondwana landmasses 195 Ma ago (Source: http://www.scotese.com/pangeanim.htmUH)

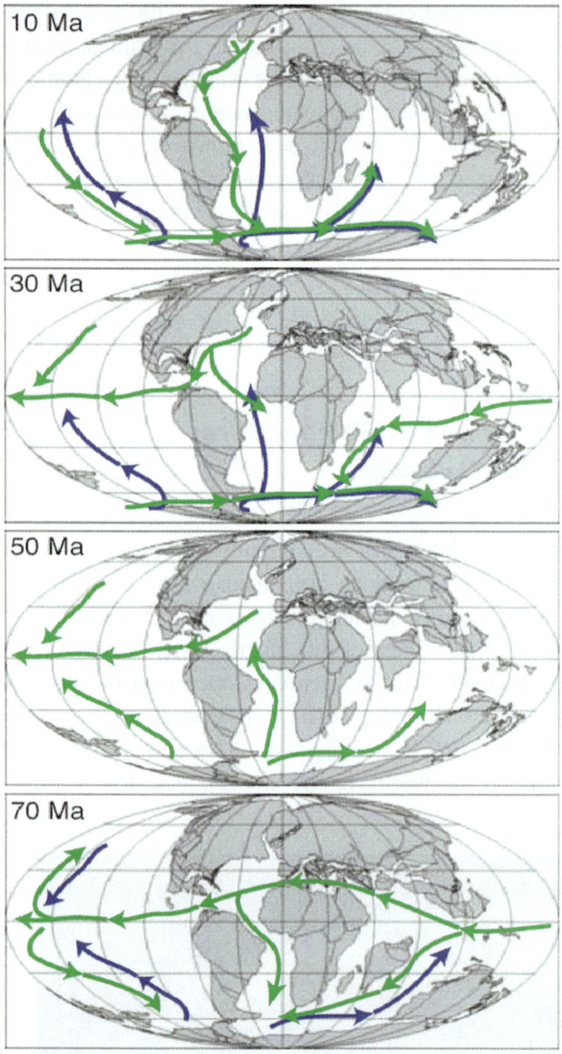

Fig. 5 Earth's changing geography of continents and oceans over the last 70 m.y. (from Ben Cramer, unpublished, with permission). The continents have largely retained their latitudinal positions, with the outstanding exception of India, which has moved from the Southern to the Northern Hemisphere and collided with Asia to form the Himalayas. The Atlantic, Indian and Southern Oceans Ocean have grown at the expense of the Pacific Ocean

new advances have made it possible to estimate temperature and atmospheric CO_2 levels from ice cores and sediments for time periods long before instruments became available (Ruddiman, 2001).

Ice cores provide continuous and at times subannually resolved records of temperature when the snow fell thousands of years ago. This thermometer can be read through measurements of the ratio of light to heavy hydrogen and also light to heavy oxygen atoms in the ice (Alley, 2000b; Mayewski and White, 2002). The atmospheric CO_2 levels can be measured directly from air bubbles trapped in the ice. However, ice in the low and mid latitudes is only a few decades to hundreds of years old (and shrinking fast) and even Greenland ice goes back only 130,000 years. The Antarctic ice record extends back at least 850,000 years, and scientists are seeking locations where the ice may be over 1 million years old. However, to obtain temperatures from lower latitudes or to look back beyond even this relatively brief period of geological time, we have to rely on data from sediments and fossils.

Beyond ice cores, there are three main ways of estimating past temperatures:

i. Geochemical analysis of fossil material that preserve isotopic (e.g. $\delta^{18}O$), elemental (e.g. Mg/Ca), or compound (e.g. Alkenone) ratios reflecting water temperature at the time of the organism's growth.
ii. Identifying fossils with known temperature ranges, e.g. reptiles, plants and many marine microfossils. (These two approaches can provide actual temperature estimates if they can be calibrated using present day measurements or species.)
iii. Recognising sediments that form in under particular climates, e.g. glacial deposits, desert sand or coal.

Estimating past temperature history is still difficult because the most continuous and readily dated records come from the ocean floor. However, hundreds of cores from the world's oceans provide an extensive coverage in space and time over the last 100 million years for a more or less continuous record of the ocean floor temperature. This turns out to be useful because the modern ocean circulation is driven largely by cold salty dense bottom water flowing from the poles towards the equator, so the ocean floor temperatures are widely regarded as a reflection of high latitude surface water temperatures (Miller et al., 2005). The best established and most widespread recorders are calcareous microfossils, some living near the sea surface and others on the ocean floor. A component of the oxygen isotopes measured from the shells results from the effect of isotopically light ice building up on land and depleting the ocean. Current research is

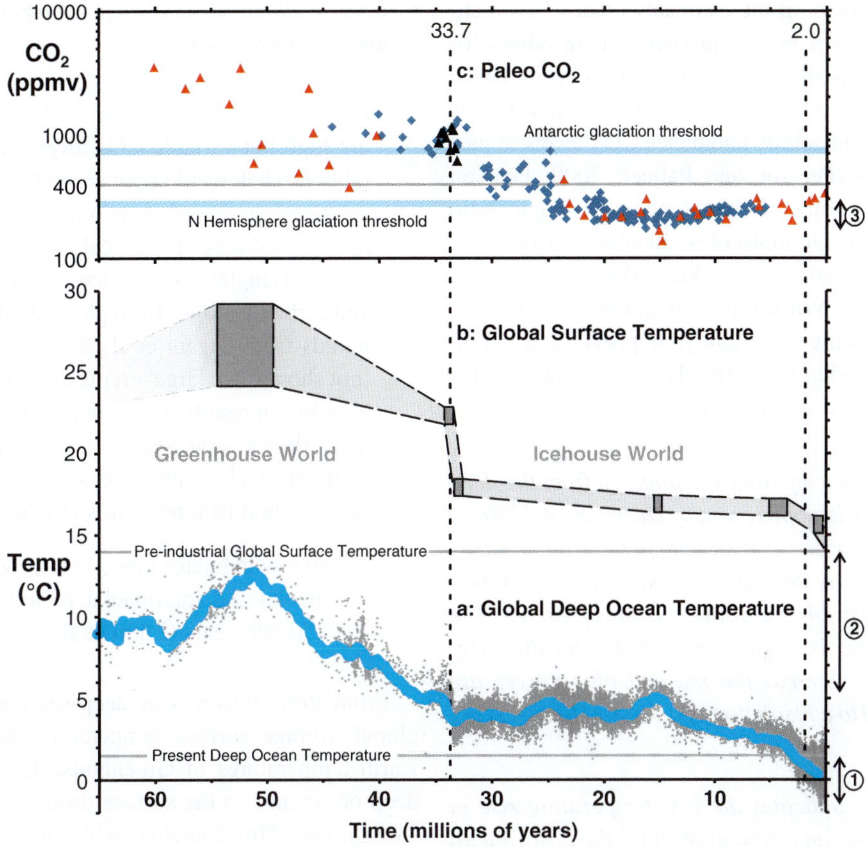

Fig. 6 History of deep ocean and global surface temperature (**a** and **b** on same arithmetic scale) and paleo CO_2 estimates (**c** on a log scale) over the last 65 m.y. See italicised section in the text

aimed at a reliable way for correcting for this "ice volume" effect. For the moment, however, we will follow Hansen et al. (2008) and make a simple adjustment assuming that half of the isotope shift results from this effect when Antarctica has large ice sheets. This can be then subtracted to give a sea floor temperature history (Fig. 6a).

Getting a record of land surface temperature over the same time period is much more difficult because there are far fewer sites with continuous records, typically less than 100 sites to represent the 140 million square km of their earth's land surface. Nevertheless there are a small though growing number of studies that can provide a global average temperature for a few key periods in the past, as shown in Fig. 6b. These use Earth System models to simulate Earth's surface processes and can also incorporate past changes in geography. The models can re-create ocean currents, climate, vegetation and ice cover for particular periods in the past and for various greenhouse gas levels, but they are crucially dependent on geochemical and palaeontological estimates of past temperature from a range of latitudes for a particular time period as an independent guide for checking model results, along with indicators of past CO_2 levels.

CO_2 is the most abundant greenhouse gas (after water vapour) and one that appears to have had a large influence on earth's history (Royer, 2006). Atmospheric CO_2 levels in the past can be estimated rather crudely using several different techniques: carbon isotopes from carbonate concretions in palaeosols (ancient soils), the density of stomata (pores for gas exchange) in fossil leaves, carbon isotopes from marine phytoplankton and boron levels in marine sediments. Because atmospheric gases are constantly mixed CO_2 levels around the world are very similar, and, unlike temperature, sample location does not matter. However, each technique has its complexities that result in errors in the estimates made. Two techniques have shown more consistency than others for the last few tens of millions of years. One is based on boron isotope ratios from calcareous microfossils to estimate ocean acidity and hence CO_2 levels in past surface waters (Pearson and Palmer, 2000; Pearson et al. 2009). Another is based on the isotopic ratio of carbon in organic molecules (alkenones) from phytoplankton (Pagani et al., 2005; 2009). Results are shown in Fig. 6c. While this body of data is still small, and open to some uncertainty, it provides a useful basis for comparing with other key proxy data and for constraining earth system models.

(A) *Deep ocean temperature, estimated from the deep sea oxygen isotope record of Zachos et al. (2001a) but with the ice volume component removed by assuming it is the source for half of the variation for the last 34 m.y. (blue section of curve). The black curve is slightly smoothed from the original data, whereas the red and blue curves are smoothed to a resolution of 0.5 m.y. (Hansen et al., 2008).*

① *Arrow indicates the 2°C temperature rise in deep ocean temperature since the Last Glacial Maximum around 20,000 years ago (Adkins et al., 2002).*

(B) *Global average surface temperature for key time slices determined from climate models checked against geological palaeotemperature proxies.*

② *Arrow indicates 6°C temperature rise in global surface temperature since the Last Glacial Maximum (Schneider von Diemling et al., 2006).*

Estimates for each time slice came from the following sources:

- *Last Interglacial period 120,000–130,000 years ago (Clark and Huybers, 2009)*
- *Mid-Pliocene global surface temperature (Haywood and Valdes, 2004)*
- *Mid-Miocene global surface temperature (You et al., 2009)*
- *Shift in global surface temperature from ~33.8 to 33.5 m.y. (Liu et al., 2009)*
- *Eocene global surface temperatures (Huber, 2008), and references therein.*

(C) Palaeo CO_2 estimates over the last 65 m.y., showing high but variable CO_2 levels in earlier times, falling to below 400 ppm around 24 m.y. ago. The estimates are based on two different techniques – boron isotopes from calcareous microfossils (green triangles from Pearson and Palmer, 2000) and carbon isotopes from phytoplankton (blue diamonds from Pagani et al., 2005). Error estimates (not shown) are large – typically +40% and –20%. New boron results from a detailed study of the climate shift at ~34 Ma (dark green triangles from Pearson et al., 2009) are similar to those from older carbon isotope results (Pagani et al., 2005).

③ Arrow indicates range in CO_2 values between glacial and interglacial periods over the last 850,000 years (Lüthi et al., 2008)

From these histories of deep-sea temperature and global average surface temperature we can see the warm temperatures of Greenhouse Earth in both the deep ocean and on the surface drop sharply at around 34 m.y. ago. This coincides with the formation of the first big Antarctic ice sheets (Zachos et al., 1992) and initiated the Icehouse Earth climate we evolved in. We can also see that during Greenhouse Earth CO_2 levels were mostly high though variable (700–1,500 ppm) until around 30 m.y. ago. Recent new data have provided more detail and better dating for the co-incidence of falling temperature and CO_2 levels below ~750 ppm at 33.7 m.y. ago (Pearson et al., 2009). This is the CO_2 threshold for Antarctic ice sheet growth indicated by a recent modelling study (DeConto et al., 2008). It seems the early ice sheets grew large enough during the fall in CO_2 at around 34 m.y. to survive the return to higher CO_2 levels until around 24 m.y. ago when they sank below ~400 ppm, ensuring the ice sheets persisted until the present day.

Green Antarctica

Even though Antarctica moved onto the pole 120 m.y. ago, fossil records of diverse plant and animal ecosystems show that Antarctica continued to experience a

Fig. 7 Forests in the Antarctic Peninsula 70–100 m.y. ago. Artistic reconstruction based on the work of J. Howe, J.E. Francis, and other British Antarctic Survey geologists. Painted by Robert Nicholls (*upper*). Fossil log showing annual growth rings and *circular burrows* formed and filled after the tree died (*lower*). Photo: Jane Francis

warm climate for many tens of millions of years, supporting large forests, dinosaurs and mammals (Francis et al., 2008) (Fig. 7). This is astonishing, as the polar position of Antarctica caused the continent to experience polar nights, requiring its inhabiting plants and animals to adapt to complete darkness through the winter months, just as they do in the present day, north of the arctic circle. Fossil tree trunks from Antarctica; growing about 60 Ma years ago, contain clear seasonal tree rings, providing further evidence that these trees were well adapted to polar nights.

This period of sustained warmth, Greenhouse Earth, reached its peak at around 50 Ma with global average temperatures 10–15°C higher than present (Huber, 2008) and CO_2 levels ranging from 700 to 1,500 ppm based on phytoplankton and boron estimates. This Greenhouse World was characterised by a much lower temperature gradient between tropics and poles. The evidence is strongest in the Arctic, where fossil palm trees and crocodiles from this period have been found at 80°N (Francis et al., 2008). Most of Antarctica is inaccessible beneath its present ice cover, but some argue there may have been cool periods when it hosted interior ice sheets large enough to change global sea level as much as 20 m (Miller et al., 2005). This issue is not yet resolved.

A remarkable feature of the Greenhouse world is the occasional temperature spike, or hyperthermal event (Bowen, 2006). The largest and best known of these occurred 55 Ma ago, clearly recognised by a sharp shift in both carbon and oxygen isotopes in less than 6,000 years (Kennett and Stott, 1991). Recent estimates indicate deep-ocean temperature rose by ~5°C above an already warm background. The cause has been ascribed to the catastrophic release of around 2,000 Gt of carbon in the form of methane from sub-sea-floor gas hydrates into the atmosphere (Zachos et al., 2005). Methane is many times more effective as a greenhouse gas than CO_2 (IPCC, 2007c). While methane lasts in the atmosphere is only ~10 years, it oxidises to CO_2, a much longer lived gas, perhaps explaining why the core records show that Earth took over 100,000 years for ocean temperature to return to its former level.

Icy Antarctica – Big Dynamic Ice Sheets

As we saw earlier in Fig. 6a, high latitude ocean warmth peaked at ~12°C around 50 m.y. ago, and then declined steadily towards the climate-transforming shift just after 34 m.y. ago. Both north and southern high latitudes cooled the same time (Liu et al., 2009) consistent with a global cause. The most obvious reason and one consistent with climate modelling is a drop in atmospheric CO_2 levels at a time when the Earth's pathway and alignment with the sun caused a reduction in solar energy reaching Earth, inducing a temporary cooling (DeConto et al., 2008). This led to the most significant climatic tipping point in the last 65 m.y. (Kump, 2009). A further indication of this event comes the deep-sea record in the tropical Pacific Ocean. Sediment cores show an increase in the depth below which calcium carbonate dissolves from 3.5 to 5 km, reflecting a less acidic ocean. The timing coincides with the temperature and ice volume changes between 33.9 and 33.6 m.y. (Coxall et al., 2005).

The cause of the long term decline in temperature has been comprehensively reviewed (Hay et al., 2002) but remains speculative. Hay and colleagues considered many possibilities including "(1) variations in solar energy output; (2) changes in the concentration of dust in space; (3) variations in the Earth's orbital motions; (4) continental drift; (5) the effect of vertical movements ('uplift') on the radiation balance and atmospheric circulation; (6) changes in oceanic circulation; (7) changes in sea-ice cover; (8) variations in the concentration of greenhouse gasses in the atmosphere; and (9) increased aerosol content of the atmosphere." In the end, they concluded that the decline in CO_2 levels over the last ~50 m.y. was most likely a factor. Whatever the cause(s), the combination of both direct and circumstantial evidence for a coupled decline in temperature and CO_2 levels, with a sharp drop at around 33.7 m.y. ago, is becoming increasingly robust.

Coincidentally, the last two fragments of Gondwana (Australia and South America) finally drifted northward away from Antarctica; Australia at about 35.5–32 Ma ago (Lawver and Gahagan, 2003; Stickley et al., 2004), and South America over a longer period between 40 and 8 Ma ago (Barker et al., 2007). This allowed the Southern Ocean to surround the continent creating the Antarctic Circumpolar Current (ACC). Today, the ACC is the largest ocean current in the world, extending to the sea floor in places (Böning et al., 2008; Turner et al., 2009). Due to its zonal flow, the development of the ACC further thermally isolated Antarctica from the rest of the world, though even if Drake Passage was opening at 34 Ma this could have brought about only a fifth of the cooling achieved from declining CO_2 levels (DeConto and Pollard, 2009).

The first big Antarctic ice sheets were similar in extent to modern Antarctic ice sheets, but warmer and thinner, and highly dynamic. The most complete record of their behaviour comes from drill cores taken at Cape Roberts on the edge of the continent through a 1,500-m-thick sequence of shallow marine sediments just seaward of the Transantarctic Mountains. The record shown in Fig. 8 spans the period from 33 to 17 m.y. ago, preserving 55 glacial-interglacial cycles, but with many time breaks represented by the unconformities between them. Still, the cycles that have been preserved provide windows into the history of the ice sheet margin as it flowed through the rising Transantarctic Mountains, expanding and contracting in response to changing climate (Naish et al., 2009, 2001; Pollard and DeConto, 2009), with sea-level rising and falling by several tens of meters each time (Dunbar et al., 2008) (Fig. 9).

Despite ice as extensive as today and the new geographic isolation of the Antarctic continent, none of the ice sheets in this period were large enough to destroy the Southern Beech forests. The fossil pollen in the Cape Roberts cores reflects a slight cooling from cool temperate to sub-polar, with tundra and a

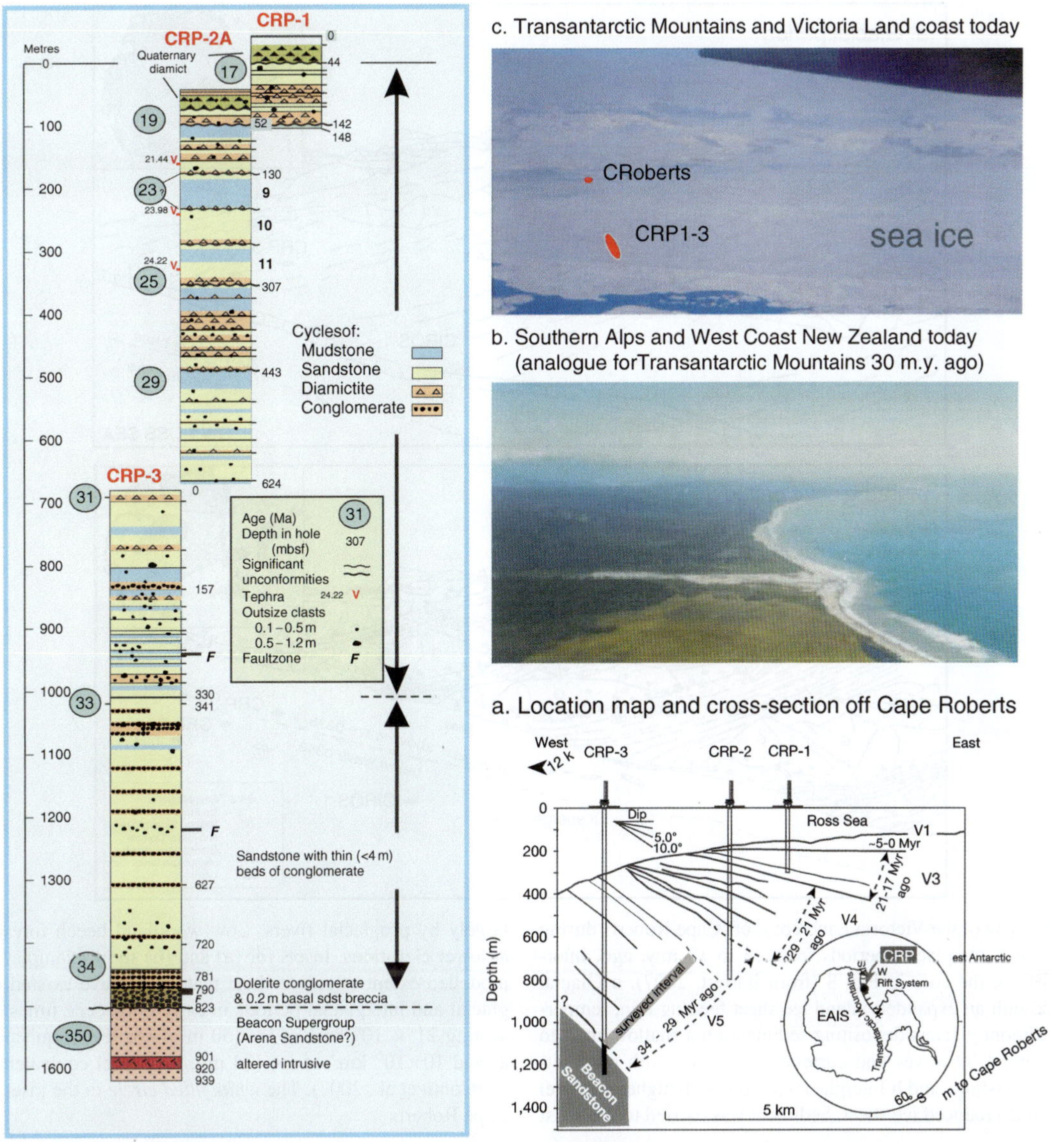

Fig. 8 A log of the Cape Roberts drill core, showing the history of ice sheet advance (diamictite-glacial sediment) and retreat (sand/mud) as well as sea level fall and rise (sand and mud) off the Antarctic Victoria Land coast from 34 to 17 m.y. ago (Barrett, 2007). *Insets* show: **a** Location and setting for the Cape Roberts Project (Naish et al., 2001). **b** Modern climatic analogue for the Victoria Land coast 30 m.y. ago – West Coast of New Zealand's South Island. Photo: Tim Naish, and **c** the Victoria Land coast off Cape Roberts in 1999

low woodland (Prebble et al., 2006), and more recent drilling in McMurdo Sound records the flora persisting and even flourishing in a warm period with summer temperatures of up to 10°C on the adjacent coast around 15 m.y.ago (Warny et al., 2009). Around 100 km inland in the Olympus Range an even younger vegetation record of moss and aquatic life has been preserved in ice-marginal lake deposits beneath a shallow layer of glacial debris and lag gravel (Lewis et al., 2008). However, the beech pollen here are of just one species close to the limit for survival. The deposit is dated by volcanic ash at 14 m.y. old and is currently regarded as the last record of forest vegetation on the Antarctic continent.

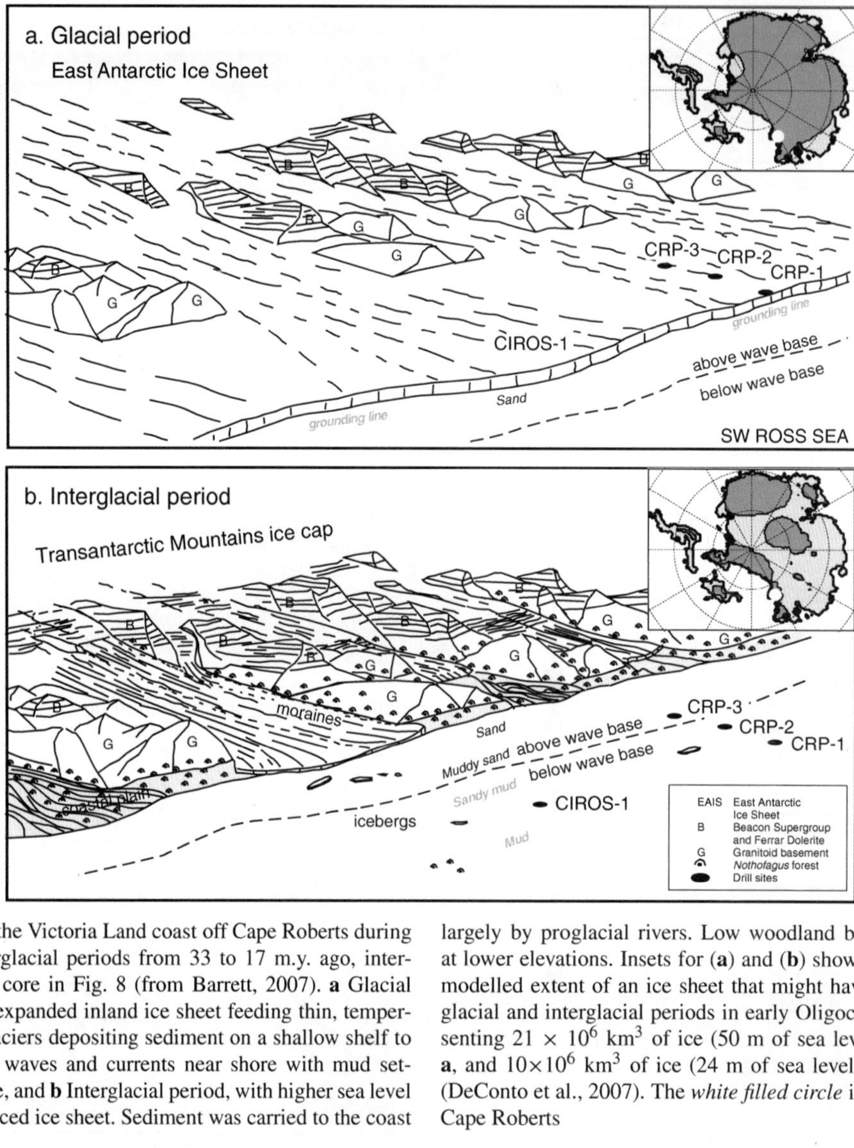

Fig. 9 View of the Victoria Land coast off Cape Roberts during glacial and interglacial periods from 33 to 17 m.y. ago, interpreted from the core in Fig. 8 (from Barrett, 2007). **a** Glacial period, with an expanded inland ice sheet feeding thin, temperate piedmont glaciers depositing sediment on a shallow shelf to be reworked by waves and currents near shore with mud settling out offshore, and **b** Interglacial period, with higher sea level and a much reduced ice sheet. Sediment was carried to the coast largely by proglacial rivers. Low woodland beech forest grew at lower elevations. Insets for (**a**) and (**b**) show examples of the modelled extent of an ice sheet that might have existed during glacial and interglacial periods in early Oligocene times, representing 21 × 10^6 km^3 of ice (50 m of sea level equivalent) in **a**, and 10×10^6 km^3 of ice (24 m of sea level equivalent) in **b** (DeConto et al., 2007). The *white filled circle* in the inset marks Cape Roberts

EAIS Freezes – WAIS Stays Dynamic

As the Cape Roberts core showed, for almost 20 m.y. huge Antarctic ice sheets reached the continental margin regularly every 40,000 years and then retreated. However, from 14.2 to 13.8 m.y. ago another cooling event increased their volume and stabilised their behaviour. There is as yet no satisfactory explanation for this because there was little change in average global temperature or atmospheric CO_2 levels, by this time tracking consistently below 400 ppm (Pagani et al., 2010). However, there was certainly a profound cooling in the high south latitudes in both deep and surface waters of the Southern Ocean (Shevenell et al., 2004), and this is seen globally in deep-sea records also (Fig. 5a). In the McMurdo Dry Valleys, the event is recorded in the transition from wet-based to dry-based lacustrine and glacial deposits remarkably preserved at the edge of the Antarctic ice sheet, and are dated at around 13.8 m.y. (Lewis et al., 2007).

The rock surfaces in this region have eroded extremely slowly over millions of years (Summerfield et al., 1999) and indicate the persistence of a dry, cold polar climate from this time to the present. Hence, the prevailing view is that the cooling ~14 m.y. ago established the East Antarctic Ice Sheet pretty much in the form we see now. However, recent studies reveal numerous subglacial lakes and rivers under the ice sheet (Siegert et al., 2005; Wingham et al., 2006), as well as areas below sea level around the East Antarctic margin, caution us to expect surprises to come.

The geological history of the WAIS is less well known, but it seems likely that West Antarctica was ice-covered from the time of the first big ice sheets. The reasoning for this comes both from the large ice volumes implied by the deep-sea isotopes (Miller et al., 2005), and also because 34 m.y. ago West Antarctica was mostly above sea level. Wilson and Luyendyk (2009) showed that as the region stretched and cooled over millions of years, it turned from subcontinent to archipelago, reducing the volume of ice it could carry. The WAIS is vulnerable to melting from warmer ocean waters not only because its outlet glaciers discharge into the surrounding ocean but also because tectonic subsidence and glacial erosion have formed troughs that deepen into the interior to almost 2,000 m below sea level. As a consequence warm ocean currents melting the ice from underneath can lead to both thinning and accelerating flow of the ice coastwards. This was first recognised by Mercer (1978), who also warned of the potential loss of the ice sheet once the ice shelves that buttress most ice streams disappeared.

This has motivated much research and speculation on the past behaviour of the WAIS in the recent past, especially between 5 and 3 m.y. ago, a period called the "Early Pliocene" and the last time Earth was persistently warmer than today. The greater warmth of the Early Pliocene globally has long been known (Kennett, 1982), and indeed the prospect of the loss of two-thirds or more of the East Antarctic Ice Sheet has been proposed (Webb et al., 1984; Barrett et al., 1992). However, significant countervailing evidence was offered (e.g. Sugden et al., 1993, and many others), and the primary evidence of rare Pliocene diatoms from debris deposited by glaciers from the Antarctic interior is now considered more likely contamination from atmospheric transport (McKay et al., 2008).

However, curiosity over the response of the WAIS (and perhaps part of the EAIS) has continued on account of persistent estimates of Pliocene sea level rise of 25 m (Dowsett and Cronin, 1990; Rohling et al., 2009). This time period is especially relevant because, as Pagani et al. (2010) have reported, atmospheric CO_2 levels were between 365 and 415 ppm, i.e. similar to those of today. Also global temperatures were 3–4°C warmer than today, the middle of the range expected by 2100 with "business as usual." So what did happen to the WAIS over the last 5 million years?

Drilling the sea floor beneath the McMurdo Ice Shelf in 2007 has captured this time period in the upper part of a 1,285-m-long core as part of the ANDRILL project (Fig. 10; Naish et al., 2009). The nature of the

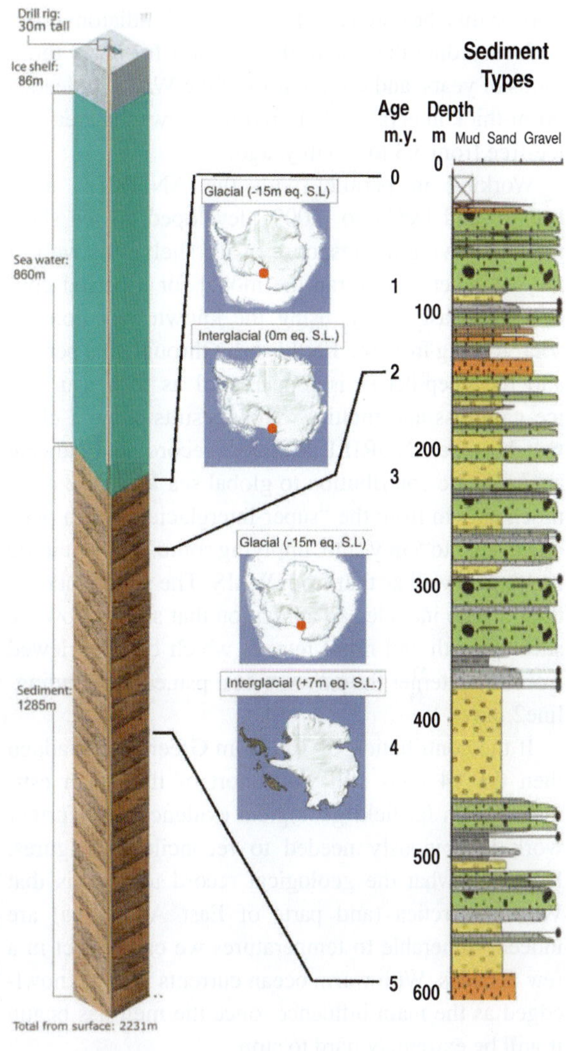

Fig. 10 Diagram of the ANDRILL drill on the McMurdo Ice Shelf and the cored strata beneath (*left*), with a log of the upper 600 m of strata, covering the last 5 m.y. (*right*). Cyclic variations in the rock types represent ice sheet expansion (grounded glacial deposits, *green*), ice shelf conditions like today (marine mud, *grey*) and ice sheet retreat with extensive open ocean (diatomite – algae from blooms in surface waters, *yellow*). The pairs of Antarctic maps (*centre*) show ice extent during glacial and interglacial times in the last 2 million years (*upper*) and from 2 to 5 m.y. (*lower*)

sediment layers records glacial advances (when the ice grounded, depositing glacial debris from the south), interglacial periods like today (when the fine mud and some microplankton settled to the sea floor) and surprisingly periods of open ocean (when virtually all of the sediment deposited came from diatom blooms in the warmer waters). The last 2 million years comprises mostly glacial debris from more extensive ice and some thin mud layers like sediments today. But prior to this the core revealed 13 layers of diatom ooze, each recording the loss of the ice shelf for many thousands of years, and with it most of the WAIS. Indeed, a 90-m-thick interval of diatom ooze shows the area was ice-free from 3.3 to 3.6 m.y. ago.

Working in parallel with the ANDRILL team Pollard and DeConto (2009) developed an ice sheet model showing the response of ice shelves as well as the ice sheets. They ran the model for a period covering the last 5 m.y. using the known variations in solar forcing from the Earth's orbit through this period, and the deep-ocean isotope record as a measure of ice gain/loss and melting. Their results show a close match to the ANDRILL MIS core record, and indicate an Antarctic contribution to global sea level rise of as much as 7 m from the "super-interglacial" warm periods from 3 to 5 m.y. ago, implying some East Antarctic melting as well as that from WAIS. The supplement to their article includes an animation that shows slow ice sheet growth and rapid retreat, which can be viewed from the internet (http://www.eesi.psu.edu/groundingline2.mov).

If the contribution of 7 m from Greenland is added then the 14 m is still well short of the 25 m estimated from far-field geological evidence, and further work is obviously needed to reconcile the figures. However, what the geological record tells us is that West Antarctica (and parts of East Antarctica) are indeed vulnerable to temperatures we can expect in a few decades. With warm ocean currents now acknowledged as the main influence, once the melt has begun it will be extremely hard to stop.

Both the sediment record and the modelling indicate that the disintegration of the WAIS occurred with a remarkable speed in geological terms, but more research is needed to determine whether this can be measured in decades or centuries. Still in Pliocene times each period of ice sheet retreat was followed by slow ice sheet growth – until the next significant warming.

Northern Hemisphere Ice Sheets Take Over

As the world began to cool further, leading to stabilisation of the Antarctic sheets, the Northern Hemisphere ice sheets began to form. While there are indications of sea ice in the Arctic from 45 m.y. and increased sediment from sea ice and possibly ice bergs from glaciers on land around 14 m.y. ago the first clear indications of significant ice sheet development on Greenland are at around 3.2 m.y. (Moran et al., 2006). By 2.7 Ma, however, vast regions of the Northern Hemisphere were in the icy grip of huge ice sheets that covered extensive areas of the Eurasian Arctic, northeast Asia and Alaska. Just 200,000 years later, North America was also covered (Denton and Hughes, 1981; Maslin et al., 1998). While the Antarctic Ice Sheets were limited in their ability to grow by the surrounding ocean basins, the Northern Hemisphere ice sheets had no such constraints, being based on continents that extended from the Arctic Ocean to mid latitudes.

About a million years ago, Earth's glacial-interglacial cycles changed tune. For the past 34 m.y., glacial–interglacial cycles virtually always came and went every 40,000 years. However, around 1 m.y. ago, their frequency changed to every 100,000 years. At the same time glacial periods became significantly colder and interglacials warmer (Imbrie and Imbrie, 1979; Zachos et al., 2001b). The coincidence between the geological dating of the cycles and the astronomical dating of Earth's orbital variations indicate these are very likely the primary cause, as we will discuss in the next section.

From Wobbles to Cycles

The "pacemaker" of these cycles is now seen to lie in the variations in the Earth's orbit around the sun (Imbrie and Imbrie, 1979), and the cycles themselves are called Milankovitch Cycles, after the Serbian mathematician who first recognised their significance (Milankovitch, 1941). As the Earth revolves around the sun, it is also under the influence of the gravitational pull of other planets, stars and distant galaxies. For this reason, the rotation of the Earth around the Sun is not constant but instead varies regularly with three main frequencies: (a) 100,000 years (related to orbital eccentricity), (b) 41,000 years (related to orbital obliquity) and (c) 21,000 years (related to the precession of the Earth's axis). These combine their influence to vary the amount of solar radiation Earth receives from the sun in a regular way. The variation is small, around ~1.3 Watts/square metre or about 0.1% of the radiation averaged over decades (IPCC, 2001). Nevertheless the changes turn out to be significant because they are amplified by greenhouse gases and other elements in the Earth's climate system, leading to the regular climate changes found by geoscientists in ice and sediment records. Weedon (2003) provides a comprehensive review of theory and applications.

The variation in eccentricity changes our orbit around the sun from a more circular to a more elliptic passage on a 100,000-year cycle. When Earth is on a more elliptical orbit, one hemisphere will be closer to the sun during its summer (more energy per m^2), and further away during its winter (less energy per m^2), while the other hemisphere has warmer winters and cooler summers – causing a climate forcing that affects the northern and southern hemisphere differently. Obliquity is the change in the tilt of the Earth's axis from 22.2° to 24.5°. The greater the tilt (e.g. 24.5°), the greater the difference between the seasons on Earth, with colder winters and warmer summers. This forcing affects both hemispheres equally. Finally, precession changes on an ~21,000-year cycle to where the axis points. Imagine the axis of the Earth represented by a rod sticking out of the North and South Pole. Precession changes the orientation of the tip of the rod from pointing towards the sun to pointing away from the sun. This changes the timing of seasons. For example, at present the Northern Hemisphere experiences its summer during June, July and August when the Southern Hemisphere is in winter. However, about ~10,500 years ago, the rod pointed in the opposite direction and the Northern Hemisphere experienced its summer during December, January and February, while the Southern Hemisphere had its winter.

The easiest way to grow an ice sheet in the Northern Hemisphere, and so push Earth into another ice age, is to reduce melt and hence allow more snow to accumulate (Huybers, 2006). Until recently it was thought that this could be done through cold summers (Bard, 2004), which is controlled by precession on a 21,000-year cycle. However, new research suggests that the short rather than cool Northern Hemispheric summers are critical, and the length of summer is governed by obliquity on a 41,000-year cycle (Huybers, 2006). Eccentricity, obliquity and precession change concurrently but on different frequencies, hence sometimes the effects enhance each other, and sometimes they neutralise each other. For example during low eccentricity, when Earth's path around the sun is almost circular, precession has very little influence, as there will be very little difference between the point that is furthest away and closest to the sun and hence it is of little consequence where the Earth's axis points. However, during high eccentricity this effect becomes very prominent.

For the past 30 or so million years the pacemaker for these glaciations seems largely to have been obliquity, the ice sheets coming and going with an ~41,000-year cyclicity. However, a million years ago the ice age rhythm changed to 100,000-year cycles, associated with eccentricity cycles (Zachos et al., 2001a). The extent, temperature and sea-level changes associated with the 41,000-year glaciations were substantially smaller than the 100,000-year glaciations (Zachos et al., 2008, 2001b). The reasons for this large shift are yet unknown, but for the past 1 Ma the Earth has experienced some of the largest climatic changes in its long history.

Swings of the Past 1 Million Years

In the last million years Earth's climate has been largely glacial, interrupted by interglacial episodes, lasting for between 10,000 and 30,000 years in each 100,000-year cycle (Fig. 11). Global sea level fell during glacial episodes to between 100 and 120 m below present day (Rohling et al., 2009) and returned to roughly its present level during interglacials. About 90% of the change came from the Northern Hemisphere ice sheets. A particularly curious characteristic of these glacial/interglacial cycles is that it took many tens of thousands of years to cool the globe to full glacial conditions but warming once started was rapid and the ice was quickly lost, giving both temperature and sea level curves a saw-tooth shape (Petit et al., 1999). The saw-tooth pattern contrasts strongly with the smooth sine wave for the variation in solar radiation from the Milankovitch cycles. In the next section we will see why, but first how we can measure Earth's past atmospheric composition and the temperature variations on the ice sheets.

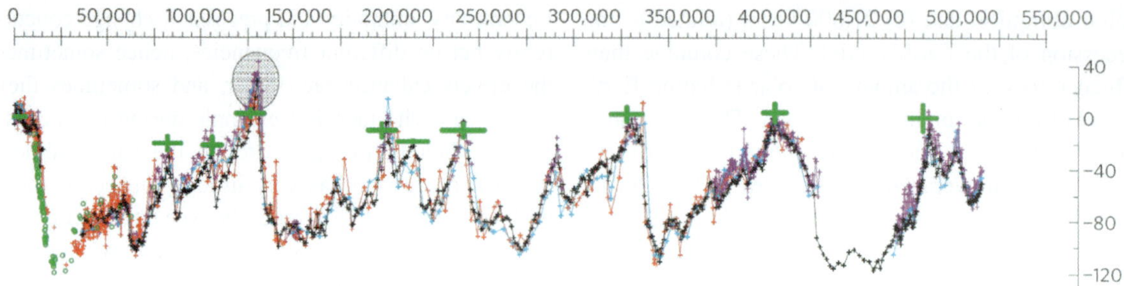

Fig. 11 Sea level variation from ice-volume change over the last five glacial cycles (from Rohling et al., 2009). The data are from carbonate ^{18}O measurements on central Red Sea cores. The *green symbols* are coral and speleothem-based high sea-level markers

Stories from Tiny Bubbles

Ice cores provide a formidable archive for studying the recent climate history of the polar regions. As snow accumulates it captures information of atmospheric conditions and composition, temperature, precipitation, storminess and even sea-ice extent (Alley, 2000b; Mayewski and White, 2002). Like tree rings, these layers of annual snow accumulate on top of each other and can be read by scientists like pages of a detailed diary of past climate. Ice cores from Greenland cover a continuous record of the past 130,000 years (Bender et al., 1994; Steffensen et al., 2008), while ice cores from Antarctica, where snow accumulation is much less, and the ice is much thicker, spans 850,000 years (EPICA Community Members, 2004). As the snow turns into firn and eventually ice from the weight of new snow accumulating above, the ice crystals enclose small bubbles that contain samples of the actual atmosphere of times past. It is from these tiny bubbles that we can measure past concentrations of greenhouse gases in the atmosphere for almost the last million years (Petit et al., 1999; Siegenthaler et al., 2005).

To reconstruct temperature, ice core scientists use isotopic ratios in water (oxygen and deuterium). This can be done because the Earth's oceans, the principal moisture source, provides a uniform mixture of these isotopes over millennial time scales (Jouzel et al., 1997). Somewhat simplified, we can assume that during warmer time periods (these can be summer periods, warm days, or interglacials), there is more energy available and hence more of the heavier isotopes are evaporated from the ocean and transported by clouds to the ice sheet to precipitate out as snow. During cooler time periods, less energy is available and hence less of the heavy isotopes are uplifted and transported to the ice sheets. There, heavier isotopes precipitate preferentially depending on the air temperature. The relationship between temperature and the amount of heavy isotopes in polar snow is very robust and can therefore be used as a good basic thermometer (Jouzel, 1999). Over longer time periods such as glacial–interglacial oscillations, the ice sheets accumulate so much ice on the continent that the oceans become proportionally depleted in the lighter isotopes. This requires the long stable isotope records in ice to be corrected for ice volume in order to calculate temperature (Jouzel, 1999; Jouzel et al., 1997).

In Fig. 12, the records for temperature (black curve), expressed as departure in ^2H (deuterium)/^1H (hydrogen) isotope ratio and greenhouse gases (red – methane, blue CO_2) for the past 650,000 years is shown from the EPICA Dome C ice core (Siegenthaler et al., 2005). At first glance it is apparent that the three curves are very similar. When temperature is warmer (upwards), greenhouse gas concentrations both for methane and CO_2 are higher and vice versa. Furthermore, as we move through time (from left to right) we observe that interglacials start with an abrupt warming, together with an abrupt increase in greenhouse gas concentration. The warm time period persists only for a few thousand years before a prolonged step-like decrease in temperature and greenhouse gas concentrations leads the planet into the next ice age over the course of about 80,000 years (Broecker, 2000a). This stands in strong contrast to the orbital forcing (eccentricity, obliquity and precession) which create smooth, sinusoidal curves of energy received on the Earth's surface. So how is it that the ice age climate shows a saw tooth pattern, with an abrupt warming leading into an interglacial?

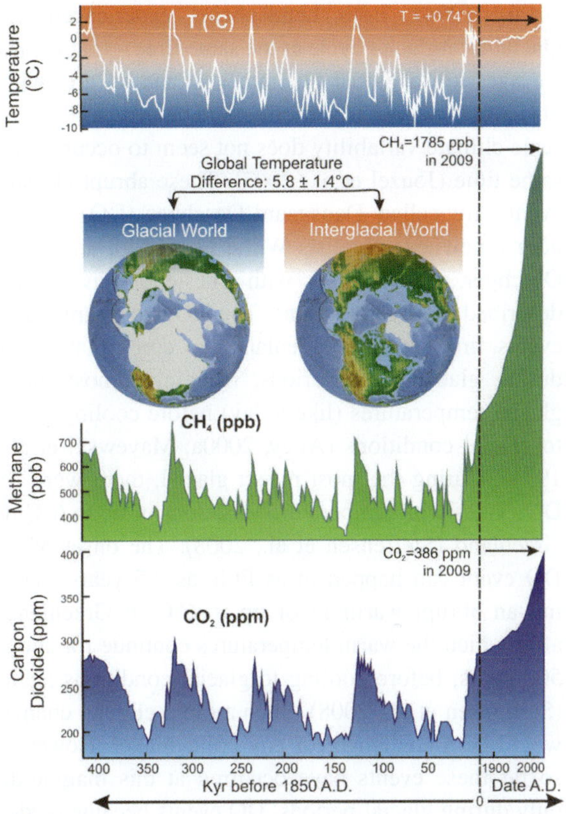

Fig. 12 Record of atmospheric CO_2, CH_4 and temperature extracted from Antarctic ice core by Petit et al. (1999) and from in situ and other data for the past century. The temperature change for the past century compared with the ice core record for earlier times is twice the global mean temperature change of Hansen et al. (2001). The temperature zero-point is the mean for 1880–1899. Difference between glacial and interglacial global temperatures calculated from mean after Schneider von Diemling et al. (2006). Modified after Hansen et al. (2005)

The answer lies in feedbacks and here in particular in the greenhouse gas feedback. A positive feedback describes a mechanism that enhances an effect. For example, interest paid on savings could be seen as a positive feedback. As the interest accumulates, the amount saved increases and so does the interest paid, which means even more is saved. A negative feedback is a mechanism that reduces an effect. A practical example would be a thermostat. As the temperature of a room reaches the temperature set on the heater's thermostat, it switches off the heater until the temperature falls below the setting and the thermostat switches the heater on again. Negative feedbacks tend to encourage stable situations (steady temperature of the living room), while positive feedbacks create runaway processes (ever increasing savings).

It is a positive feedback that lifts Earth out of an ice age. It begins with an orbital configuration that slightly increases the energy received from the sun (Imbrie and Imbrie, 1979), which triggers feedbacks that change the greenhouse budget. These feedbacks are not yet fully understood but include a number of key events (Broecker, 2000b). One of the first feedbacks is a decrease in wind strength, leading to less dust and hence nutrients deposited to the oceans (Broecker, 2000b; Lambert et al., 2008). The Southern Ocean is particularly deprived of iron (Boyd et al., 2000). As winds of the continent weaken, less iron is transported from the continents into the ocean, starving algae of nutrients and hence reduce their activity and ability to fixate CO_2, (Boyd et al., 2000). This increases atmospheric concentrations of CO_2, which further enhances warming (Boyd et al., 2000; Lambert et al., 2008), which leads to further decreased wind, and so on. Another feedback from warming over the continents comes with the melting of permafrost regions, which releases methane, which further increases greenhouse gas concentrations and hence temperature, leading to more melt of permafrost regions.

Furthermore, as the atmosphere and ocean warm, the sea-ice and other ice starts to melt. As ice melts, it reduces quickly the bright white blanket with a dark ocean or continent. While up to 90% of incoming solar radiation is reflected back into space from white surfaces such as sea-ice, the dark surfaces(such as oceans or soils) reflect less than 50% back into space (IPCC, 2007c). The additional energy absorbed by the Earth further increases warming.

Another major feedback comes from atmospheric water vapour, the most influential of all gases (IPCC, 2007c). As temperatures increase, the warmer air is capable of holding more moisture. Water vapour in the upper troposphere (e.g. 5–10 km height) is particularly effective as a greenhouse gas (IPCC, 2007c; Pierrehumbert, 2002). It is important to note that water vapour is not a forcing of the climate system but a response. So while it reinforces the warming (or cooling), it is only responding to a warming or cooling caused by other forces, e.g. solar radiation or greenhouse gases.

This list of feedback processes represents only examples and is by no means complete. In addition, there are many feedbacks that are not yet understood and perhaps even discovered. However, an argument that is at times confused in the public discussion is that in the past, temperature rose before greenhouse gas concentrations increased and hence greenhouse gas concentration cannot be responsible for temperature increases. Past records show us that this line of reasoning is not correct. Instead, while the initial temperature increase was caused by orbital forcing (increase in energy from the sun), it is the strong response in the greenhouse gas budget that quickly led to further warming, and with that even higher greenhouse gas concentrations, leading to further warming yet again (Boyd et al., 2000). It is these positive feedbacks that led to the abrupt warming that terminated glaciations in the past (EPICA Community Members, 2004). Today, we initiate the same feedback mechanisms, but instead of starting with a warming with orbital forcing, now it originates from our own release of greenhouse gases through burning fossil fuels and deforestation.

Looking at past greenhouse gas concentrations gives us the opportunity to compare today's concentration to a "natural" background (Fig. 12). Over the past 650,000 years, we can observe that greenhouse gas concentrations fluctuated within very narrow band, varying from about 180 ppm for CO_2 during cold glacials to about 280 ppm during very warm interglacials (Siegenthaler et al., 2005). Today's CO_2 concentrations are at 386 ppm and rising ever faster. This is also the case for methane, with a natural range from 320 to 790 ppb, reaching 1,785 ppb in 2009 and nitrous oxide, which rose from 270 to 319 ppb. Plainly the direct measurements of past and modern greenhouse gas concentrations show they are higher now than during any time of the past 650,000 years (IPCC, 2007c). And as we saw earlier, they are likely higher than at any time in the last 20 million years. As greenhouse gases increase, what response can we expect of the Greenland and Antarctic ice sheets?

The Polar See-Saw

For most of the last 100,000 years, Earth has been in the grip of ice age conditions, only emerging from it about 10,000 years ago. For the past ~13 years, high resolution ice core records from both polar regions provide us with an opportunity to study the expression of global events on both hemispheres (Bender et al., 1994). Both hemispheres experienced the glacial/interglacial cycle at the same time – that means ice ages are a global phenomenon and so are interglacials. However, if we look more closely, millennia-scale climate variability does not seem to occur at the same time (Jouzel et al., 2007). These abrupt climate events are called Dansgaard-Oeschger (DO) Events after renowned scientists Willie Dansgaard and Hans Oeschger, who together with Chester Langway first described these fluctuations. The DO cycles are warm events first seen in Greenland ice cores that occur during glacial time periods, reaching almost interglacial temperatures (like today) before cooling again to glacial conditions (Alley, 2000a; Mayewski et al., 1994). During the most recent glacial, there were 25 DO cycles with temperature variations of up to 6°C in Greenland (Steffensen et al., 2008). The onset of an DO event can happen in as little as 2.5 years, causing an abrupt warming of up to 6°C in Greenland, after which the warm temperatures continue for about 500 years, before cooling to glacial conditions again (Steffensen et al., 2008). Such a rapid climate change would be a challenge for any policymaker. However, so far these events have occurred at this magnitude only during glacial periods. DO events became colder with each event. The coldest DO event is usually accompanied by extensive iceberg rafting in the North Atlantic as seen from deep sea cores (Heinrich, 1988). These events were named Heinrich events after scientist Hartmut Heinrich. A total of six Heinrich events occurred during the past glacial time period (Heinrich, 1988) with far reaching effects, such as increasing the runoff of the Amazon River (Broecker, 1995).

As higher resolution Antarctic ice core records became available, it emerged that the hemispheres were out of phase (Fig. 13) (Jouzel et al., 2007; Luthi et al., 2008; Stocker, 1999, 2003), with Antarctica being the first to warm. However, the warming is gradual. When Antarctica reaches its warmest conditions, Greenland starts to warm abruptly, after which Antarctica starts to cool and Greenland follows. However, Antarctica's temperature changes are symmetrical (both warming and cooling are gradual), while Greenland's changes begin with an abrupt warming followed by a gradual cooling (Jouzel et al., 2007; Stocker, 1999, 2003). The mechanism for abrupt changes is not yet fully understood, but they could be caused by changes in the thermohaline circulation (Broecker, 2003; Stocker, 1999, 2003; Wolff et al.,

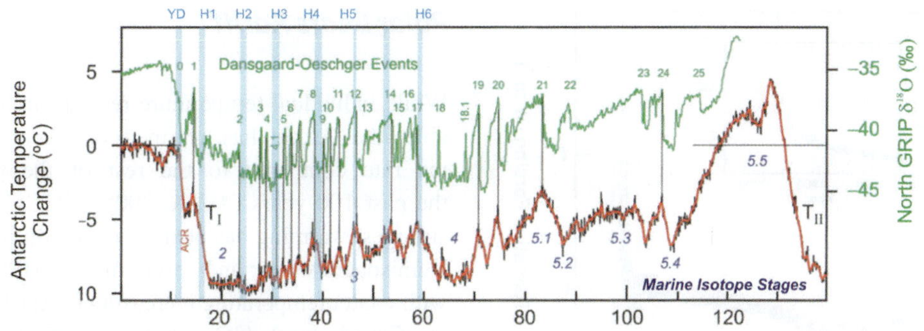

Fig. 13 Seesaw climate behaviour of the two hemispheres over the past 140,000 years, as shown from ice core records from Greenland (*green*, North GRIP) and Antarctica (*red*, EPICA Dome C). Modified after Jouzel et al. (2007)

2009). During Heinrich events, the large input of fresh water from iceberg armadas could slow or shut down the thermohaline circulation. This would increase Antarctica's sea surface temperature and reduce sea ice cover in the North Atlantic. In response, this could kick-start the thermohaline circulation again, leading to warm waters being transported from low latitudes to Greenland, causing the observed abrupt increase (Broecker, 2003; 2006; Steffensen et al., 2008; Stocker, 2003).

The observations indicate that slow, gradual changes in climate forcing can lead to abrupt climate shifts both within and beyond major regions of the Earth, which shows why we need to better understand processes and interactions between all parts of Earth's climate system (Barker et al., 2009; Severinghaus, 2009; Stocker, 2003). The 0.7°C rise in temperature in both hemispheres synchronously since 1850 AD is a dramatic change to past behaviour, indicating that the primary cause must have also changed (Turner et al., 2009).

Modern Rise of Greenhouse Gases

For the past 10,000 years, our planet enjoyed warm temperatures of an interglacial. Yet a closer look at the past 10,000 years is useful as we compare past atmospheric CO_2 levels from ice core data with historical observations from Mauna Loa in Hawaii (IPCC, 2007c; Keeling et al., 1995). The direct overlap of the two data sets shows that ice core, while difficult to process and analyse, are reliable means of reconstructing past greenhouse gas concentrations. It is plain that the rise of greenhouse gas concentrations in the last century is unprecedented over the past 10,000 years, which has been a period of relative stability in greenhouse gas concentration and temperature (Fig. 14). This increase coincides with the industrial revolution and the rapid increase in human population (IPCC, 2007c), and has led to substantially different boundary conditions for Earth's climate.

Climate models run with only natural forcing (solar + volcanic gases) cannot produce the warming observed in the last few decades. In contrast, climate models that add anthropogenic forcing to the natural forcing shows a significant temperature increase for the last few decades that compares closely with observed temperature (Fig. 15).

Feeling the Heat

The last decade was the warmest decade in the instrumental record of global surface temperature since the onset of the industrial evolution in 1850 AD (IPCC, 2007c). Since 1906 AD the Earth's surface warmed by 0.74°C in response to emissions of greenhouse gases from fossil fuels and changes in agriculture and land use (IPCC, 2007c). However, over the last 50 years the warming has accelerated twice as fast as the warming of the previous 100 years (0.13°C per decade) (IPCC, 2007c). Moreover, the Arctic is expected to warm faster than the rest of the globe, due to polar amplification (Masson-Delmotte et al., 2006). The Antarctic is responding more slowly on account of the loss of ozone and thermal inertia from its huge ice sheet (Thompson and Solomon, 2002), but the surrounding oceans are warming (Böning et al., 2008). So how are the polar ice sheets responding to the heat?

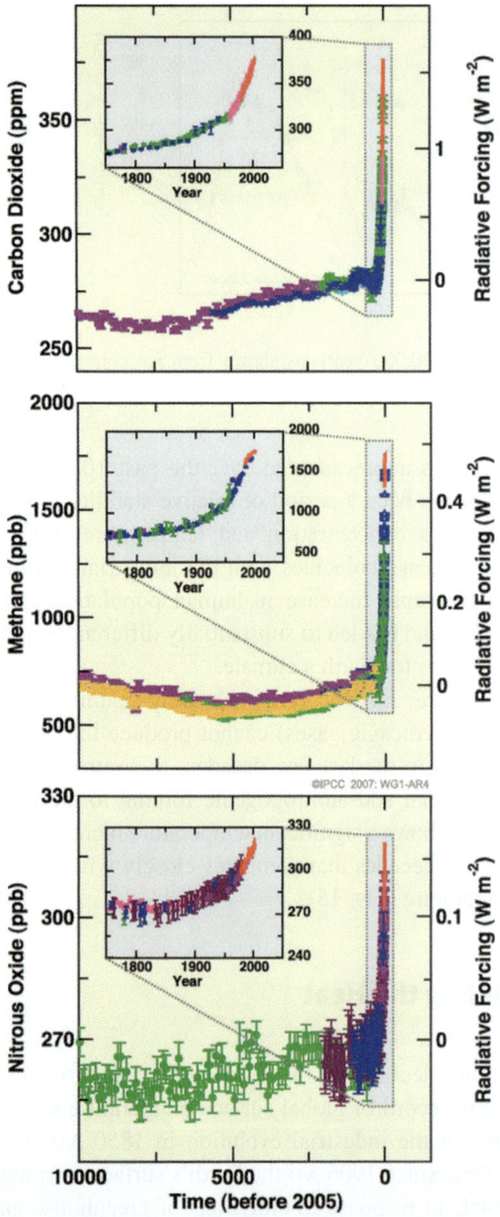

Fig. 14 Atmospheric concentrations of carbon dioxide, methane and nitrous oxide over the last 10,000 years (*large panels*) and since 1750 (*inset panels*). Measurements are shown from ice cores (symbols with different *colours* for different studies) and atmospheric samples (*red lines*). The corresponding radiative forcing is shown on the right hand axes of the large panels. Source: Climate Change 2007: The Physical Science Basis. Working Group I Contribution to the Fourth Assessment Report of the Intergovernmental Panel on Climate Change. Figure SPM1. Cambridge University Press (IPCC, 2007c)

Polar Bears Adrift

While individual temperature records are highly variable, the Arctic has warmed on average at twice the rate compared to the rest of the planet over the past 100 years (ACIA, 2005; IPCC, 2007c). The strongest warming has been observed during Northern Hemisphere winter and spring months (IPCC, 2007c), with winter temperature increasing in Alaska and western Canada by 3–4°C over the past 50 years (ACIA, 2005). In addition, precipitation has increased by about 8% (ACIA, 2005). More importantly though, most of the precipitation increase now occurs as rain and not snow with a bias towards winter precipitation (ACIA, 2005). Pouring liquid water onto to snow and ice is one of the most efficient methods of melting it. For this reason with the increase in rainfall we can expect greater and faster snow melt, and as a result occasional flash flooding (ACIA, 2005). In response to warmer Arctic temperatures, snow cover has decreased by about 10% over the past 30 years (ACIA, 2005) while snow precipitation in Greenland has increased by about 10% as warmer air can carry more moisture (IPCC, 2007c). Also permafrost regions of the Arctic are thawing at around 4 cm/year (IPCC, 2007c), ever increasing the depth to which the permafrost melts (ACIA, 2005).

One of the most dramatic changes in the Arctic, however, concerns sea ice extent. This has declined by about by 3% per decade over the last 30 years until 2005 when in 2 years it shrank by 23% (Kwok et al., 2009). It has recovered a little in the last couple of years but the decline is expected to continue in the next few years. However, even more severe changes are observed, when the sea ice is distinguished by type. First year ice is thin and fragile and easy to disintegrate in comparison to multi year ice, which is thicker and hence more stable. Until very recently it posed a major challenge to establish ice thickness short of walking across the sea ice and drilling it. Due to the vast area (albeit rapidly shrinking) in situ measurements can only provide a snapshot of the health of the Arctic sea ice. However, a new satellite "ICESat" – Ice, Cloud and land Elevation Satellite has now the ability to provide these important measurements for a 5-year time period since its launch (Kwok et al., 2009). In the time between 2003 and 2008, an astonishing 42% reduction and ~0.6 m thinning of multiyear sea ice was observed. Over this brief time period, the prominent ice type changed from multiyear ice to first year ice

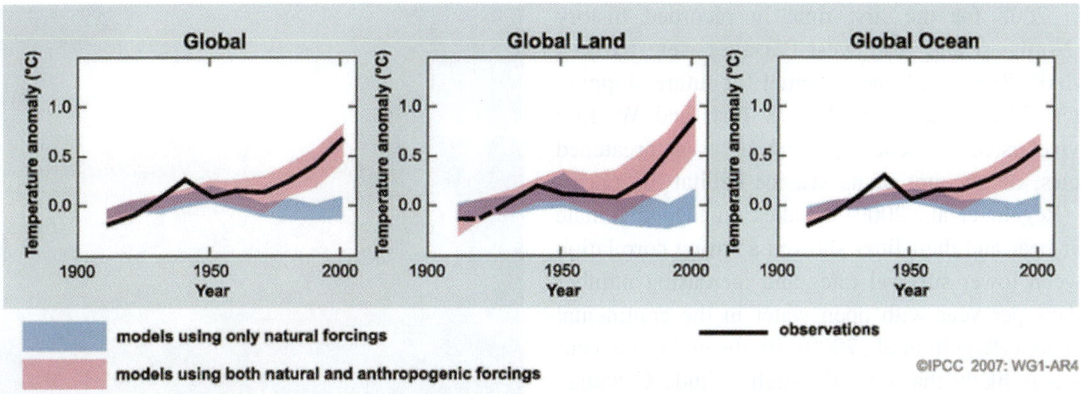

Fig. 15 Modelled average global surface temperature (total-land-ocean) comparing natural and all forcings with observations (*black line*) (Hegerl et al., 2007). Source: Climate Change 2007: The Physical Science Basis. Working Group I Contribution to the Fourth Assessment Report of the Intergovernmental Panel on Climate Change. Figure SPM4 (*lower panel*). Cambridge University Press (IPCC, 2007b)

(Fig. 16). This is particularly important in the view of new research that found that the strength of cyclones in the Arctic basin increases with reduced September sea ice extent and thickness, reducing sea ice still further (Simmonds and Keay, 2009). In addition, the Arctic sea surface temperatures are increasing. In 2007, ice free areas were around 5°C warmer than the long term average (AMAP, 2009).

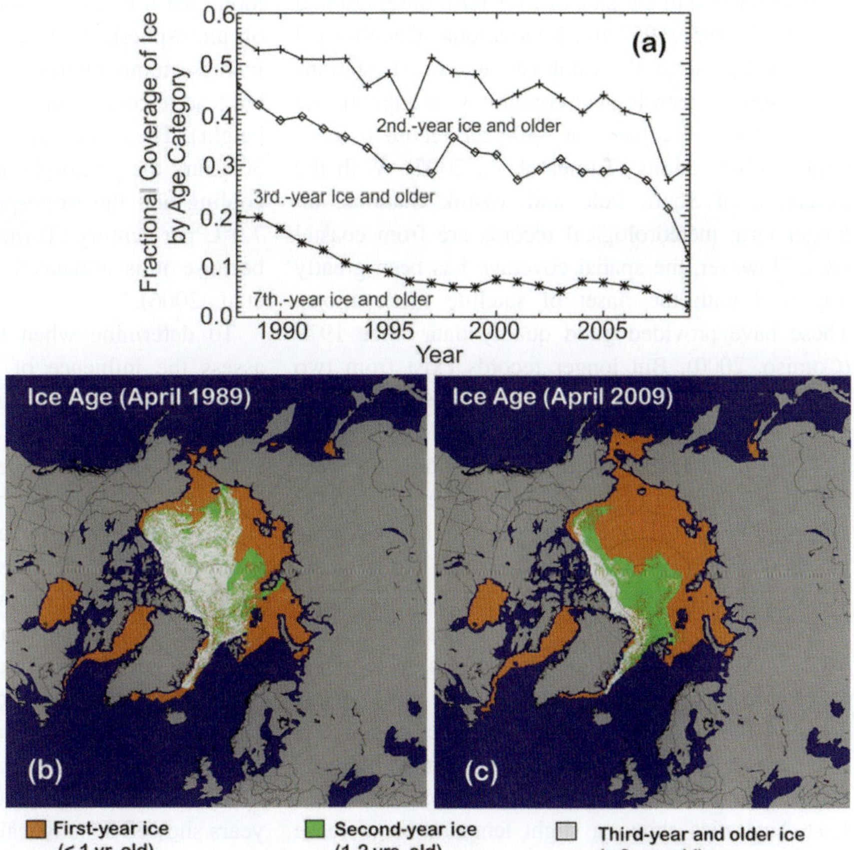

Fig. 16 Median sea ice conditions shown as average between 1981 and 2000 (*left upper panel*) and compared with 2009 conditions. Colour coding distinguishes between first year ice (*orange*), second year ice (*green*) and older ice (*white*). Numerical visualisation shows relative contribution of the three sea-ice types by percentage (*top panel*). Source: Courtesy of C. Fowler and J. Maslanik, University of Colorado (Maslanik et al., 2007)

In 2008 for the first time in recorded history the Northeast and Northwest Passages were ice free (AMAP, 2009), with the potential for future shipping routes. That same year, the US Fish and Wildlife Service listed the iconic polar bear as a threatened species, on account of its sea ice habitat disappearing (Regehr et al., 2009). Studies on tagged female polar bear and their litter showed a strong correlation between lower survival rates and increasing number of days per year with open water in the continental shelf area (Regehr et al., 2009). By the end of the century it is likely that only the high latitude Canadian archipelago and Greenland offer viable habitats for polar bears (ACIA, 2005). Many more species are affected (IPCC, 2007a), perhaps their decline will be less noticed as they lack the iconic presence of the polar bear.

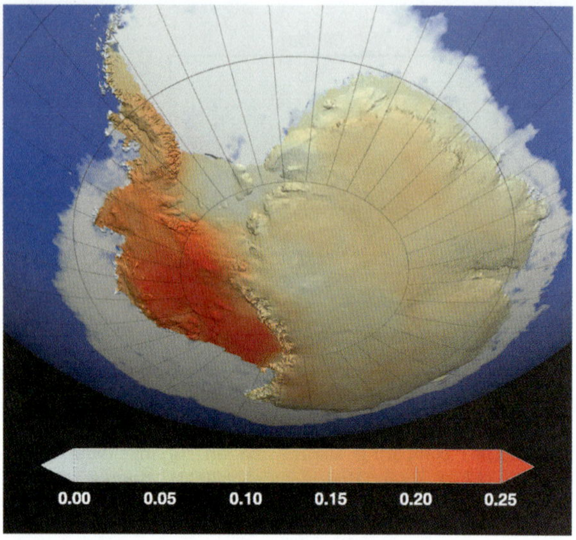

Fig. 17 Warming over Antarctica since 1957 AD (Source: Steig et al., 2009)

Antarctica Loses Chill

Assessing Antarctica's current temperature trend is challenging due to the lack of long-term observational records. In 1956/1957 the International Geophysical Year (IGY) led to the establishment of 16 stations to measure meteorological conditions throughout the year, and now we have an ~50-year record of continuous observations (Turner et al., 2005). With the exception of South Pole and Vostok Stations, all longer term meteorological records are from coastal sites. However, the spatial coverage has been greatly improved with the onset of satellite observations. These have provided good quality data since 1979 (Comiso, 2000). But longer records exist from two sub-Antarctic Islands: South Orkney Islands (since 1903) and Faraday Station, Argentine Islands (since 1947) (Turner et al., 2009). These records are very important because they lie in the region of the Antarctic Peninsula, one of the most rapidly warming places on Earth (+0.53°C per decade) (Turner et al., 2005).

In a recent study, Steig and colleagues (2009) used climate-field reconstructions and a climate model to establish temperature trends in the data sparse areas of East and West Antarctica. They found that large regions of West Antarctica might warm even faster than the Antarctic Peninsula and that even East Antarctica shows a slight temperature increase (Fig. 17). In addition to surface temperatures, we can now also study temperatures throughout the atmosphere with the help of satellite and weather balloon measurements. This is important, as an increase of atmospheric CO_2 concentrations is expected to increase temperatures of the troposphere (0–~8 km), but lead to cooling in the stratosphere (from ~8–50 km height). These measurements show that over the past 30 years the stratosphere above Antarctica has been cooling and the troposphere warming at about 5.0–7.0°C per century (Turner et al., 2006). This is likely because it is enhanced by ozone depletion (Turner et al., 2006).

To determine when the warming started and to assess the influence of oscillating climate variability, we need to look further back in time beyond meteorological records. The International Trans Antarctic Scientific Expedition (ITASE) is a consortium of 29 nations that have been collecting shallow ice cores (~200 m) from across the Antarctic continent for the past decade or so to reconstruct its climate for the past 200 years (Mayewski et al., 2005). This time period was chosen as it covers the transition from pre-industrial (natural) background to today's anthropogenically altered climate. Also shallow ice cores are logistically easier to recover than deep ice cores. A first attempt to combine available well dated and continuous records covering the past 100 years showed a statistically significant continent-wide increase of temperatures of 0.2°C since the late nineteenth century (Schneider et al., 2006). It further showed large inter-annual to decadal variability, with a strong antiphase behaviour between the Antarctic Peninsula (warming) and the rest of the Antarctic continent (less warming or cooling) (Schneider et al., 2006).

Antarctic climate is dominated by two oscillating climate drivers: (a) the Southern Annular Mode (SAM), which provides a measure of the strength of the Westerly winds sweeping around the continent between latitudes 40 and 60°S (popularly known as the "roaring forties" and "furious fifties") (Thompson and Solomon, 2002) and (b) the El Niño Southern Oscillation, which changes the location of the dominating pressure centres in the Southern Ocean (Bertler et al., 2004; Turner, 2004). The stronger these winds, the further south they move. As the cyclones spin faster around Antarctica, they thermally isolate and hence cool most of the continent (Mayewski et al., 2009). Depending on their average locations, their position can have strong regional impact (Bertler et al., 2006). The Antarctic Peninsula, however, reaches into this band of cyclones. As they move further south and become stronger, relatively warm, moist air is forced over the mountainous spine, creating a föhn effect, with very warm, dry air descending into the Weddell Sea, warming the Antarctic Peninsula (Marshall et al., 2006; Thompson and Solomon, 2002). Over the past 50 years, SAM has steadily and significantly intensified (Thompson and Solomon, 2002). This is largely due to the development of the Antarctic ozone hole and the rise of global temperature (Arblaster and Meehl, 2006; Thompson and Solomon, 2002). The cooling effect is now outweighed by the background warming of the globe, and Antarctica is warming. Can the Southern Ocean keep Antarctica cool enough?

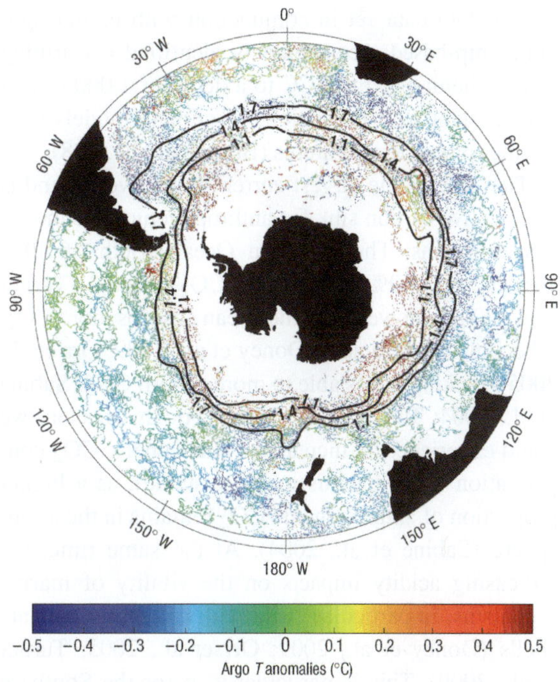

Fig. 18 Spatial pattern of changes in the ACC. The *dots* show the distribution of the 52,447 Argo profiles used in this study. The *colour* indicates the deviations of Argo potential temperatures averaged over the neutral-density layer 26.9–27.7 from the climatological mean (Böning et al., 2008)

Southern Ocean Warming

It takes about 4.2 joules to heat 1 cm^3 of water by 1°C. This means, it takes 4,200,000 joules (or ~11,700 Wh) to heat 1 m^3 of water by 1°C. In short, it requires a lot of energy to heat an ocean, yet the oceans have begun to warm from the surface down, reaching depths of between 1,000 m (Böning et al., 2008) (Fig. 18) and 3,000 m (IPCC, 2007c). As a consequence, the oceans have absorbed over 80% of the heat added to the Earth's system through the increase of greenhouse gas emissions (IPCC, 2007c). Since the 1970s, this is the equivalent of 190,000 average nuclear power plants heating directly the oceans.

Determining that the ocean is warming has been a challenge for a number of reasons: it is a vast area, it is difficult to reach the deeper basins of the ocean, oceans are made up of different water masses moving at various levels and different speeds, and these vary between seasons and from year to year, with phenomena such as the El Niño Southern Oscillation and seasonal sea-ice cover. However, since 2001 current measurements from a range of depths and locations world-wide have been taken through the ARGO programme (Böning et al., 2008). The ARGO floats are smart devices that drift with ocean currents, and rise and sink to measure temperature and salinity from the surface to depths of 2,000 m. Every few days these floats surface and transmit their data via satellite. A particular important water mass that has been studied now with the ARGO network is the Antarctic Circumpolar Current (ACC).

The ARGO data set in conjunction with earlier sporadic ship-based measurements identified a warming and freshening of the ACC to a depth of 1,000 m that is consistent with predictions of climate models with warmer global temperatures (Böning et al., 2008).

The ACC is the largest current in the world, and is an effective carbon sink for anthropogenic greenhouse gas emissions. The Southern Ocean south of 40°S absorbs over 40% of the total CO_2 emitted (Turner et al., 2009). However, as the ocean absorbs more CO_2, it becomes more acidic (Doney et al., 2009; Orr et al., 2005), making it less able to more absorb CO_2 (Sabine et al., 2004; Turner et al., 2009). For this reason, we could experience an increase in atmospheric CO_2 concentration (even if emissions are stable), as a higher proportion of anthropogenic CO_2 remains in the atmosphere (Sabine et al., 2004). At the same time, the increasing acidity impacts on the vitality of marine organisms, in particular those with calcium carbonate shells (Doney et al., 2009; Orr et al., 2005; Turner et al., 2009). This is particularly so for the Southern Ocean, which has low saturation levels of $CaCO_3$ and because it is a primary sink for atmospheric CO_2 (Orr et al., 2005).

However, the polar oceans are also warming (Böning et al., 2008), reducing the power of the thermohaline circulation. The world's ocean current system works like a global conveyor belt (Fig. 19) transporting vast amounts of energy from the warmer tropics to the cooler polar regions. This conveyor is driven by the sinking of cold (*thermo*), salty (*haline*) water in the polar oceans, which is caused largely by the formation of sea-ice (Broecker, 1997). The flowing of these cold dense water masses from polar continental shelves to the deep mid-low latitude ocean basins drives the currents that make up the entire ocean circulation system. The sinking occurs predominantly in three areas: in the Weddell and Ross Seas around Antarctica and in the North Atlantic off the Greenland ice sheet (Fig. 19). In the tropical regions, these water masses surface due to upwelling initiated by extensive evaporation. This allows warm currents to flow towards the high latitudes again, where the water cools and sinks (Clark et al., 2002; Rahmstorf, 2002). As the ocean warms, and less sea-ice forms, less salty brines are produced. The decrease in sinking water weakens the circulation and with that reduces the heat transfer around the globe. It is those critical areas where heavy

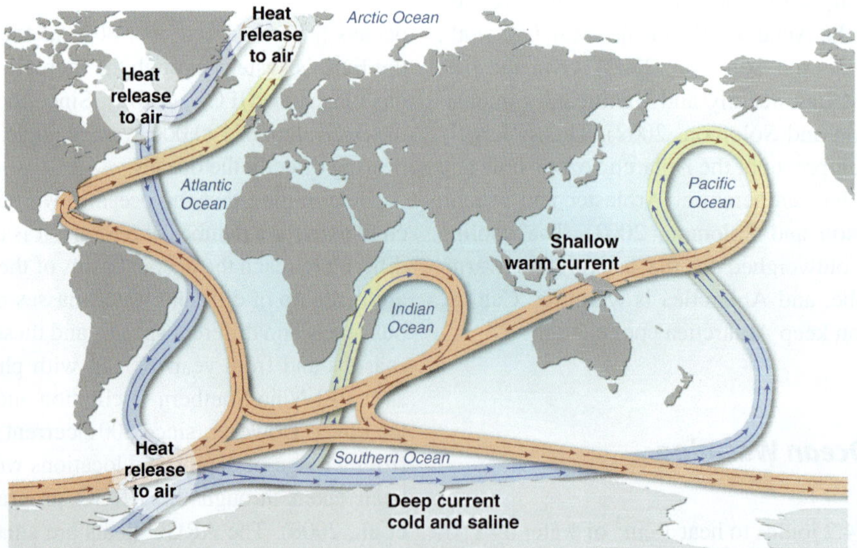

Fig. 19 Image of the Ocean Thermohaline Circulation. *Blue* currents depict cold, salty, deep currents, while *red* currents depict warm surface currents. *Yellow* currents depict vertical motion – transiting from deep to shallow currents and vice versa. Source: World ocean thermohaline circulation (June 2007). In UNEP/GRID-Arendal Maps and Graphics Library. Retrieved from http://maps.grida.no/go/graphic/world-ocean-thermohaline-circulation1

water sinks, where CO_2 is drawn into the deep ocean, storing it for hundreds of years. As the current slows, less CO_2 is drawn into the ocean and instead will remain in the atmosphere, where it warms the atmosphere and hence slows ocean currents further (Clark et al., 2002; Mayewski et al., 2009; Rahmstorf, 2002; Sabine et al., 2004; Turner et al., 2009).

As the ocean warms, another problems worries scientists – could a warmer polar ocean melt the ice sheets faster?

Ice Shelf Collapse

One of the most spectacular responses to warming temperatures has been the recent disintegration of ice shelves in the Antarctic. They are called "collapses" because of the speed with which sections covering hundreds of square kilometres have broken out. While these events have been observed only from satellite images and subsequent images can be separated by days or weeks due to cloud cover, and the break outs could well have taken just a few days.

But what is an ice shelf and how does it collapse? An ice shelf is a floating platform of ice tens to hundreds of metres thick that forms where a glacier or ice sheet flows down to and beyond the coast, When it extends far enough offshore to float (i.e. beyond the point where water depth is greater than 90% of the ice thickness), the ice begins to thin, the surface becomes flat and pieces break off around the edge to form ice bergs. This is normal and shelves are generally sustained by the glaciers feeding them, along with some surface snow accumulation. Ice shelves become vulnerable to collapse when either summer air temperatures are above freezing, leading to surface melt water, or the ocean waters beneath them are warm enough to melt and thin them faster than they are gathering snow and ice.

So far, all 13 Antarctic ice shelf collapses in recent history have occurred in the Antarctica Peninsula, an area that experiences the world's steepest warming trend with 0.53°C per decade (Mayewski et al., 2009; Turner et al., 2005). In 2002, the Larsen B ice shelf collapsed over the course of 3 weeks (or perhaps less) (Fig. 20). From marine sediment cores recovered from there area, it is evident that the Larsen B ice shelf existed at least for the past 10,000 years (Domack et al., 2005). Shortly before the Larsen B ice shelf disintegrated, melt pools were observed on its surface. It is now understood that these were symptoms of a structural weakening which might have started 20 years earlier with less visible processes (Glasser and Scambos, 2008).

Fig. 20 Satellite image of the Larsen B ice shelf collapse as seen on 7 March 2002. (Source: Image courtesy of the NASA/GSFC/LaRC/JPL, MISR Team)

When an ice shelf collapses it does not contribute to a rise in sea-level, as the ice is already floating in the ocean. However, ice shelves have an important buttressing effect, slowing the flow of glaciers and ice sheets feeding into the shelf. Their removal allows glacier ice to flow faster into the ocean, and this increases sea-level (Rignot et al., 2004; Scambos et al., 2004). The Larsen B ice shelf was a particularly interesting event, as glacial flow had been measured prior to its disintegration. With the ice shelf gone, these measurements were repeated and a two- to sixfold increase in the centreline speed of at least four glaciers draining into the Larsen B region was observed (Scambos et al., 2004). The warming of the Peninsula has a distinct North-South gradient, which is reflected in the ice shelf response as the warming reaches further and further south. From the Larsen ice shelf groups, the most northern ice shelf – Larsen A – disintegrated in 1995, then followed Larsen B in 2003, while its most southern ice shelf, Larsen C, still exists, but shows significant thinning (Shepherd et al., 2003).

The rapid collapse of ice shelves in the Antarctic Peninsula concerns scientists, especially as they look to the larger ice shelves. The Ross Ice Shelf is roughly the size of France and the largest of its kind. The ice moves at between 500 and 700 m/year, and the glaciers feeding it drain about a third of the WAIS (Rignot et al., 2008). As we saw in the "Green Anatarctica" section, core studies from the ANDRILL show that in the past, the Ross Ice Shelf collapsed many times between 1 and 5 million years ago, when global temperatures were around 3°C warmer than today and CO_2 concentrations were less than 400 ppm (Naish et al., 2009). Modelling studies indicate that these conditions not only collapsed the Ross Ice Shelf but also led to a huge loss of ice in West Antarctica and some from East Antarctica, contributing to a rise of 7 m in global sea level (Naish et al., 2009). What do the latest mass balance estimates for the polar ice sheets indicate?

Pace Picking Up

Global sea level on time scales of decades to centuries varies primarily as a consequence of expansion or contraction of ocean water due to temperature changes and loss or gain of ice on land. The Third IPCC Assessment Report (IPCC, 2001) noted that the sea level changes had been slight for the last 3,000 years, rising at around 0.1 mm/year. In the twentieth century, however, sea level rose between 1 and 2 mm/year. With the inclusion of satellite data IPCC (2007b) noted sea level in the last decade of the twentieth century was rising at 3 mm/year. The data showed greatest increase between the last three decades and the last to be from thermal expansion (Table 1). However, the contribution of ice sheets to sea level rise has also increased and the potential from this source is much greater than from warming oceans. The IPCC projected the likely rise in sea level to 2100 (IPCC, 2007a), their range being 0.18–0.59 m, the value depending on the particular emissions scenario adopted. This was below similar estimates by IPCC (2001), but specifically excluded effects from increased ice sheet melt because the processes, especially the influence of warm ocean waters beneath ice margins, were not yet understood. Might the ice sheets turn out to be greater contributors than the earlier models suggested?

Measuring the rate of gain or loss on an ice sheet 4,000 km across is simple in principle but difficult in practise. It amounts to gauging the mass balance of the ice sheet, that is whether the ice sheet is growing (positive mass balance) or shrinking (negative mass balance), and by how much. The factors determining this are the amount of snow that falls less the amount of ice lost. This might be either through calving (coastal ice margin calving into the ocean) or surface ablation (melt and evaporation – yes, ice does evaporate). If the ice sheet receives as much snowfall as it loses through calving and/or ablation then ice sheet is in balance (Pritchard et al., 2009; Rignot et al., 2008). In Antarctica, owing to its cold temperatures and margins mainly beyond the coast, almost all ice is lost due to calving (ice breaking off into the sea). In Greenland, calving and ablation (loss of ice from evaporation and/or melting) account for about equal amounts of ice loss (Pritchard et al., 2009). Mass balance is calculated using three different

Table 1 Rates of sea level rise separated by source for the last decades and the last decade of the twentieth century (from IPCC, 2007a, Table SPM-1). Potential sea level rise from ice from Lythe et al. (2001) and Lemke et al. (2007)

	Rate of sea level rise (mm/year)		Potential sea level rise
Source of sea level rise	1961–2003	1993–2003	
Thermal expansion	0.42 ± 0.12	1.6 ± 0.5	0.3 m for 3°C rise
Glaciers and ice caps	0.50 + 0.18	0.77 + .22	~0.4 m
Greenland ice sheet	0.05 + 0.12	0.21 + 0.07	7 m
Antarctic ice sheet	0.14 + 0.41	0.21 + 0.35	57 m
Sum of individual climate contributions to sea level rise	1.1 + 0.5	2.8 + 0.7	
Observed total sea level rise	1.8 + 0.5	3.1 + 0.7	

and independent approaches: measurement of surface elevation changes (lower surface elevation indicates loss), difference in gravitational force (an decrease in force indicates loss), and calculated balance between addition of ice from snow accumulation and loss of ice by measuring ice loss around the margin from known glacier cross-sections and velocities (Pritchard et al., 2009; Rignot et al., 2008; Turner et al., 2009). Most of these measurements are made from satellite observations with only few in situ measurements and all have their own limitations. These projects include Gravity Recovery and Climate Experiment (GRACE) satellites, interferometric synthetic-aperture radar (InSAR) data and Ice, Cloud, and land Elevation Satellite (ICESat) observations. Together these methods provide a convincing picture of rapid change (Pritchard et al., 2009). A history of these measurements is shown in Fig. 21.

Codes for Fig. 21 Greenland: B (orange; Box et al., 2006), surface mass balance, using stated trend in accumulation, ice flow discharge (assumed constant) and standard error on regression of accumulation trend, with added arrow indicating additional loss from ice flow acceleration; H (brown; Hanna et al., 2005), surface mass balance, with arrow as for B; T (dark green; Thomas et al., 2006), laser altimetry, showing new results and revision of Krabill et al. (2000) to include firn densification changes; Z (violet; Zwally et al., 2005), primarily radar altimetry, with uncertainty reflecting the difference between a thickness change due to ice everywhere and that due to low-density firn in the accumulation zone; R (red; Rignot and Kanagaratnam, 2006), ice discharge combined with surface mass balance; V (blue; Velicogna and Wahr, 2005) GRACE gravity; RL (blue; Ramillien et al., 2006) GRACE gravity; J (magenta dashed; Johannessen et al., 2005), radar altimetry without firn densification correction and applying only to central regions that are thickening but omitting thinning of coastal regions, V2 (light blue; Velicogna, 2009).

Codes for Antarctica: Z (violet; Zwally et al., 2005), radar altimetry, with uncertainty reflecting the difference between a thickness change due to ice everywhere and that due to low-density firn everywhere; RT (dark green; Rignot and Thomas, 2002), ice discharge and surface mass balance, with dashed end line because some of the accumulation rate data extend beyond the time limits shown; RT2 (dark green; Rignot and Thomas, 2002), updated to include additional mass losses indicated by Thomas et al. (2004) and Rignot et al. (2005) dashed because the original authors did not produce this as an estimate for the whole ice sheet nor are accumulation rates updated; V1

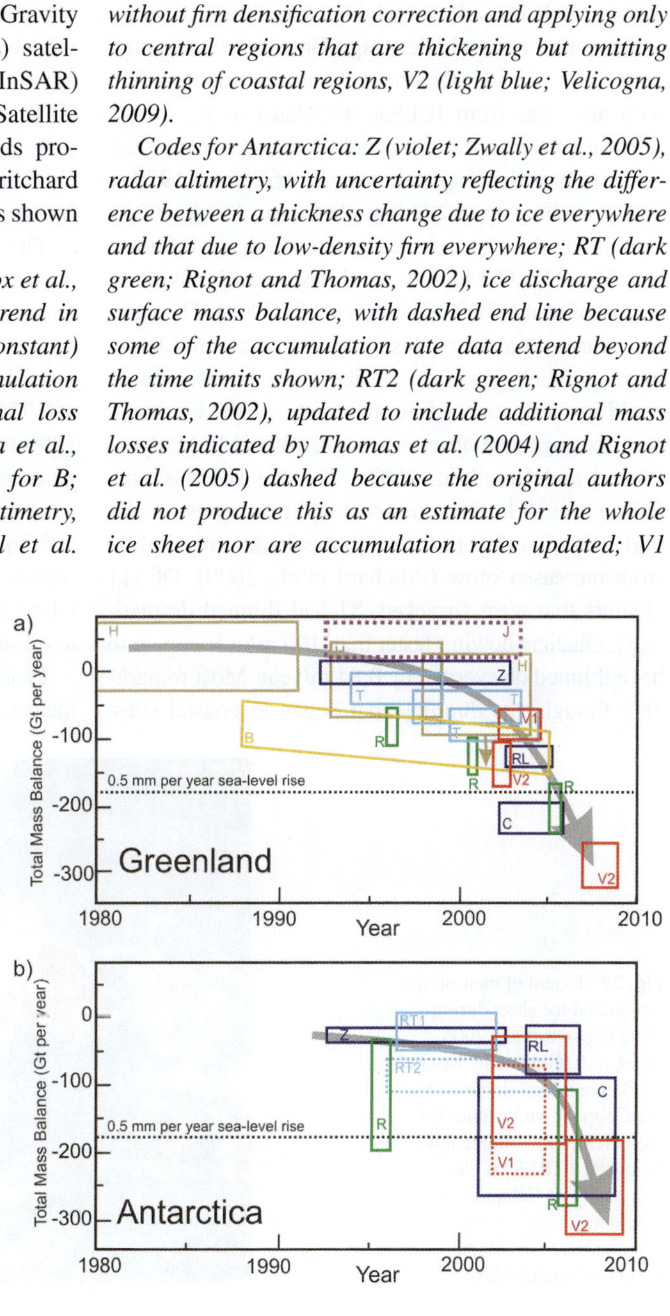

Fig. 21 Mass balance estimates for Greenland (*top*) and Antarctica (*bottom*). The coloured rectangles indicate the time span over which the measurements apply and the range, and follow Thomas et al. (2006). The range estimate is given as (mean ± uncertainty) as reported in the original papers. See italicised section in the text

(light blue; Velicogna and Wahr, 2006), V2 (light blue dashed; Velicogna, 2009), GRACE gravity; RL (blue; Ramillien et al., 2006), GRACE gravity, R (red; Rignot et al., 2008), and C (yellow; Chen et al., 2009). Figure modified and based on Climate Change 2007: The Physical Science Basis. Working Group I Contribution to the Fourth Assessment Report of the Intergovernmental Panel on Climate Change, Fig. 18 Cambridge University Press (2007c).

Because the Greenland ice sheet loses mass through surface melt as well as calving, estimating its mass balance has been challenging. However, substantial progress was made (Rignot and Kanagaratnam, 2006), with new data from ICESat (Pritchard et al., 2009). The data confirm the Greenland ice sheet is melting. The area experiencing melt grew by 16% from 1979 to 2002, reaching up to 2,000 m elevation (ACIA, 2005). Since then, the affected area has steadily grown further (see Fig. 22) (ACIA, 2005) and new record maximum melt achieved in 2007 (Wouters et al., 2008). The summer warmth produces slushy snow and created large, turquoise melt ponds on the ice sheet surface. These could instantly empty into the depths of the ice sheet, contributing to the observed loss of around 170 Gt/year (Sneed and Hamilton, 2007). The ice sheet is now thinning most in the south-east and north west, while high elevation sites in the south are thickening slightly from increased snow (Pritchard et al., 2009). Of 111 glaciers that were surveyed, 81 had thinned dramatically. Glaciers flowing faster than 100 m/year appear to have thinned on average by 0.84 m/year. Most remarkably though, the thinning that began at coastal sites, penetrated 120 km into the centre of the ice sheet and to elevations of 2,000 m (Pritchard et al., 2009).

In Antarctica, the small average annual accumulation of snow in the interior has in the past made it difficult to calculate the mass balance. However, the new satellite data, in particular ICESat and InSAR data, allow elevation changes to be measured more accurately (Pritchard et al., 2009; Rignot et al., 2008). Between 1992 and 2006, WAIS ice loss increased by 59%, (132 ± 60 Gt/year; red circles Fig. 23). The rate of ice loss was even greater in the Antarctic Peninsula, increasing by 140% between 1996 and 2006, reaching 60 ± 46 Gt/year (Rignot et al., 2008). In contrast, EAIS mass balance showed essentially no change (−4 ± 61 Gt/year). While some areas gained (blue circles, Fig. 23), others lost (red circles) (Rignot et al., 2008).

The largest ice loss is seen in the Amundsen sector (Rignot et al., 2008), where some glaciers thinned by as much as 9 m/year (Pritchard et al., 2009). The area around the Pine Island Glacier shows the largest ice loss of all. The glaciers have accelerated between 20 and 75%, doubling their loss of ice in the decade from 1996 to 2006 (Rignot et al., 2008). As a result these glaciers have thinned and become buoyant at their seaward end, as the grounding line has retreated to within 100 km of the ice divide (Pritchard et al., 2009). At the seaward end, the Getz ice shelf, the largest in the Pine Island region, shows thinning of 2–3 m/year, likely due to ocean-driven warming.

From both the sea-level record and recent satellite measurements of ice mass balance, it seems that the

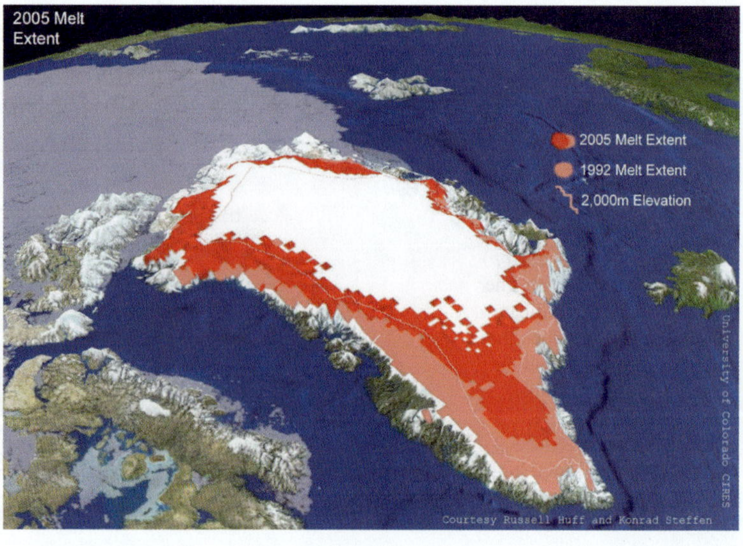

Fig. 22 Extent of melt on the Greenland ice sheet during 1992 (*light red*) and 2006 (*dark red*). Graphic by Konrad Steffen and Russell Huff of the Cooperative Institute for Research in Environmental Sciences (CIRES) at the University of Colorado at Boulder.
Source: HUcires.colorado.edu/steffen/greenland/melt2005/U

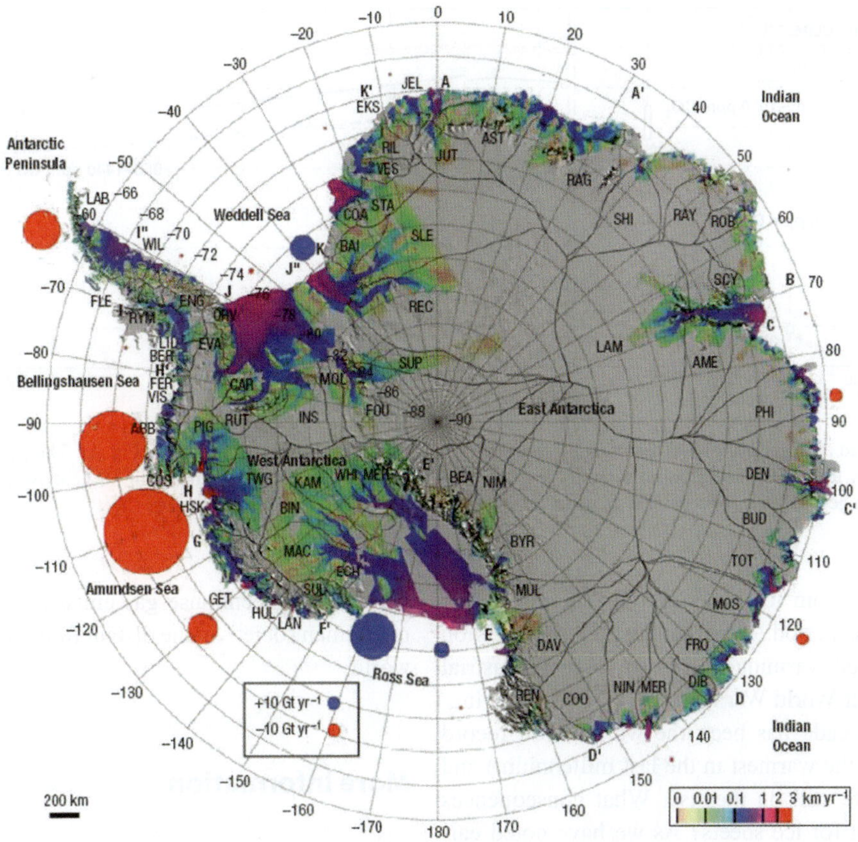

Fig. 23 Mass balance measurements for Antarctica from satellite measurements. *Red circles* indicate net loss, *blue circles* indicate net gain. The size of the circles is proportional to the amount lost or gained. *Colours* on the map indicated flow velocities of ice and ice streams with *blue* and *red* indicating high velocities (>0.1 km/year) and *yellow* and *green* indicating slower movement (<0.1 km/year).
Source: Rignot (2008)

response of the ice sheets to warming is both real and accelerating. The errors associated with the measurements are still quite high (20–40%), but the different methods are showing similar trends. These observations along with our knowledge of past level rise from palaeoclimate studies indicates we should be planning to respond now, both to reduce the ice loss and deal with the consequences of sea level rise "in the pipeline."

Bracing for the Future – It Is Our Choice

In this last section we summarise what we currently know of both long term and recent history of the ice sheets and consider their future prospects in the light of climate change. The geological record tells us that the first big ice sheets formed on Antarctica around 34 m.y. ago, when global temperature dropped ~4° as part of a steady decline over the last 50 m.y. The big Northern Hemisphere ice sheets developed just 2.5 m.y. ago as temperature dropped a further 2–3°C Estimates of CO_2 levels in the distant past are more speculative because of greater uncertainties in the techniques, but they indicate levels ranging from two to eight times pre-industrial in the Greenhouse world, and values less than two times pre-industrial (likely less than 400 ppm) for the last 24 million years of the Icehouse world.

Basic physics shows we can expect a causal relationship between CO_2 and temperature as a consequence of the "greenhouse effect," and climate modelling reveals the extent of the influence of CO_2 on temperature. The close connection is evident from Antarctic ice core data which show both in "lock-step" through the eight glacial cycles of the last 850,000 years. The connection is also clear from direct temperature observations over the last 150 years and CO_2

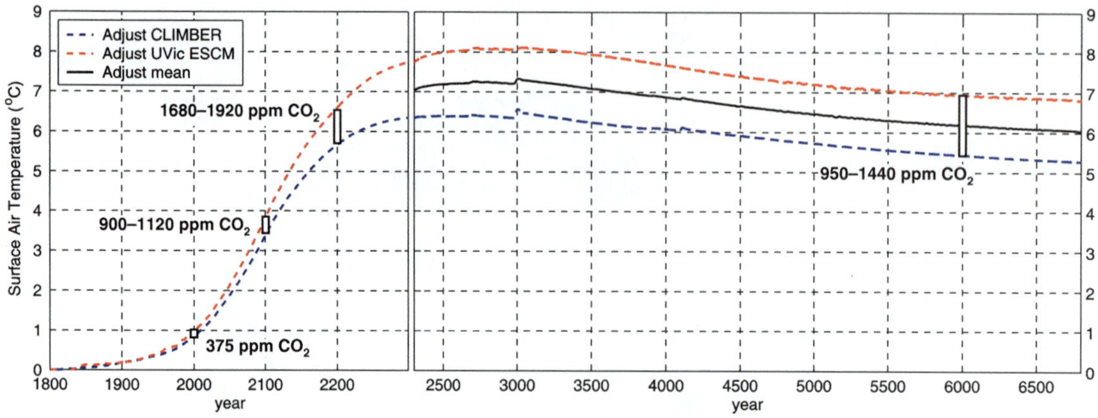

Fig. 24 Projected atmospheric global temperature rise over the next 6,000 years following the carbon discharge from all fossil fuels over the next 200 years ("business as usual") (modified from Montenegro et al., 2007, Fig. 1). The persistence of global average temperatures more than ~6°C above pre-industrial risks the loss of even the largest ice sheet

measurements from ice cores and air sampling over the same time period. Here though, the suppression of CO_2-induced warming by aerosols from industrial expansion after World War II needs to be factored in.

The last decade has been the warmest on record (and possibly the warmest in the last millennium) and CO_2 levels are still on the rise. What consequences can we expect for ice sheets? As we have noted earlier satellite-based measurements indicate that both the Greenland and the West Antarctic Ice Sheets have in the last decade or so now begun to lose mass. Although current estimates indicate the contribution of ice sheets to sea level rise is still small (~0.4 + 0.2 mm; IPCC, 2007a) recent estimate indicates the rate has doubled in the last 5 years (Velicogna, 2009).

Montonegro et al. (2007) provide a worst case scenario with results from two climate models that simulate Earth's global surface temperature response to atmospheric CO_2 if all available fossil energy were used over the next two centuries. They indicate the persistence of CO_2 in the atmosphere was such that even after 6,000 years CO_2 levels were still 3.3–5 times pre-industrial levels (Fig. 24). Ice sheet modelling studies, though still at an early stage, indicate much higher levels of CO_2 are needed to remove ice sheets than form them. DeConto et al. (2007) found that 4–8 times pre-industrial CO_2 levels was required to remove the East Antarctic Ice Sheet, whereas the threshold for its growth was just 750 ppm (~2.5 times pre-industrial). Further model development is plainly needed, but for the moment there remains the risk that our current greenhouse gas emissions path will commit human society in the distant future to a Greenhouse world.

More Information

Review reports:
For further information we would like to point to a few key references:

ACCE (2009) Antarctic Climate Change and the Environment (529p). Scientific Committee on Antarctic Research, Cambridge.

ACIA (2005) Arctic Climate Impact Assessment (1042p). Cambridge: Cambridge University Press.

AMAP (2009) 2009 Update on Selected Climate Issues of Concern (p. 15). Oslo: Arctic Monitoring and Assessment Programme.

IPCC (2007) Climate Change 2007: The Physical Science Basis (996p). Contribution of Working Group I to the third assessment report of the Intergovernmental Panel on Climate Change. Cambridge, UK and New York, NY: Cambridge University Press.

Koç N, Njåstad B, Armstrong R, Corell RW, Jensen DD, Leslie KR, Rivera A, Tandong Y, Winther J-G (Eds) (2009) Melting Snow and Ice: A Call for Action. Centre for Ice, Climate and Ecosystems, Norwegian Polar Institute.

Mayewski PA, Meredith MP, Summerhayes CP, Turner J, Worby A, Barrett PJ, Casassa G, Bertler NAN, Bracegirdle T, Naveira-Garabato AC, et al. (2009) State of the Antarctic and southern ocean climate system (SASOCS). Rev Geophys 47:1–38.

Turner J, Bindschandler R, Convey P, di Prisco G, Fahrbach E, Gutt J, Hodgson D, Mayewski PA, Summerhayes CP (2009) Antarctic Climate Change and the Environment (529p). Cambridge: Scientific Committee on Antarctic Research.

Scientific webpage debate on recent findings:
http://www.realclimate.org
http://www.ipcc.ch

References

ACIA (2005) Arctic Climate Impact Assessment (p. 1042). Cambridge: Cambridge University Press.

Adkins F, McIntyre K, Schrag DP (2002) The salinity, temperature, and 18O of the glacial deep ocean. Science 298:1769–1773.

Alley RB (2000a) Ice-core evidence of abrupt climate changes. Proc Natl Acad Sci USA 97:1331–1334.

Alley RB (2000b) The Two-Mile Time Machine (229p). Princeton, NJ: Princeton University Press.

AMAP (2009) AMAP 2009 Update on Selected Climate Issues of Concern (p. 15). Oslo: Arctic Monitoring and Assessment Programme.

Arblaster JM, Meehl GA (2006) Contributions of external forcings to southern annular mode trends. J Clim 19:2896–2905.

Bamber J, Riva R, Vermeersen B, LeBrocq A (2009) Reassessment of the Potential Sea-Level Rise from a Collapse of the West Antarctic Ice Sheet. Science 324(5929): 901–903. doi: 10.1126/science.1169335.

Bard E (2004) Greenhouse effect and ice ages: Historical perspective. C R Geosci 336:603–638.

Barker PF, Filippelli GM, Florindo F, Martin EE, Scher HD (2007) Onset and role of the Antarctic circumpolar current. Deep Sea Res IIp:2388–2398.

Barker S, Diz P, Vautravers MJ, Pike J, Knorr G, Hall IR, et al. (2009) Interhemispheric Atlantic seesaw response during the last deglaciation. Nature 457:1097–1102.

Barrett PJ, Adams CJ, McLntosh WC, Swisher CC, Wilson GS (1992) Geochronological evidence supporting Antarctic deglaciation three million years ago. Nature 359:816–818.

Barrett PJ (2007) Cenozoic climate and sea-level history from glacimarine strata off the Victoria Land Coast, Cape Roberts Project, Antarctica. In MJ Hambrey, P Christoffersen, NF Glasser, B Hubbart (Eds), Glacial Processes and Products, International Association of Sedimentologists Special Publication. 39:259–287.

Bender M, Sowers T, Dickson M-L, Orchardo J, Grootes P, Mayewski PA, et al. (1994) Climate correlations between Greenland and Antarctica during the past 100,000 years. Nature 372:663–666.

Bertler NAN, Barrett PJ, Mayewski PA, Fogt RL, Kreutz KJ, Shulmeister J (2004) El Niño suppresses Antarctic warming. Geophys Res Lett 31:L15207. doi: 10.1029/2004GL020749.

Bertler NAN, Naish TR, Oerter H, Kipfstuhl S, Barrett PJ, Mayewski PA, et al. (2006) The effects of joint ENSO-Antarctic oscillation forcing on the McMurdo Dry Valleys, Antarctica. Antarct Sci 18:507–514.

Böning CW, Dispert A, Visbeck M, Rintoul SR, Schwarzkopf FU (2008) The response of the Antarctic circumpolar current to recent climate change. Geosci Nat 1:864–869.

Bowen GJ (2006) Eocene hyperthermal event offers insight into greenhouse warming. EOS 87:165 and 169.

Box JE, Bromwich DH, Veenhuis BA, Bai L-S, Stroeve JC, Rogers JC, et al. (2006) Greenland ice sheet surface mass balance variability (1988–2004) from calibrated polar MM5 Output. J Clim 19:2783–2800.

Boyd PW, Watson AJ, Law CS, Abraham ER, Trull T, Murdoch R, et al. (2000) A mesoscale phytoplankton bloom in the polar Southern Ocean stimulated by iron fertilization. Nature 407:695–702.

Broecker WS (1995) Chaotic climate. Sci Am 273:62–68.

Broecker WS (1997) Thermohaline circulation, the Achilles Heel of our climate system: Will man-made CO_2 upset the current balance? Science 278:1582–1588.

Broecker WS (2000a) Abrupt climate change: Causal constraints provided by the paleoclimate record. Earth Sci Rev 51: 137–154.

Broecker WS (2000b) Abrupt climate change: Causal constraints provided by the paleoclimate record. Earth Sci Rev 51: 137–154.

Broecker WS (2003) Does the trigger for abrupt climate change reside in the ocean or in the atmosphere? Science 300: 1519–1522.

Broecker WS (2006) Abrupt climate change revisited. Glob Planet Change 54:211–215.

Buchardt SL, Dahl-Jensen D (2008) At what depth is the Eemian layer expected to be found at NEEM? Ann Glaciol 48: 100–102.

Chen JL, Wilson CR, Blankenship DD, Tapley BD (2009) Accelerated Antarctic ice loss from satellite gravity measurements. Nat Geosci 2:859–862.

Clark PU, Pisias NG, Stocker TF, Weaver AJ (2002) The role of the thermohaline circulation in abrupt climate change. Nature 415:863–869.

Clark PU, Huybers P (2009) Global change: Interglacial and future sea level. Nature 462, 856–857.

Comiso JC (2000) Variability and trends in antarctic surface temperatures from in situ and satellite infrared measurements. J Clim 13.1674–1696.

Coxall HK, Wilson PA, Palike H, Lear CH, Backman J (2005) Rapid stepwise onset of Antarctic glaciation and deeper calcite compensation in the Pacific Ocean. Nature 433: 53–57.

DeConto R, Pollard D, Harwood D (2007) Sea ice feedback and Cenozoic evolution of Antarctic climate and ice sheets. Paleoceanography 22:PA3214. doi: 10.1029/2006PA001350.

DeConto RM, Pollard D, Wilson PA, Palike H, Lear CH, Pagani M (2008) Thresholds for Cenozoic bipolar glaciation. Nature 455:652–656.

DeConto RM, Pollard D (2009) Modeling West Antarctic ice sheet growth and collapse through the last 5 million years. Nature 458:329–332.

Denton GH, Hughes TJ (1981) The Last Great Ice Sheets. New York, NY: Wiley.

Domack E, Duran D, Leventer A, Ishman S, Doane S, McCallum S, et al. (2005) Stabitity of the Larsen B ice shelf on the Antarctic Peninsula during the Holocene epoch. Nature 436:681–685.

Doney SC, Fabry VJ, Feely RA, Kleypas JA (2009) Ocean acidification: The other CO_2 problem. Annu Rev Marine Sci 1:169–192.

Dowsett HJ, Cronin TM (1990) High eustatic sea level during the middle Pliocene: Evidence from the southern U.S. Atlantic Coastal Plain. Geology 18:435–438.

Dunbar GB, Naish TR, Barrett PJ, Fielding CR, Powell RD (2008) Contraining the amplitude of late Oligocene bathymetric changes in Western Ross Sea during orbitally-induced oscillation in the East Antarctic Ice Sheet: (1) Implications for glacimarine sequence stratigraphic models. Palaeogeogr Palaeoclimatol Palaeoecol 260:50–65.

Community Members EPICA (2004) Eight glacial cycles from an Antarctic ice core. Nature 429:623–628.

Francis JE, Ashworth A, Cantrill DJ, Crame JA, Howe J, Stephens R, et al. (2008) 100 Million years of antarctic climate evolution: Evidence from fossil plants. In AK Cooper, PJ Barrett, H Stagg, B Storey, EStump, W Wise, the 10th ISAES editorial team (Eds), Antarctica: A Keystone in a Changing World (pp. 19–28). Proceedings of the 10th International Symposium on Antarctic Earth Sciences. Washington, DC: The National Academies Press.

Glasser NF, Scambos TA (2008) A structural analysis of the 2002 Larsen B ice shelf collapse. J Glaciol 54:3–16.

Hanna E, Huybrechts P, Janssens I, Cappelen J, Steffen K, Stephens A (2005) Runoff and mass balance of the Greenland ice sheet: 1958–2003. J Geophys Res 110:D13108. doi: 10.1029/2004JD005641.

Hansen J, Ruedy R, Sato M, Imhoff M, Lawrence W, Easterling D, et al. (2001) A closer look at United States and global surface temperature change. J Geophys Res 106: 23947–23963.

Hansen J, Sato M, Ruedy R, Nazarenko L, Lacis A, Schmidt GA, et al. (2005) Efficacy of climate forcings. J Geophys Res 110.

Hansen J, Sato M, Kharecha P, Beerling D, Masson-Delmotte V, Pagani M, et al. (2008) Target atmospheric CO_2: Where should humanity aim? Open Atmos Sci J 2: 217–231.

Hay WW, Soeding E, DeConto RM, Wold CN (2002) The Late Cenozoic uplift – Climate change paradox. Int J Earth Sci (Geol Rundsch) 91:746–774.

Haywood AM, Valdes PJ (2004) Modelling Pliocene warmth: Contribution of atmosphere, oceans and cryosphere. Earth Planet Sci Lett 218:363–377.

Hegerl GC, Zwiers FW, Braconnot P, Gillett NP, Luo Y, Marengo Orsini JA, et al. (2007) Understanding and attributing climate change. In S Solomon, D Qin, M Manning, Z Chen, M Marquis, KB Averyt, M Tignor, HL Miller (Eds), Climate Change 2007: The Physical Science Basis. Contribution of Working Group I to the fourth assessment report of the Intergovernmental Panel on Climate Change. Cambridge, UK and New York, NY: Cambridge University Press.

Heinrich H (1988) Origin and consequences of cyclic ice rafting in the Northeast Atlantic Ocean during the past 130,000 years. Quat Res 29:142–152.

Huber M (2008) Climate change – A hotter greenhouse? Science 321:353–354.

Huybers P (2006) Early pleistocene glacial cycles and the integrated summer insolation forcing. Science 313:508–511.

Imbrie JA, Imbrie KP (1979) Ice Ages – Solving the Mystery. Cambridge, MA: Harvard University Press.

IPCC (2001) Climate Change 2001: The Scientific Basis: Cambridge, United Kingdom and New York, NY, USA, Cambridge University Press.

IPCC (2007a) Climate Change 2007: Impacts, Adaptation and Vulnerability. Cambridge, UK and New York, NY: Cambridge University Press.

IPCC (2007b) Climate Change 2007. The Physical Science Basis. Contribution of Working Group I to the Third Assessment Report of the Intergovernmental Panel on Climate Change (996p). Cambridge, UK and New York, NY: Cambridge University Press.

IPCC (2007c) Climate Change 2007. The Physical Science Basis. Contribution of Working Group I to the Third Assessment Report of the Intergovernmental Panel on Climate Change. Summary for Policymakers (21p). Cambridge, UK and New York, NY: Cambridge University Press.

IPICS (2008) The oldest ice core: A 1.5 million year record of climate and greenhouse gases from Antarctica, Science and outline implementation plan as approved by IPICS SC: 1st June 2008, p. 30.

Johannessen OM, Khvorostovsky K, Miles MW, Bobylev LP (2005) Recent ice-sheet growth in the interior of greenland. Science 310:1013–1016.

Jouzel J, Alley RB, Cuffey KM, Dansgaard W, Grootes P, Hoffmann G, et al. (1997) Validity of the temperature reconstruction from water isotopes in ice cores. J Geophys Res 102:26471–26487.

Jouzel J (1999) Calibrating the isotopic paleothermometer. Science 286:910–911.

Jouzel J, Masson-Delmotte V, Cattani O, Dreyfus G, Falourd S, Hoffmann G, et al. (2007) Orbital and millennial Antarctic climate variability over the past 800,000 years. Science 317:793–796.

Keeling CD, Whorf TP, Wahlen M, van der Plichtt J (1995) Interannual extremes in the rate of rise of atmospheric carbon dioxide since 1980. Nature 375:666–670.

Kennett JP (1982) Marine Geology. Englewood Cliffs, NJ: Prentice-Hall.

Kennett JP, Stott LD (1991) Abrupt deep-sea warming, palaeoceanographic changes and benthic extinctions at the end of the Palaeocene. Nature 353:225–229.

King JC, Turner J (1997) Antarctic Meteorology and Climatology: Atmospheric and Space Science Series (409p). Cambridge: Cambridge University Press.

Krabill W, Abdalati W, Frederick E, Manizade S, Martin C, Sonntag J, et al. (2000) Greenland ice sheet: High-elevation balance and peripheral thinning. Science 289:428–430.

Kump LR (2009) ATMOSPHERIC SCIENCE: Tipping pointedly colder. Science 323:1175–1176.

Kwok R, Cunningham GF, Wensnahan M, Rigor I, Zwally HJ, Yi D (2009) Thinning and volume loss of the

Arctic Ocean sea ice cover: 2003–2008. J Geophys Res 114.

Lambert F, Delmonte B, Petit JR, Bigler M, Kaufmann PR, Hutterli MA, et al. (2008) Dust-climate couplings over the past 800,000 years from the EPICA Dome C ice core. Nature 452:616–619.

Lawver LA, Gahagan LM (2003) Evolution of Cenozoic seaways in the circum-Antarctic region. Palaeogeogr Palaeoclimatol Palaeoecol 198:11–37.

Lemke P, Ren J, Alley RB, Allison I, Carrasco J, Flato G, Fujii Y, Kaser G, Mote P, Thomas RH, Zhang T (2007) Observations: Changes in snow, ice and frozen ground, in Climate Change 2007: The Physical Science Basis. Contribution of Working Group I to the Fourth Assessment Report of the Intergovernmental Panel on Climate Change, edited by S. Solomon et al., pp. 337–383, Cambridge University Press, New York.

Lewis AR, Marchant DR, Ashworth AC, Hemming SR, Machlus ML (2007) Major middle Miocene global change: Evidence from East Antarctica and the transantarctic mountains. Geol Soc Am Bull 119:1449–1461.

Lewis AR, Marchant DR, Ashworth AC, Heden L, Hemming SR, Johnson JV, et al. (2008) Mid-Miocene cooling and the extinction of tundra in continental Antarctica. Proc Natl Acad Sci 105:10676–10680.

Liu Z, Pagani M, Zinniker D (2009) Global cooling during the Eocene-Oligocene climate transition. Science 323:1187–1190.

Lunt DJ, Foster GL, Haywood AM, Stone EJ (2008) Late Pliocene Greenland glaciation controlled by a decline in atmospheric CO_2 levels. Nature 454:1102–1105.

Luthi D, Le Floch M, Bereiter B, Blunier T, Barnola J-M, Siegenthaler U, et al. (2008) High-resolution carbon dioxide concentration record 650,000–800,000 years before present. Nature 453:379–382.

Lythe MB, Vaughan DG, the BEDMAP Consortium (2001) BEDMAP: A new ice thickness and subglacial topographic model of Antarctica. J Geophys Res 106:11335–11351.

Marshall GJ, Orr A, van Lipzig NPM, King JC (2006) The impact of a changing southern hemisphere annular mode on Antarctic peninsula summer temperatures. J Clim 19:5388–5404.

Maslanik JA, Fowler C, Stroeve J, Drobot S, Zwally J, Yi D, et al. (2007) A younger, thinner Arctic ice cover: Increased potential for rapid, extensive sea-ice loss. Geophys Res Lett 34.

Maslin MA, Li XS, Loutre MF, Berger A (1998) The contribution of orbital forcing to the progressive intensification of northern hemisphere glaciation. Quat Sci Rev 17:411–426.

Masson-Delmotte V, Kageyama M, Braconnot P, Charbit S, Krinner G, Ritz C, et al. (2006) Past and future polar amplification of climate change: Climate model intercomparisons and ice-core constraints. Clim Dyn 26:513–529.

Mayewski PA, Meeker LD, Whitlow S, Twickler MS, Morrison MC, Bloomfield P, et al. (1994) Changes in atmospheric circulation and ocean ice cover over the North Atlantic during the last 41,000 years. Science 263:1747–1751.

Mayewski PA, White F (2002) The Ice Chronicles (233p). Hanover, NH: University Press of New England.

Mayewski PA, Frezzotti M, Bertler NAN, van Ommen T, Hamilton GS, Jacka TH, et al. (2005) The International Trans-Antarctic Scientifc Expedition (ITASE) – An overview. Ann Glaciol 41:180–185.

Mayewski PA, Meredith MP, Summerhayes CP, Turner J, Worby A, Barrett PJ, et al. (2009) State of the Antarctic and Southern Ocean Climate System (SASOCS). Rev Geophys 47:1–38.

McKay RM, Barrett PJ, Harper MA, Hannah MJ (2008) Atmospheric transport and concentration of diatoms in surficial and glacial sediments of the Allan Hills, Transantarctic Mountains: Palaeogeogr Palaeoclimatol Palaeoecol 260:168–183.

Meinshausen M, Meinshausen N, Hare W, Raper SCB, Frieler K, Knutti R, et al. (2009) Greenhouse-gas emission targets for limiting global warming to 2°C. Nature 458:1158–1162.

Mercer JH (1978) West Antarctic ice sheet and CO_2 greenhouse effect: a threat of disaster. Nature 271:321–325.

Milankovitch M (1941) Canon of insolation and the ice age problem. Koniglich Serbische Akademie, Belgrade 132.

Miller KG, Wright JD, Browning JV (2005) Visions of ice sheets in a greenhouse world. Marine Geol 217:215–231.

Montenegro A, Brovkin V, Eby M, Archer D, Weaver AJ (2007) Long term fate of anthropogenic carbon. Geophys Res Lett 34:L19707. doi: 10.1029/2007GL030905.

Moran K, Backman J, Brinkhuis H, Clemens SC, Cronin T, Dickens GR, et al. (2006) The Cenozoic palaeoenvironment of the Arctic Ocean. Nature 441:601–605.

Naish TR, Woolfe KJ, Barrett PJ, Wilson GS, Atkins C, Bohaty SM, et al. (2001) Orbitally induced oscillations in the East Antarctic ice sheet at the Oligocene/Miocene boundary. Nature 413:719–723.

Naish T, Powell R, Levy R, Wilson G, Scherer R, Talarico F, et al. (2009) Obliquity-paced Pliocene West Antarctic ice sheet oscillations. Nature 458:322–328.

Orr JC, Fabry VJ, Aumont O, Bopp L, Doney SC, Feely RA, et al. (2005) Anthropogenic ocean acidification over the twenty-first century and its impact on calcifying organisms. Nature 437:681–686.

Pagani M, Zachos JC, Freeman KH, Tipple B, Bohaty S (2005) Marked decline in atmospheric carbon dioxide concentrations during the paleogene. Science 309:600–603.

Pagani M, Caldeira K, Berner R, Beerling DJ (2009) The role of terrestrial plants in limiting atmospheric CO_2 decline over the past 24 million years. Nature 460:85–88.

Pagani M, Liu Z, LaRiviere J, Ravelo AC (2010) High Earth-system climate sensitivity determined from Pliocene carbon dioxide concentrations. Nature 3:27–30.

Palmer S, Shepherd A, Bjornsson H, Palsson F (2009) Ice velocity measurements of Langjokull, Iceland, from interferometric synthetic aperture radar (InSAR): J Glaciol 55:834–838.

Pearson PN, Foster GL, Wade BS (2009) Atmospheric carbon dioxide through the Eocene-Oligocene climate transition. Nature 461:1110–1113.

Pearson PN, Palmer MR (2000) Atmospheric carbon dioxide concentrations over the past 60 million years. Nature 406:695–699.

Petit J-R, Jouzel J, Raynaud D, Barkov NI, Barnola JM, Basile I, et al. (1999) Climate and atmospheric history of the past 420,000 years from the Vostok ice core, Antarctica. Nature 399:429–436.

Pierrehumbert RT (2002) The hydrologic cycle in deep-time climate problems. Nature 419:191–198.

Pollard D, DeConto RM (2009) Modelling West Antarctic ice sheet growth and collapse through the past five million years. Nature 458:329–332.

Prebble JG, Raine JI, Barrett PJ, Hannah MJ (2006) Vegetation and climate from two Oligocene glacioeustatic sedimentary cycles (31 and 24 Ma) cored by the Cape Roberts Project, Victoria Land Basin, Antarctica: Palaeogeogr Palaeoclimatol Palaeoecol 231:41–57.

Pritchard HD, Arthern RJ, Vaughan DG, Edwards LA (2009) Extensive dynamic thinning on the margins of the Greenland and Antarctic ice sheets. Nature (advance online publication).

Rahmstorf S (2002) Ocean circulation and climate during the past 120,000 years. Nature 419:207–214.

Ramillien G, Lombard A, Cazenave A, Ivins ER, Llubes M, Remy F, et al. (2006) Interannual variations of the mass balance of the Antarctica and Greenland ice sheets from GRACE. Glob Planet Change 53:198–208.

Regehr EV, Hunter CM, Caswell H, Amstrup SC, Stirling I (2009) Survival and breeding of polar bears in the southern Beaufort Sea in relation to sea ice. J Anim Ecol 9999.

Rignot E, Thomas RH (2002) Mass balance of polar ice sheets. Science 297:1502–1506.

Rignot E, Casassa G, Gogineni P, Krabill W, Rivera A, Thomas R (2004) Accelerated ice discharge from the Antarctic Peninsula following the collapse of the Larsen B ice shelf. Geophys Res Lett 31:L18401. doi: 10.1029/2004GL020697.

Rignot E, Casassa G, Gogineni S, Kanagaratnam P, Krabill W, Pritchard H, et al. (2005) Recent ice loss from the Fleming and other glaciers, Wordie Bay, West Antarctic Peninsula. Geophys Res Lett 32.

Rignot E, Kanagaratnam P (2006) Changes in the velocity structure of the greenland ice sheet. Science 311: 986–990.

Rignot E, Bamber JL, van den Broeke MR, Davis C, Li Y, van de Berg WJ, et al. (2008) Recent Antarctic ice mass loss from radar interferometry and regional climate modelling. Nat Geosci 1:106–110.

Rohling EJ, Grant K, Bolshaw M, Roberts AP, Siddall M, Hemleben C, et al. (2009) Antarctic temperature and global sea level closely coupled over the past five glacial cycles. Nat Geosci 2:500–504.

Royer DL (2006) CO_2-forced climate thresholds during the Phanerozoic. Geochim Cosmochim Acta 70:5665–5675.

Ruddiman WF (2001) Earth's Climate: Past and Future (465p). New York, NY: W.H. Freeman.

Sabine CL, Feely RA, Gruber N, Key RM, Lee K, Bullister JL, et al. (2004) The oceanic sink for anthropogenic CO_2. Science 305:367–371.

Scambos TA, Bohlander JA, Shuman CA, Skvarca P (2004) Glacier acceleration and thinning after ice shelf collapse in the Larsen B embayment, Antarctica. Geophys Res Lett 31:L18402. doi: 10.1029/2004GL020670.

Schneider DP, Steig EJ, van Ommen TD, Dixon DA, Mayewski PA, Jones JM, et al. (2006) Antarctic temperatures over the past two centuries from ice cores. Geophys Res Lett 33:L16707. doi: 10.1029/2006GL027057.

Schneider von Diemling T, Ganopolski A, Held H, Rahmstorf S (2006) How cold was the Last Glacial Maximum? Geophys Res Lett 33:L14709.

Severinghaus JP (2009) Climate change: Southern see-saw seen. Nature 457:1093–1094.

Shepherd A, Wingham D, Payne T, Skvarca P (2003) Larsen ice shelf has progresively thinned. Science 302:856–859.

Shevenell AE, Kennett JP, Lea DW (2004) Middle Miocene southern ocean cooling and antarctic cryosphere expansion. Science 305:1766–1770.

Siegenthaler U, Stocker TF, Monnin E, Lüthi D, Schwander J, Stauffer B, et al. (2005) Stable carbon cycle-climate relationship during the Late Pleistocene. Science 310: 1313–1317.

Siegert MJ, Carter S, Tabacco I, Popov S, Blankenship DD (2005) A revised inventory of Antarctic subglacial lakes. Antarct Sci 17:453–460.

Simmonds I, Keay K (2009) Extraordinary September Arctic sea ice reductions and their relationships with storm behavior over 1979–2008. Geophys Res Lett 36.

Sneed WA, Hamilton GS (2007) Evolution of melt pond volume on the surface of the Greenland Ice Sheet. Geophys Res Lett 34:L03501. doi: 10.1029/2006GL028697.

Steffensen JP, Andersen KK, Bigler M, Clausen HB, Dahl-Jensen D, Fischer H, et al. (2008) High-resolution greenland ice core data show abrupt climate change happens in few years. Science 321:680–684.

Steig EJ, Schneider DP, Rutherford SD, Mann ME, Comiso JC, Shindell DT (2009) Warming of the Antarctic ice-sheet surface since the 1957 international geophysical year. Nature 457:459–462.

Stickley CE, Brinkhuis H, Schellenberg S, Sluijs A, Roehl U, Fuller M, et al. (2004) Timing and nature of the deepening of the Tasmanian Gateway. Paleoceanography 19:PA4027.

Stocker TF (1999) Abrupt climate changes: From the past to the future – A review. Int J Earth Sci 88:365–374.

Stocker TF (2003) South dials north. Nature 424:496–497.

Sugden DE, Marchant DR, Denton GH (1993) The case for a stable Antarctic ice sheet. Geogr Annal 75A(4): 155–351.

Sugden DE, Marchant DR, Potter N Jr, Souchez RA, Denton GH, Swisher CC III, et al. (2002) Preservation of Miocene glacier ice in East Antarctica. Nature 376:412–414.

Summerfield MA, Stuart FM, Cockburn HAP, Sugden DE, Denton GH, Dunai T, et al. (1999) Long-term rates of denudation in the Dry Valleys, Transantarctic Mountains, southern Victoria Land, Antarctica based on in situ produced cosmogenic 21Ne. Geomorphology 27:113–129.

Thomas R, Rignot E, Casassa G, Kanagaratnam P, Acuna C, Akins T, et al. (2004) Accelerated sea-level rise from West Antarctica. Science 306:255–258.

Thomas R, Frederick E, Krabill W, Manizade S, Martin C (2006) Progressive increase in ice loss from Greenland. Geophys Res Lett 33:L10503. doi: 10.1029/2006GL026075.

Thompson DWJ, Solomon S (2002) Interpretation of recent southern hemisphere climate change. Science 296: 895–899.

Turner J (2004) The El Niño-southern oscillation and Antarctica. Int J Climatol 24:1–31.

Turner J, Colwell SR, Marshall GJ, Lachlan-Cope TA, Carleton AM, Jones PD, et al. (2005) Antarctic climate change during the last 50 years. Int J Climatol 25:279–294.

Turner J, Lachlan-Cope TA, Colwell S, Marshall GJ, Connolley WM (2006) Significant warming of the Antarctic winter troposphere. Science 311:1914–1917.

Turner J, Bindschandler R, Convey P, di Prisco G, Fahrbach E, Gutt J, et al. (2009) Antarctic Climate Change and the

Environment. Cambridge: Scientific Committee on Antarctic Research.

United Nations Framework Convention on Climate Change (1992) The United Nations Framework Convention on Climate Change (p. 25). Accessed on http://unfccc.int/resource/docs/convkp/conveng.pdf

Velicogna I, Wahr J (2005) Greenland mass balance from GRACE. Geophys Res Lett 32(8):L18505. doi: 10.1029/2005GL023955.

Velicogna I, Wahr J (2006) Acceleration of Greenland ice mass loss in spring 2004. Nature 443:329–331.

Velicogna I (2009) Increasing rates of ice mass loss from the Greenland and Antarctic ice sheets revealed by GRACE. Geophys Res Lett 36:L19503, 4 PP., doi: 10.1029/2009GL040222.

Warny S, Askin R, Hannah MJ, Mohr BAR, Raine JI, Harwood DM, et al., the SMS Science Team (2009) Palynomorphs from a sediment core reveal a sudden remarkably warm Antarctica during the middle Miocene. Geology 37: 955–958.

Webb PN, Harwood DM, McKelvey BC, Mercer JH, Stott LD (1984) Cenozoic marine sedimentation and ice-volume variation on the East Antarctic craton. Geology 12:287–291.

Weedon G (2003) Time-Series Analysis and Cyclostratigraphy. Cambridge: Cambridge University Press.

Wilson DS, Luyendyk BP (2009) West Antarctic paleotopography estimated at the Eocene-Oligocene climate transition: Geophys Res Lett 36:L16302.

Wingham DJ, Siegert MJ, Shepherd A, Muir AS (2006) Rapid discharge connects Antarctic subglacial lakes. Nature 440:1033–1036.

Wolff EW, Fischer H, Rothlisberger R (2009) Glacial terminations as southern warmings without northern control. Nat Geosci 2:206–209.

Wouters B, Chambers D, Schrama EJO (2008) GRACE observes small-scale mass loss in Greenland. Geophys Res Lett 35:L20501, 5 PP., doi: 10.1029/2008GL034816.

You Y, Huber M, Muller RD, Poulsen CJ, Ribbe J (2009) Simulation of the middle Miocene climate optimum. Geophys Res Lett 36.

Zachos JC, Breza J, Wise SW (1992) Early Oligocene ice sheet expansion on Antarctica, sedimentological and isotopic evidence from the Kergulen Plateau. Geology 20: 569–573.

Zachos J, Pagani M, Sloan L, Thomas E, Billups K (2001a) Trends, rhythms, and aberrations in global climate 65 Ma to present. Science 292:686–693.

Zachos JC, Shackleton NJ, Revenaugh JS, Palike H, Flower BP (2001b) Climate response to orbital forcing across the Oligocene-Miocene boundary. Science 292: 274–278.

Zachos JC, Rohl U, Schellenberg SA, Sluijs A, Hodell DA, Kelly DC, et al. (2005) Rapid acidification of the ocean during the paleocene-eocene thermal maximum. Science 308:1611–1615.

Zachos JC, Dickens GR, Zeebe RE (2008) An early Cenozoic perspective on greenhouse warming and carbon-cycle dynamics. Nature 451:279–283.

Zwally HJ, Giovinetto MB, Li J, Cornejo HG, Beckley MA, Brenner AC, et al. (2005) Mass changes of the Greenland and Antarctic ice sheets and shelves and contributions to sea-level rise: 1992–2002. J Glaciol 51:509–527.

Climate and Peatlands

Rixt de Jong, Maarten Blaauw, Frank M. Chambers, Torben R. Christensen, François de Vleeschouwer, Walter Finsinger, Stefan Fronzek, Margareta Johansson, Ulla Kokfelt, Mariusz Lamentowicz, Gaël Le Roux, Dmitri Mauquoy, Edward A.D. Mitchell, Jonathan E. Nichols, Emanuela Samaritani, and Bas van Geel

Abstract Peatlands are an important natural archive for past climatic changes, primarily due to their sensitivity to changes in the water balance and the dating possibilities of peat sediments. In addition, peatlands are an important sink as well as potential source of greenhouse gases. The first part of this chapter discusses a range of well-established and novel proxies studied in peat cores (peat humification, macrofossils, testate amoebae, stomatal records from subfossil leaves, organic biomarkers and stable isotope ratios, aeolian sediment influx and geochemistry) that are used for climatic and environmental reconstructions, as well as recent developments in the dating of these sediments. The second part focuses on the role that peatland ecosystems may play as a source or sink of greenhouse gases. Emphasis is placed on the past and future development of peatlands in the discontinuous permafrost areas of northern Scandinavia, and the role of regenerating mined peatlands in north-western Europe as a carbon sink or source.

Keywords Bog surface wetness variations · ^{14}C dating · Plants' responses to CO_2 · Stable isotopes · Methane emissions from peatlands · Palsas

Introduction

Following the pioneering work of Blytt (1876), Sernander (1908) and Von Post and Sernander (1910), peatlands are regarded as important archives of climatic change (Blackford, 2000; Chambers and Charman, 2004). Peatlands cover around 4.3 million km^2 of the world's surface, particularly in the cooler regions of the Northern Hemisphere (Chambers and Charman, 2004). Climatic changes throughout the Holocene have been reconstructed from peat, using a wide array of biological, geochemical and mineral proxies. Many different proxy-indicators can be derived from peat cores, allowing for a multi-proxy approach for climatic reconstructions (Chambers and Charman, 2004). Peat-based climatic and environmental reconstructions are currently available from many sites in Europe (see, e.g. Hughes et al., 2000; Barber et al., 2003; Chambers and Charman, 2004), North America (e.g. Jones et al., 2009), South America (e.g. Chambers et al., 2007a), New Zealand (e.g. Hajdas et al., 2006), South Atlantic Islands (e.g. Van der Putten et al., 2004) and Australia (e.g. Fletcher and Thomas, 2007). Such studies complement other reconstructions (based on, e.g. treerings or lake sediments) and thereby provide a more complete reconstruction of past climatic and environmental changes (e.g. Kokfelt et al., 2010).

Ombrotrophic peatlands, in particular, are frequently used for climatic reconstructions since their water balance and nutrient status depends on the input of atmospheric water and evapotranspiration from the bog surface (see Charman, 2002). In peatland systems that function close to their climatic thresholds, small changes in external factors may cause large alterations in ecosystem functioning, making these areas sensitive indicators of past as well as current climatic changes. For example, this was observed in monitoring studies on peatlands in the Alpine region (e.g. Bragazza, 2008) and in the discontinuous permafrost regions of northern Fennoscandia (e.g. Christensen et al., 2004).

R. de Jong (✉)
Institute of Geography, University of Bern, CH-3012 Bern, Switzerland
e-mail: dejong@giub.unibe.ch

J. Dodson (ed.), *Changing Climates, Earth Systems and Society*, International Year of Planet Earth,
DOI 10.1007/978-90-481-8716-4_5,

Currently, much research is focused on the role of peatland areas as potentially large sources and sinks of carbon dioxide and methane (Christensen et al., 2004; Johansson et al., 2006). High-latitude wetland ecosystems play an important role in the global carbon budget as they store up to one-third of global soil carbon (Beilman et al., 2008). In general, these ecosystems are net sinks for atmospheric CO_2 owing to the prevailing waterlogged, anoxic and cool conditions that effectively reduce decomposition rates, thereby favouring the formation of peat. However, these conditions also favour anaerobic decomposition, making wetlands significant sources of atmospheric methane (Ström and Christensen, 2007). In a warming climate, the release of methane due to permafrost decay may have large feedbacks on the global climate (e.g. Zona et al., 2009), and improved estimates of the amount of greenhouse gases and their rate of release are required. Although peatlands in permafrost areas receive most attention in the greenhouse gas debate, the large number of mined peatlands in northwest Europe, when restored, could potentially also act as a considerable carbon store (Moore, 2002; Samaritani et al., 2010).

Here we review recent developments in a selection of climate proxy-indicators, as well as advances in dating methods. Reconstructions of atmospheric CO_2 from subfossil leaves as well as the potential impact of CO_2 changes on plant physiology are also discussed. We then briefly review recent findings on climatic changes reconstructed from peatlands. In the third section we focus on the role of peatlands as sources and stores for carbon dioxide and as sources for methane.

Peatlands as Archives of Past Climate Variability

With an increasing number of proxy reconstructions from peat bogs becoming available from many regions in the world, a direct comparison between records from different sites and different archives can be made. In theory, a compilation of multi-proxy reconstructions from different terrestrial archives in one region covers more climatic factors and seasons than reconstructions based on only one type of archive. However, for a reliable comparison two factors are of crucial importance: chronological errors must be minimal and the 'climatic' meaning of each different proxy must be clear. Below we discuss a selection of frequently used proxies. An overview of additional proxy methods can be found in Blackford (2000), Charman (2002) and Chambers and Charman (2004).

Chronologies of Wetlands

Peat stratigraphy was applied early to estimate the ages of layers in peat (von Post, 1946). Layers of different peat composition were thought to reflect past climate changes, and the superposition of these layers, when linked to independent dating evidence (e.g. the Swedish varve chronology, links of pollen zones in peat with archaeological zones), allowed for chronologies at millennial scale precision. These chronologies were relative, at best, because they depended on inferred temporal links. With the application of radiocarbon dating from the 1950s, absolute ages could be obtained, resulting in precisely and independently dated wetland archives (as long as the carbon in the organic matter is younger than the limit of ^{14}C dating at 50,000 ^{14}C years). Indeed, ^{14}C dating of pollen zones across the British Isles led Smith and Pilcher (1973) to reject the previous paradigm that these zones were simultaneous over large areas.

For ^{14}C dating of a peat core, generally organic material is sampled at regular intervals or at depths with major proxy changes across the core. As only milligrams of carbon are needed for an AMS (accelerator mass spectrometer) ^{14}C date, a cubic centimetre of bulk peat (or several milligrams of selected fossil plant remains such as leaves, seeds or pollen) often contains sufficient carbon. However, small amounts of contamination with modern carbon can result in unreliable dates.

Radiocarbon dating of tree rings with dendrochronologically obtained calendar ages allows calibration of the ^{14}C time scale (Reimer et al., 2004). Resulting calibrated age confidence intervals are often considerably larger than the original lab-provided measurement uncertainties, especially during periods where the calibration curve shows wiggles or plateaus. For peat deposits spanning the last few decades to centuries, other types of dating can additionally be applied, e.g. bomb-pulse ^{14}C dating (van der Linden et al., 2008) or lead-210 (^{210}Pb) dating (Turetsky et al.,

2004). Moreover, well-dated volcanic ash layers can provide additional precise dating points, and can also be used to link together proxy archives across a region (Pilcher et al., 1995).

Reliability of dateable material and funding limitations often prohibit dating at dense sediment intervals, and additionally ^{14}C dates themselves can have measurement errors (Scott, 2007). Therefore some form of interpolation is needed to provide age models for each level of a core. The most basic method for such age-depth models is linear interpolation between the depths and ages of the dated levels (Bennett, 1994). More ecologically realistic age-depth models include linear accumulation (e.g. Belyea and Clymo, 2001; Blaauw and Christen, 2005), convex models (as deposits reach a height limit; Belyea and Baird, 2006), concave models (owing to decomposition of fossil matter; Yu et al., 2001), and flexible Bayesian 'Poisson' models (Bronk Ramsey, 2007). Each type of model has its merits and limitations, and the choice of age-depth model will affect the age estimates assigned to depths in a peat core (Bennett and Fuller, 2002; Telford et al., 2004; Yeloff et al., 2006; Blockley et al., 2007).

A special case of age-model which has been studied intensively in peat deposits is ^{14}C wiggle-match dating, where wiggles in high-density ^{14}C dated sections of a core are matched to those of the ^{14}C calibration curve by assuming linear accumulation rates (van Geel and Mook, 1989; Christen et al., 1995; Kilian et al., 1995; 2000; Mauquoy et al., 2002b; Blaauw et al., 2003; Blaauw and Christen, 2005). This works best during periods where the calibration curve shows major wiggles, e.g. during the Little Ice Age (Mauquoy et al., 2002b) or around ca. 3,000–2,000 years ago (Kilian et al., 1995, 2000). Although ^{14}C wiggle-match dating is a time-consuming and expensive method as dozens of ^{14}C dates are needed, it can provide age-models at (multi-) decadal precision. Moreover, using Bayesian methods, additional information such as prior knowledge of accumulation rates or hiatus sizes can be included to obtain more realistic age-models (e.g. Blaauw and Christen, 2005).

Determination of Peat Humification

Peat humification is a measure of the degree of peat breakdown or decomposition, most of which is assumed to take place while the plant (and other) material is in the upper part of the peat – the aerated acrotelm, the assumption being that far less breakdown occurs below in the permanently saturated catotelm (Clymo, 1984; Belyea and Clymo, 2001). The degree of peat humification is therefore taken to indicate the prevailing environmental conditions at (or shortly after) the time of peat accumulation (Aaby and Tauber, 1975), with dark-coloured highly humified peat indicating dry or warm conditions (and so either a longer time in the acrotelm, or faster decomposition), and light-coloured unhumified peat accumulating under wet or cool conditions (with a short transfer time to the catotelm, or a slow rate of decomposition). It can be assessed visually in the field using von Post's 10-point scale (von Post and Granlund, 1926), or Troels-Smith's (1955) 5-point scale, or by using a range of measures such as colour and fibrosity. The most widely used laboratory technique is colorimetry, based on light-absorbance or light transmission measured by colorimeter or spectrophotometer at a fixed wavelength through an alkali extract of 0.2 g dried and ground peat (see Blackford and Chambers, 1993, for review; http://www2.glos.ac.uk/accrotelm/humproto.html for laboratory protocol).

The colorimetric technique was first applied on *Sphagnum*-rich raised mire peats in Denmark (Bahnson, 1968; Aaby and Tauber, 1975; Aaby, 1976). Its potential for providing a proxy-climate signal from highly humified blanket peat was then explored by Chambers (1984). It has since been applied in a range of peat types, especially in Northwest Europe (e.g. Blackford and Chambers, 1991; Nilssen and Vorren, 1991; Borgmark, 2005) but also in North America (e.g. Booth and Jackson, 2003) and more recently, in South America (Chambers et al., 2007a). It has been regarded as one of three standard principal proxy-climate indicators from peats (cf. Hughes et al., 2006), the other two being macrofossil analysis and analysis of testate amoebae.

The strengths of the technique are that it is easy to use, can be applied on slim (0.5 or 1 cm) contiguous slices of peat, 0 is largely operator-independent (provided the batches of samples adhere to a strict time schedule) and the data are easily graphed for visual inspection and interpretation. It can also provide proxy-climate information in those parts of a *Sphagnum* peat core in which the peat macrofossil composition is largely uniform, being 'locked in' to a species dominant, such as *Sphagnum austinii (S. imbricatum)*, or *S. magellanicum*, and so not able to provide any indication of climate change.

Practitioners have focussed on identifying 'wet shifts', when relatively unhumified peat succeeds well-humified peat, as these are thought to be less susceptible to secondary decomposition and were believed to be easier to date using conventional bulk-sample radiocarbon dating, as the peat may be accumulating faster. In moisture-receiving (as opposed to moisture-shedding) peat macrotopes, such as valley or basin mires, the degree of peat humification may instead indicate the balance between the amount of woodland (with moisture interception and high evapotranspiration in the catchment) and the degree of woodland clearance and agricultural land-use (and so higher run-off) from surrounding land (Chambers, 1988).

The weaknesses of the technique include the question as to what compounds are being measured and what the data really indicate; that there is likely to be a plant-species signal in the data (Chambers et al., 1997; Yeloff and Mauquoy, 2006); that (theoretically) there may be secondary decomposition if a very dry spell of climate succeeds a wet period; and that the technique can in some peat horizons give contrary indications from other peat proxy-climate techniques, especially testate amoebae data, with no clear explanation. It is problematic when used in sedge-rich peat, as very different readings can arise from adjacent samples if one sample mainly comprises decay-resistant sedge macrofossils (e.g. *Eriophorum*, with the appearance and texture of frayed string), and the next comprises more humified material. The assumption that the use of alkali will extract mainly humic (and fulvic) acids has been questioned by Caseldine et al. (2000) and explored using size-exclusion chromatography by Morgan et al. (2005).

In recent years, multi-proxy approaches, particularly using the three principal proxy-climate measures referred to above, have become more common, but how to combine the data to derive a single 'bog-wetness climate proxy' (e.g. Hughes et al., 2006) remains problematic, especially when the temporal resolution of macrofossils and testate amoebae often does not match that of peat humification: of the three, it was the technique that is likely to be conducted on the slimmest samples, and always contiguously, so providing a higher-resolution proxy-record of climate change. Even in cores in which all three are conducted contiguously on the same thickness of samples, the contra-indications may be important, albeit difficult to interpret or explain, and so it may be more appropriate to present all three supposed climate-proxies separately. Despite reservations expressed by Yeloff and Mauquoy (2006), which were based on a comparison of peat humification and macrofossil data from a single profile, the peat humification colorimetric technique has much to recommend it, and may be the only one of the three that is feasible in highly humified peats containing few testate amoebae. Nevertheless, chronological control is important, and it is recommended that radiocarbon dating be conducted on small samples of above-ground plant components (preferably *Sphagnum* leaves). In view of the potential role of solar forcing on both climate and cosmogenic isotope production (Beer and van Geel, 2008), it may be no coincidence that 'wet-shifts', identified in peat cores by marked changes in peat humification, provide the best periods for potential wiggle-matches (cf. Mauquoy et al., 2004), although a brief dry shift featured in a successful wiggle-match in Tierra del Fuego (Chambers et al., 2007a), in which both the peat humification and plant macrofossil data suggested a dry interlude at precisely the same time (2,650 ^{14}C years BP; 2,800 cal BP) as a pronounced wet shift in continental Europe (e.g. Speranza et al., 2000, 2002).

Peat Macrofossils

Sub-fossil plant macrofossils of mosses, dwarf shrubs and graminoids preserved in peat samples have been used to reconstruct the temporal variability of local peat bog vegetation in hummock/hollow complexes at the microform scale (McMullen et al., 2004; Blundell et al., 2007). The preservation of the macrofossils in these samples can be excellent and species level identification of taxa is frequently possible. Multiple profiles have also been recovered from single raised peat bogs to reconstruct the spatial and temporal variability of peat bog vegetation at the mesoform scale (Barber et al., 1998; Hughes et al., 2000; Mauquoy et al., 2002a). Deposits collected from ombrotrophic (rain-fed) peat bogs have largely been used to make palaeoclimate reconstructions, since a range of *Sphagnum* mosses, Cyperaceae and dwarf shrubs are sensitive to changes in mire surface wetness which is in turn dependent upon changes in precipitation and temperature (Barber et al., 2004; Mauquoy et al., 2008). The macrofossil data produced from this research have also been used to reconstruct peatland carbon accumulation rate changes (Oldfield et al., 1997; Mauquoy et al., 2002a; Heijmans et al., 2008) and the identification of successional processes in peatlands, for example, the nature and timing of the fen-bog transition (Hughes and Barber, 2004). Macrofossil data generated using plant macrofossil data are also valuable in applied palaeoecological research, since long-term base line data for the origin, nature and potential causes of local vegetation changes can be generated. This information can then be used to guide conservation of threatened blanket peat and raised peat bog ecosystems (Chambers et al., 1999, 2007b). The nature and timing of recent human disturbance to ombrotrophic bogs, for example burning and drainage, has also been assessed (Pellerin and Lavoie, 2003). Macrofossil analyses can help to ensure the generation of precise and accurate ^{14}C chronologies since above-ground parts of identified plants can be selected for ^{14}C AMS dating (preferably *Sphagnum* stems and leaves free of fungal contamination), and cyperaceous roots and the stem bases and attached lower parts of *Eriophorum vaginatum* leaves can be avoided (Mauquoy et al., 2004). These have been shown to be younger (because of downward growth) than bryophyte fragments in the same ^{14}C dated stratigraphic levels (Kilian et al., 2000). Accurate and precise chronologies generated using ^{14}C Bayesian chronologies on peat sequences have served to pinpoint rapid environmental changes, and possible climate forcing factors, for example changes in solar activity (Mauquoy et al., 2004). Peat sequences are also capable of providing high-resolution studies as a ~1310-year-old peat profile from Denmark recorded rapid accumulation rates up to 2–3 year/cm (Mauquoy et al., 2008).

Selective preservation is a potential limitation with the analyses of mire and peat macrofossils, particularly for the above-ground parts of dwarf shrubs and Cyperaceae. Regarding the identification of mire and peat macrofossils, an excellent overview is presented by Birks (2007), whilst identification plates are available in Mauquoy and van Geel (2007, see Fig. 1). Decay-resistant *Sphagnum* mosses have different decomposition rates (Johnson and Damman, 1991). It is not always possible to identify *Sphagnum* species to the lowest taxonomic level, particularly where their stem leaves are absent. Cyperaceous remains are also difficult to identify consistently, since in many instances seeds and diagnostic epidermal tissues are simply not present. Owing to these problems, successional and/or palaeoclimate reconstructions produced from mire and peat macrofossils need to be interpreted cautiously and ideally should be combined with other techniques, such as pollen, non-pollen and testate amoebae analyses (Swindles et al., 2007; Yeloff et al., 2007).

Quantitative reconstructions of mire surface wetness which use the relationship between the modern peat-forming vegetation and water-table depth are now possible (Väliranta et al., 2007). Transfer functions have been generated using plant macrofossils in the same way as those applied to testate amoebae data (Charman et al., 2007), and offer the prospect of more refined water-table depth reconstructions using plant macrofossil data. A key question regarding raised peat bogs is what ultimately drives local water table depths: summer precipitation or temperature? It is highly important to identify this in order to provide accurate palaeoclimate data for the climate modelling community. Correlations between instrumental measurements of water table depths and climate variables presented in Charman et al. (2009) suggest that water-tables in Northwest European peat bogs are controlled by warm season moisture deficits, with precipitation more important than temperature. More analyses of the relationships between climate variables and water table depths in other regions will serve to confirm the importance of warm season moisture deficit.

Since northern peatlands store a large proportion of global soil carbon (Beilman et al., 2008), there is an important need to understand how carbon sequestration rates in peatlands in the boreal and subarctic regions have responded to previous periods of climate change. This may give clues to their likely future response to projected climate warming scenarios. Given that the rate of carbon accumulation has been shown to be dependent upon the botanical composition of peat deposits (Oldfield et al., 1997; Mauquoy et al., 2002a), more of these studies are needed to identify the spatial and temporal response of northern peatlands to former episodes of climate change. In the Mauquoy et al. (2002a) study, peat accumulation rates decreased during Little Ice Age climatic deteriorations following the Medieval Warm

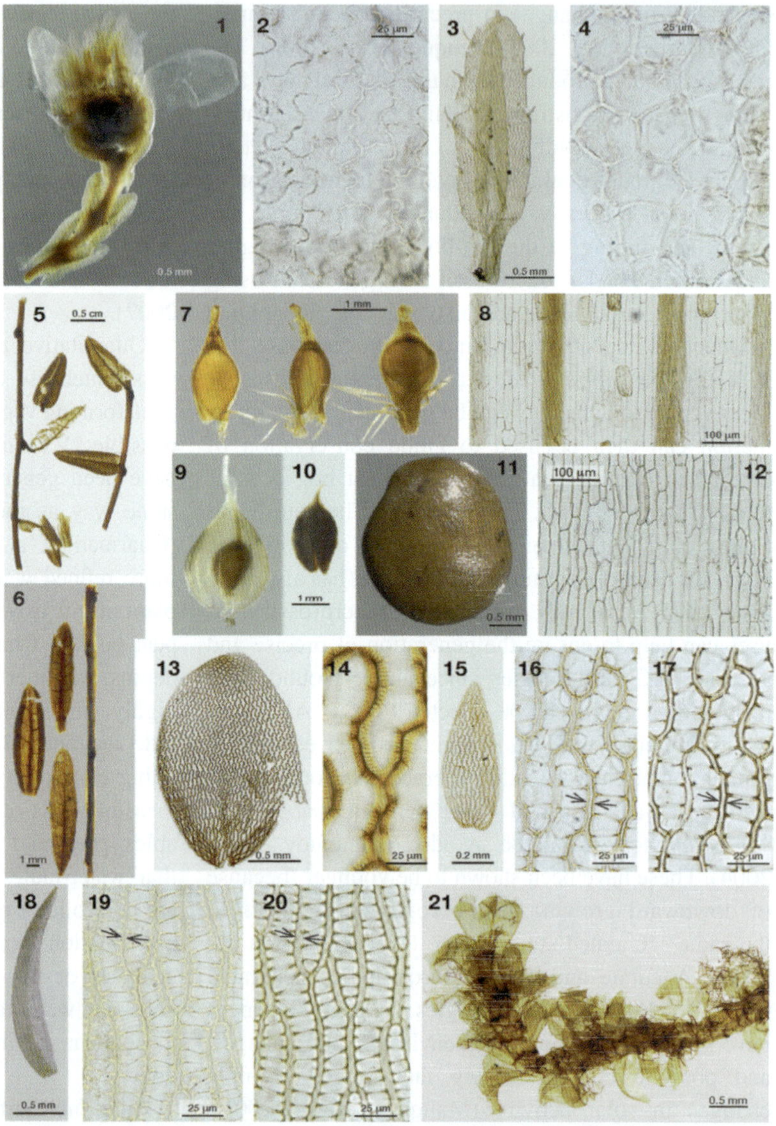

Fig. 1 1 and 2: *Calluna vulgaris*, stem with flower and leaf epidermis. 3 and 4: *Erica tetralix*, leaf and leaf epidermis. 5: *Vaccinium oxycoccus*, twigs with leaves. 6: *Andromeda polifolia*, leaves and twig. 7: *Rhynchospora alba*, seeds. 8: *Eriophorum vaginatum*, epidermis. 9 and 10: *Carex rostrata*, perigynium and achenes. 11 and 12: *Menyanthes trifoliata*, seed and leaf epidermis. 13 and 14: *Sphagnum austinii*, leaf. 15–17: *Sphagnum* sect. Acutifolia, leaf and ventral high and low focus. 18–20: *Sphagnum* sect. Cuspidata, leaf and ventral high and low focus. 21: *Paludella squarrosa*

Period, owing to the presence of aquatic *S. cuspidatum* leaves. These have lower decay resistance than *Sphagnum* section *Acutifolia* and *S. imbricatum* leaves. Future research will serve to confirm the importance of species composition upon reconstructed carbon accumulation rate changes.

Testate Amoebae

Testate amoebae are unicellular organisms that live mainly in fresh water, soils, mosses, sand and lake sediments and are especially abundant and diverse in peatlands. They are divided into two taxonomic groups, characterised by either lobopodia (Arcellinida) or filopodia (Euglyphida and related Cercozoan testate amoebae) (Meisterfeld, 2001a, 2001b). They form shells (tests) of different types (agglutinating, idiosomic, and proteinaceous) that are used for identification of living as well as fossil testate amoebae (Fig. 2). Testate amoebae play an important functional role in the cycling of elements in terrestrial ecosystems (Schröter et al., 2003; Wilkinson, 2008).

The structure of testate amoeba communities in peatland varies in relation to ecological gradients such as moisture and water chemistry (Lamentowicz

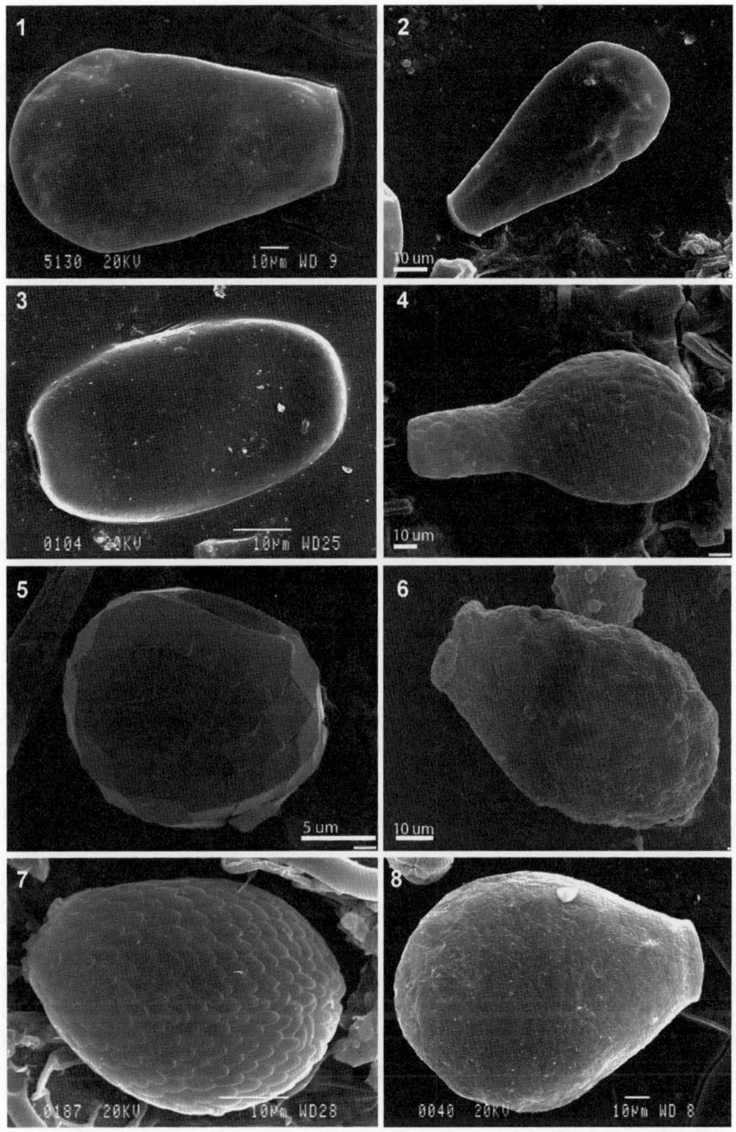

Fig. 2 Scanning electron microscope pictures of selected testate amoebae species from various peatland habitats: (1) *Hyalosphenia papilio* Leidy, 1879 (Photo by M. Lamentowicz); (2) *Hyalosphenia elegans* Leidy, 1879 (Photo by M. Lamentowicz); (3) *Archerella flavum* Archer, 1877 (Photo by M. Lamentowicz); (4) *Nebela lageniformis* Penard (Photo by L. Lamentowicz), 1890; (5) *Paraquadrula irregularis* (Archer, 1877) Deflandre, 1932 (Photo by L. Lamentowicz); (6) *Physochila griseola* Penard, 1911 (Photo by L. Lamentowicz); (7) *Assulina muscorum* Greeff (Photo by M. Lamentowicz), 1888; (8) *Nebela tincta* (Leidy, 1879), Awerintzew, 1906 (Photo by M. Lamentowicz)

et al., 2008a; Mitchell et al., 2008a). Their shells are well preserved in peat deposits, thus allowing palaeoenvironmental reconstructions (Charman, 2001) and management of peatlands (Buttler et al., 1996; Davis and Wilkinson, 2004; Mitchell et al., 2008a). The link between climate and testate amoeba assemblages is strong as they depend directly on micro-macroenvironmental conditions that are strongly affected by climate. They provide information on the coring location as they live, die and remain in situ in the peat. This tight link between local conditions and testate amoeba assemblages allows for detailed ecological reconstructions. However, this ecology is in part also controlled by non-climatic autogenic processes in the peatland as well as by direct and indirect human impact (e.g. on local hydrology).

The taxonomy of testate amoebae is complex and, until recently, based only on morphological criteria, but molecular methods have been introduced in combination with detailed morphological analyses (e.g. by scanning electron microscope). This combination reveals a higher than expected diversity (Lara et al., 2007; 2008) and the existence of pseudo-cryptic species (i.e. genetically different but with morphological differences that are easily overlooked when observed by light microscopy or that can be seen only with a scanning electron microscope; Todorov et al., 2009; Heger et al., 2010). This has also been reported for other groups of protists (e.g. Saez et al., 2003).

As a prerequisite to their use in paleoecology, the ecology of testate amoebae in peatlands has been studied extensively, e.g. in Canada (Warner, 1987; Charman and Warner, 1997), New Zealand (Charman, 1997), the United States (Payne et al., 2006; Booth, 2008), the Jura Mountains of Switzerland and France (Mitchell et al., 1999), the Swiss Alps (Lamentowicz et al., 2009b), Russia (e.g. Mazei et al., 2009; Muller et al., 2009), the United Kingdom (Woodland et al., 1998), Finland (Tolonen et al., 1994a, 1994b), Poland (Lamentowicz et al., 2008b), Greece (Payne and Mitchell, 2007), Turkey (Payne et al., 2008) and for Europe in general (Charman et al., 2007). Most ecological studies showed a strong correlation between testate amoeba community structure and bog surface wetness (usually expressed as water-table depth) in controlling the assemblages of testate amoebae in ombrotrophic peatlands (e.g. Hendon et al., 2001; Charman et al., 2006; Blundell et al., 2007). The rationale behind these different studies is that depending on the biogeographical and climatic location, testate amoeba faunas and the responses of cosmopolitan taxa to ecological gradients may differ. For that reason, high-quality training sets from the same biogeographical and climatic settings are preferred for reliable palaeoclimatic reconstructions.

Although a relatively large number of ecological and palaeoecological studies of peatland testate amoebae have been published, the geographical distribution of the research effort is uneven. For Western Europe a large number of studies and training sets are available, in particular from the UK, where a regional curve of water table change was constructed on the basis of subfossil testate amoebae (Charman et al., 2006; Blundell et al., 2007; Swindles et al., 2010). However, in Eastern Europe testate amoebae are not yet extensively used as a tool in palaeoecological research except for Poland (e.g. Lamentowicz et al., 2008b, 2009a). A new training set for testate amoebae in Poland was recently developed (Lamentowicz et al., 2008b). These studies have provided data on climatic change (wet and dry phases) and human impact on the landscape (deforestation and peatland exploitations). In the Alpine region so far only a few studies have used testate amoebae as a tool for palaeohydrological reconstructions. Modern feedbacks between temperature, hydrology and precipitation in Alpine peatlands are not well known. The first testate amoeba training set from the Alpine region was constructed on the basis of samples collected in the Engadin Valley (south-east Switzerland) from a variety of peatland types (Lamentowicz et al., 2009b). This transfer function was applied to a finely sub-sampled core (near-annual resolution) from the Mauntschas peatland to reconstruct variations in the water table during the past 135 years (Lamentowicz et al., 2010). Comparison with measured temperature data showed a correlation to air temperatures since AD 1864.

Studies describing more environmental variables than water table depth and/or including other groups of organisms such as plants are very rare (Tolonen et al., 1994a). Minerotrophic peatlands (fens) have more plant and testate amoebae species and are characterized by a complex hydrochemistry (Heal, 1961; Hájek et al., 2006). More work is needed over broader ecological gradients than just *Sphagnum*-dominated peatlands of the temperate to boreal zones, and including several key ecological variables, as well as comparative studies of different groups of organisms. Such studies would allow for an improved assessment of the bioindicative potential of each species in neoecology (restoration ecology) and palaeoecology.

Testate amoeba analysis is now a well-established method in Quaternary science and, despite several shortcomings, its progress is very promising (Mitchell et al., 2008a). However, effort should be put into improving this tool to make it even more useful. The three priorities we see are as follows:

1. *Taxonomy*. Good taxonomy combining morphological and molecular methods is urgently needed. Paleontologists and biologists sometimes use different taxonomic approaches and this causes confusion. Ecologists often pool testate amoebae taxa in species complexes because it is difficult or impossible to identify species with any certainty (Charman et al., 2000). It is therefore unclear if these closely related taxa have different ecological requirements. Future work should aim to determine if these closely related taxa have different ecologies.
2. *Ecology*. Interpretation of testate amoeba communities data requires a sound understanding not only of species–environment correlations but also of the mechanisms responsible for these patterns. Beyond

correlative studies such as those currently used to build transfer function models, a good ecological understanding based on a combination of descriptive and experimental studies is essential to support palaeoclimatic reconstructions (Mitchell et al., 2008a). Such studies are rare (Mitchell, 2004). The integration of descriptive ecological and experimental studies on testate amoebae in field and laboratory conditions will be advantageous. Studies on the fine scale spatial variations in testate amoebae communities and their seasonal patterns are also scarce (Heal, 1964; Mitchell et al., 2000; Warner et al., 2007). Further studies are needed to understand the timing and patterns of testate amoebae response to factors such as temperature and hydrology. For the reconstruction of palaeohydrology of peatlands in general, a challenge is the application of testate amoeba analysis in minerotrophic fens, which also includes the basal part of most peatland palaeoecological records. In fens the complexity of testate amoeba communities and identification increases (Heal, 1961; Lamentowicz et al., 2010) and most transfer functions do not cover the minerotrophic end of the gradient well (Mitchell et al., 2008b; Payne and Mitchell, 2009; Wall et al., 2010).

3. *Theoretical aspects*. Recently several studies have addressed questions such as how the number of counted individuals and the mesh-size of filters used for sample preparation affect the reliability of community estimates and the quality of palaeoecological inferences (Mitchell et al., 2008b; Payne and Mitchell, 2009; Wall et al., 2010). More work is needed to optimize the protocols for sample preparation and analysis.

Biomarker Compounds and Stable Isotope Ratios as Indicators of Environmental Change in Peatlands

Biomarkers

Organic biomarkers, or molecular fossils, are recalcitrant compounds whose origin can be traced to a particular taxon of organism. These compounds can be used to create reconstructions similar to traditional plant macrofossils. In peatlands, these compounds are typically produced by plant epicuticular waxes and are especially useful when decay leaves macrofossils unidentifiable. Molecules used in reconstructions can be general indicators of broad groups of plants. Saturated hydrocarbons and their homologues (*n*-alkanes, *n*-fatty acids and *n*-fatty alcohols) are a typical example. The shorter chain varieties of these molecules (20–26 carbons) are produced mainly by *Sphagnum*, while the longer chain varieties (26–34 carbons) are produced mainly by vascular plants (Baas et al., 2000; Nott et al., 2000; Pancost et al., 2002). The relative abundance of short and long chain molecules can be used as indicators of the past vegetation assemblages (Nichols et al., 2006). Specific biomarkers, molecules that are produced by plants belonging to only one genus, have also been identified. These include 5-*n*-alkylresorcinols, a biomarker for the family Cyperaceae (Avsejs et al., 2002) and *n*-alkan-2-ones, a biomarker for the genus *Sphagnum* (Nichols and Huang, 2007).

Biomarker methods for paleoenvironmental reconstruction benefit from automation and user-independent results. Traditional methods of counting macrofossils and palynomorphs are time consuming and are dependent on the skill of the person performing the analysis. However, the library of species-specific biomarkers is incomplete; the most useful biomarker reconstructions from peatlands use general biomarkers that distinguish broadly between *Sphagnum* and vascular species.

Environmental Stable Isotopes

The natural abundance ratios of stable isotopes of several common elements (H, O, C) can be used as environmental indicators. Hydrogen and oxygen isotopes are used as indicators of hydrologic change, and carbon isotopes are indicators of carbon cycle and hydrologic cycle changes. Oxygen isotope ratios are typically measured from cellulose extracted from peat. Carbon isotopes can be measured both from cellulose and from leaf-wax compounds (Xie et al., 2004).

Hydrogen and Oxygen Isotopes

Since the ultimate source of the hydrogen and oxygen isotope ratios (δD and $\delta^{18}O$) of *Sphagnum* molecules is environmental water, these measurements are an excellent way of tracking past changes in the hydrologic cycle. The sediments of ombrotrophic peatlands are particularly well suited to this task as the only input of water to these systems is by direct precipitation. The hydrogen and oxygen isotopic composition of *Sphagnum* molecules is therefore dictated by two factors: the isotopic composition of the precipitation and any enrichment by evaporation of water inside and between the *Sphagnum* individuals. The δD and $\delta^{18}O$ of precipitation are extremely useful parameters for paleoclimate reconstructions and provides information about temperature, rainfall amount and moisture source (Clark and Fritz, 1997). The amount of evaporation from peatland surfaces is also an important palaeoclimate parameter which can be gleaned from the δD and $\delta^{18}O$ of *Sphagnum* molecules. Both precipitation isotopes and in situ evaporation must be taken into account when interpreting hydrogen and oxygen isotope data from peatlands.

Carbon Isotopes

Interpretation of carbon isotope measurements of *Sphagnum* molecules can be a complicated matter (Xie et al., 2004). Two competing factors influence the carbon isotope ratio of fossil *Sphagnum* molecules (e.g. cellulose, leaf waxes, etc.). The first is the carbon isotope ratio of the CO_2 used for photosynthesis. Though the preindustrial $\delta^{13}C$ value for atmospheric CO_2 is often assumed constant for the Holocene, not all the CO_2 used by *Sphagnum* comes from the atmosphere. Up to 15% of the CO_2 used by *Sphagnum* can come from recycled methane from lower in the peatland (Raghoebarsing et al., 2005). Because the $\delta^{13}C$ value of methane is so low (–40‰ to –60‰), it can dramatically affect the ultimate $\delta^{13}C$ of the *Sphagnum* molecules. The amount of recycled CO_2 available to *Sphagnum* is influenced mainly by the wetness of the surface of the peatland. When the peatland is wetter, more methane-derived CO_2 is available at the surface for use by *Sphagnum* ($\delta^{13}C$ decreases). A record of $\delta^{13}C$ of *Sphagnum* lipids from a Norwegian peatland has been interpreted this way to represent surface moisture (Nichols et al., 2009).Wetness of the *Sphagnum* itself also directly influences the $\delta^{13}C$ of its molecules. When the *Sphagnum* plant is more saturated, the water film over the photosynthetic cells impedes the incorporation of CO_2 and thus the plant becomes less selective against ^{13}C ($\delta^{13}C$ increases) (Williams and Flannagan, 1996). Because of these competing factors, $\delta^{13}C$ records derived from peat sequences are, at this time, rare and must be evaluated carefully.

Sand Grain and Dust Influx in Peat Bogs as Proxies of Past Circulation Strength

The interest in past atmospheric circulation changes, as well as changes in dust flux, has increased during recent decades (Goudie and Middleton, 2006). Although past changes in atmospheric circulation vigour and the position of Westerlies have been reconstructed using, for example, sea-salt spray indicators in ice cores from Greenland (e.g. Meeker and Mayewski, 2002), many reconstructions from north-west Europe are indirect; changes in effective precipitation, reconstructed from glaciers (e.g. Bakke et al., 2008), lake sediments (e.g. Seppä et al., 2005) or peat bogs (e.g. Swindles et al., 2010) are often interpreted in terms of atmospheric circulation changes. However, few records exist that can also be directly interpreted in terms of increased or decreased cyclone frequency and/or intensity. In addition, the sources and flux variability of dust storms are poorly understood; however, dust flux is of particular importance as it has strong feedbacks with global (Goudie and Middleton, 2006) and local (Painter et al., 2007) environmental and climatic changes. During the last century, an increase in the frequency of dust storms has been observed for some areas (for example, in North America; Neff et al., 2008). The origin (natural or anthropogenic) of this increase is still poorly understood. Therefore, it is necessary to study the fluctuation of atmospheric dust flux during pre-anthropogenic times in detail.

Sand Grains in Peat Deposits

Recently, detailed analysis of sand grains (>60 μm) in peat bogs has been carried out in North-West Europe (e.g. Björck and Clemmensen, 2004, De Jong et al., 2006; 2009, Sjögren, 2009). These discuss the use of the sand particle content in peat bogs as a proxy for past atmospheric circulation characteristics. In raised bogs, in the absence of inflowing streams, non-biogenic mineral particles found in peat cores are assumed to have been transported through the air. However, this is only true if the peat bog is very large, or situated in a flat or gently sloping topography, so that surface runoff onto the peat surface can be excluded (see, e.g. Caseldine et al., 2005). In addition, changes in the availability of erodible materials as well as bog microtopography have to be taken into account.

Sand grains are transported either by strong winds alone, or in combination with a snow cover on the peat surface which would facilitate sand grain transport over the bog surface (Björck and Clemmensen, 2004). Therefore, the quantity and size ranges of these sand grains can be used to reconstruct past atmospheric circulation strength (cyclone frequency and/or intensity) in a region. The total aeolian sediment influx (ASI, quartz grains > 125 μm) has been used to derive the occurrence and frequency of strong cyclonic activity in south-west Sweden during the past 6,500 years (Björck and Clemmensen, 2004; De Jong et al., 2006, 2009). Similarly, Sjögren (2009) used sand mass accumulation rates (SMAR, quartz grains > 60 μm) in peat profiles in northern Norway as a proxy for past atmospheric circulation types (Fig. 3). In the aforementioned studies, peat bogs are situated near the local shorelines and in close proximity to sandy deposits, such as coastal dunes. Therefore, sand availability was not a limiting factor for sand transport. A direct comparison between pollen records and ASI values from two sites showed that changes in land use were not a driving force behind the ASI peaks either (De Jong et al., 2009).

Dust Particles in Peat Bogs

Reconstructed soil dust flux and peat geochemistry are also used as proxies for past circulation strength (de Vleeschouwer et al., 2009). Because peat bogs are typical for humid climates where the soil would naturally be stabilized by vegetation, wind erosion of the soil is reduced to a minimum (Mattson and Koutler-Andersson, 1954). Therefore a large contribution of fine dust in peat bogs may be long-distance transported. Sahara dust, for example, is a source of mineral dust to European peat bogs (Kylander et al., 2005) and is mainly composed of quartz, clays, carbonates and feldspars (Bücher and Lucas, 1984). Ti (Titanium), like other conservative elements (Al, Sc, La, etc.), can be used as a proxy for soil dust influx. By taking into account the mean Ti concentration in soils (in the upper continental crust, Ti = 4,100 μg/g, McLennan, 2001), the mean accumulation rate of peat and the density of the peat, the amount of atmospheric soil dust flux (ASD) into the peat bog can be calculated (Shotyk et al., 1998). ASD can be used as a proxy for the combined effects of the degree of soil erosion and atmospheric circulation strength over a region. In ombrotrophic bogs, variations in Ti concentrations have been linked to variations in soil dust inputs (Görres and Frenzel, 1993; Shotyk et al., 1998) originating from either agricultural activities (Hölzer and Hölzer, 1998) or variation in natural atmospheric soil dust input (Shotyk et al., 1998). Recently, de Vleeschouwer et al. (2009) reconstructed ASD fluxes using Ti concentrations, bulk densities and accumulation rates from a peat bog in North-West Poland, which showed strong fluctuations during the Little Ice Age (Fig. 3).

Tracing the Sources of Dust Input

To discriminate between remote and local dust sources, geochemical analyses can be used. Conservative lithogenic elements in the peat column can be used as fingerprints of the dust input. The preferred elements are Titanium, Scandium and Yttrium (Ti, Sc and Y) and rare earth elements (REE). For example, several studies have shown that most of the Ti comes from natural sources (e.g. soil erosion). Most of the particles coming from soil and rock erosion are silicates and aluminosilicates. These particles have low sensitivity to dissolution, even in the acidic environment of a peat bog (Le Roux et al., 2006; Le Roux and Shotyk, 2006). Despite their anoxic and reductive conditions, Aubert et al. (2006) have recently demonstrated that the REE signal is preserved in the peat column. This has been used to reconstruct dust origin in the Southern Hemisphere: Australia (Kylander et al., 2007; Muller et al., 2008), New Zealand (Marx et al., 2009) and Indonesia (Weiss et al., 2002).

As in ice cores, the dust origin can be traced with radiogenic isotopes of strontium, lead and neodymium. However, lead only represents natural dust input earlier than ca. 6,000 BP, before lead metallurgy. Strontium is problematic since it can be mobile in the peat column.

However, the strontium isotope signature can be measured in acid insoluble particles, where strontium is assumed to be unaffected by post-mobilization processes (Sapkota, 2006). Neodymium (a REE) isotope analysis has so far not been used on peat bog sediments. Kamenov et al. (2009) used this method to trace dust input in a core from Florida Bay marsh, suggesting this method can be applied successfully to peat cores as well.

Reconstructions of Past Atmospheric Circulation Characteristics

ASI reconstructions from two peat bogs in south-west Sweden, situated ca. 60 km apart, show a nearly identical sand influx pattern throughout the past 3,000 years (De Jong et al., 2009). ASI values reconstructed from two additional bogs in south-west Sweden (Björck and Clemmensen, 2004; De Jong et al., 2009) also show generally similar trends (Fig. 3). Furthermore, comparison with reconstructed dune-reactivation phases along the west coast of Denmark (Clemmensen et al., 2009) shows broadly similar patterns, indicating the reliability of the ASI method as a proxy for regional-scale circulation strength. In south-west Sweden, increased circulation strength was recorded in particular after ca. 2,800 cal years BP (De Jong et al., 2006; 2009)

Similarly, SMAR measured in three peat profiles from an ombrotrophic bog situated on the west coast of northern Norway (Sjögren, 2009) gave consistent signals, although the magnitude of total sand accumulation fluctuated strongly between the three profiles and over time (Fig. 3). Strong increases in SMAR values around 3,150–2,800, 2,500–2,100, 1,450–1,100, 950–550 and 250 to +50 cal years BP were interpreted mainly as periods of major shifts in the wind regime. However, overgrazing and intense agricultural activities have certainly played an important role during some of these phases (Sjögren, 2009).

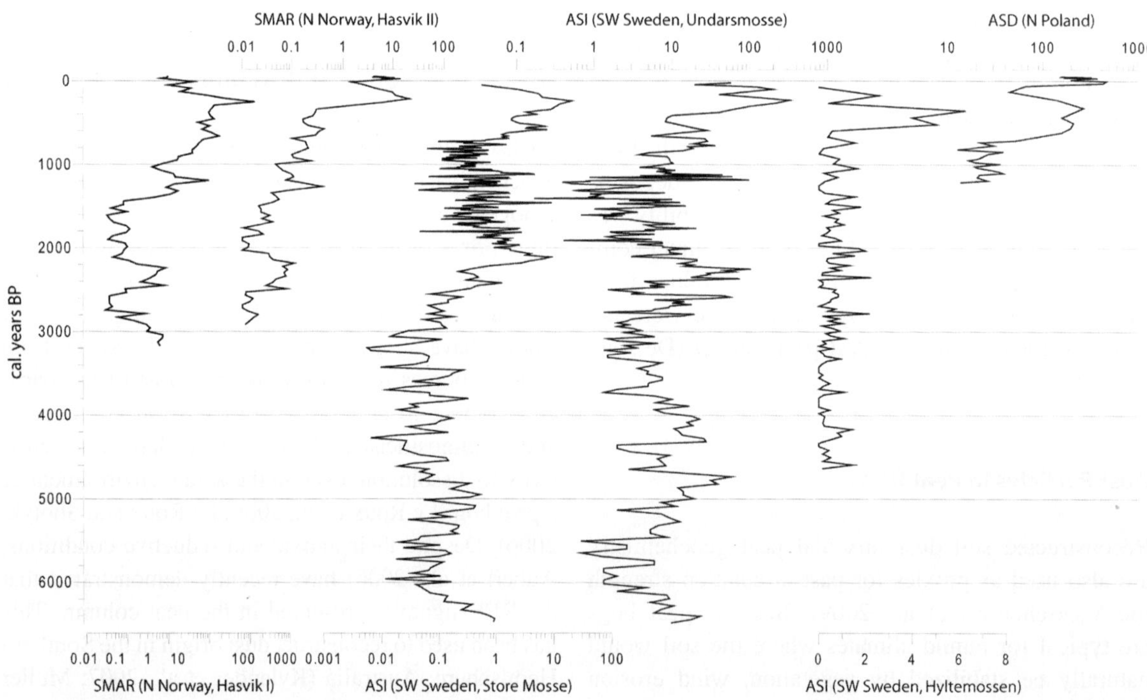

Fig. 3 Reconstructions of mineral influx into peat bogs in Northwest Europe, showing SMAR (sand mass accumulation rates, Sjögren, 2009) for two cores from Hasvik, northern Norway (notice the logarithmic scale on the *x*-axis); ASI (aeolian sediment influx) from the Store Mosse bog (De Jong et al., 2009), Undarsmosse Bog (De Jong et al., 2006) (notice the logarithmic scale on the *x*-axis) and Hyltemossen bog (Björck and Clemmensen, 2004); ASD (aeolian soil dust flux) from Slowínskie Blota bog, northern Poland (de Vleeschouwer et al., 2009). The majority of the records reflect increased mineral particle influx values around 2,700–2,000 cal years BP and around 600–100 cal years BP, which was interpreted as an increase in atmospheric circulation vigour. During the most recent phase, increased sediment availability most likely enhanced peak amplitudes

The ASD profile reconstructed from the Slowínskie Blota Bog in northern Poland (de Vleeschouwer et al., 2009) also showed increased values during LIA cold shifts (Fig. 3). Focusing on the period between AD 1000 and AD 1900, the ASD displays two main peaks around ca. AD 1370 and ca. AD 1650. These ASD peaks are also reflected in the density and the Ti profiles (de Vleeschouwer et al., 2009), indicating increased soil erosion around these times.

Strengths and Weaknesses

For the interpretation of minerogenic influx into peat bogs, it has to be kept in mind that the three mentioned proxies (ASI, SMAR and ASD) reflect the accumulation of different sizes of mineral particles, the last reflecting the smallest fraction. Therefore, ASI is likely to reflect extreme events only, whereas ASD is likely to record overall changes in the mean wind strength. Flux reconstructions should be used only in records with robust, high-resolution age-depth models since changes in peat accumulation rates strongly affect flux values. Moreover, as it involves the density of the peat, this latter parameter should be measured by taking multiple bulk peat samples of a specific volume to minimize errors. It is not possible to reconstruct whether high particle concentrations are related to single extreme events or to a series of moderately strong wind events. Moreover, the quantity and size range of particles is not just dependent on wind characteristics but also on sediment availability. Throughout the course of the Holocene, sediment availability in most regions in North-West Europe has fluctuated. This is partly due to natural causes (e.g. vegetation succession, climate change) and to an increasing extent due to human impact on the landscape. Large-scale deforestation, grazing and the development of agriculture has caused a high degree of openness of the landscape (e.g. Berglund et al., 1991). Intense agriculture has led to the nearly complete exposure of soils to wind erosion, in particular in autumn and winter, which is the season in which westerly storms are most frequent and strong in North-West Europe (Alexandersson et al., 1998). For a correct interpretation of the different mineral fractions in the peat, a reconstruction of land use and deforestation is necessary. For the fine fractions, an estimation of particle emissions from, for example, mines and cement factories would be helpful to assess the particle input into the atmosphere over time. If human impact has been accounted for, these proxies have the potential to provide detailed information on large-scale variability of past atmospheric circulation strength.

Stomatal Records from Subfossil Leaves as Proxies of Plants' Responses to Past Atmospheric CO_2 Variations

Carbon dioxide (CO_2) is the most important anthropogenic greenhouse gas and a large percentage of its recent change is attributed to anthropogenic carbon emissions (IPCC, 2007). Over the past 800,000 years atmospheric CO_2 concentrations have varied from minima of 170–200 parts per million by volume (ppmv) in glacials to maxima of 280–300 ppmv in the recent interglacials (Petit et al., 1999; Lüthi et al., 2008). Present CO_2 concentrations already exceed the natural variability of this greenhouse gas and predictions for atmospheric CO_2 concentrations over the next two centuries suggest that CO_2 concentrations will increase at a rate that will depend on the reductions of anthropogenic emissions (Meehl et al., 2007).

Photosynthesis by C_3 plants is highly sensitive to variations of CO_2 concentrations in the air. Stomata are small pores on the protective leaf cuticle around the outside of the aerial parts of land plants (Fig. 4). They are the main link between the plants' interior and the atmosphere and are necessary because land plants have internalized their gas exchange surfaces and developed waterproof coatings to avoid water loss. Stomata allow CO_2 in for photosynthesis and let water molecules out, thus carbon gain and water loss are intimately connected (McCarroll and Loader, 2004).

Plants are able to regulate water loss and carbon gain with mechanisms that respond to short-term as well as to long-term changes in CO_2. After the completion of leaf expansion, stomatal densities remain constant while the stomatal aperture changes on short timescales (in the order of minutes) in response to environmental signals, including CO_2 concentrations. These signals can trigger alterations to guard cell turgor, cytoskeleton organization, membrane transport and gene expression (Hetherington and Woodward, 2003 and references therein; Young et al., 2006). It has been shown that leaf stomata can close in response

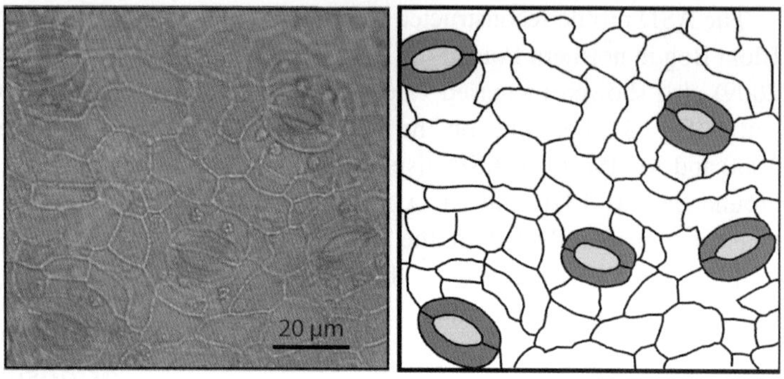

Fig. 4 *Left panel*: Image of a *Betula nana* leaf cuticle. *Right panel*: Schematic figure with guard cells of stomata (*dark grey*), stomatal pores (*light grey*), and epidermal cells (*white*)

to high CO_2 levels and open at low CO_2. For example, experiments carried out under high (double that of present) atmospheric CO_2 concentrations show that stomatal apertures are reduced by 20–40% in several plant species (e.g. Bunce, 2004). Such changes in stomatal apertures can also occur in response to changes in photosynthesis and respiration, which are influenced by the light/dark transitions (in the dark CO_2 increases in leaves by cellular respiration). Hence, stomatal apertures change also along with the diurnal cycle.

In addition to stomatal aperture, stomatal density (SD, i.e. the number of stomata per unit area) or the stomatal index (SI, i.e. the ratio of stomata versus epidermal cells per unit area) can vary. Woodward (1987) first reported changes in SD in response to changes in atmospheric CO_2 concentrations for seven species of temperate forest trees and a shrub species. Responses of SD and/or SI to an increased atmospheric CO_2 concentration have since then been reported for several C_3 species (Royer, 2001). According to Hetherington and Woodward (2003), the mean response is an 11% reduction with a doubling of the CO_2 concentration. However, several plants show an increase or no change at all of stomatal densities in response to CO_2 increases.

Although the regulating pathway is not yet completely defined, the discovery of the *HIC* gene in *Arabidopsis thaliana* indicates a genetic control for the CO_2 response of stomatal development (Gray et al., 2000). Additional support for developmental control emerges from experiments showing that the CO_2 concentration around mature leaves determines the SD of new leaves (Lake et al., 2001). However, any mechanism should be valid for the entire range of species-specific responses to CO_2 increases including increases, decrease, or no response of SD (Hetherington and Woodward, 2003). Lake and Woodward (2008) proposed a putative mechanism that involves stomatal-development control by transpiration rate and leaf abscisic acid (ABA) concentrations.

Experimental Settings to Study Stomatal Frequency Responses

Stomatal frequency (SF, i.e. SD and SI) responses to changes in atmospheric CO_2 concentrations can be studied using several techniques, including CO_2 enrichment experiments, herbarium studies and palaeorecords. CO_2 enrichment experiments (whether glasshouse, open top chambers (OTCs), and free-air CO_2 enrichment (FACE) experiments) have the advantage of allowing short-term (interannual to subdecadal scales) investigation of plants' responses to CO_2 changes while keeping other environmental variables (e.g. light intensity, temperature, soil conditions and humidity) constant or continuously monitored (Tricker et al., 2005; Körner, 2006; Ainsworth and Rogers, 2007). By contrast, to assess longer-term adaptation of plants to CO_2 changes (e.g. at decadal or centennial scales), leaves collected from herbaria as well as subfossil leaves from peat bogs or lake sediments are useful (e.g. Woodward, 1987; Wagner et al., 1996). The method involves the analysis of cuticles, which mirror the morphological structure of the leaf epidermis, including stomata. However, herbarium studies do not cover timescales longer than the past ca. 200 years and leaves analysed in such studies often have diverse provenances and indeterminable environmental conditions. Subfossil leaves, in contrast, have the advantage of providing long-term records of plant species' responses over centennial and millennial timescales. Since cuticles are made of waxes resistant to degradation under anoxic conditions (e.g. in peat or lake sediments), subfossil leaf cuticles accumulated in wetlands retain a record of long-term stomatal responses and adaptation to CO_2 changes. This approach allows the assessment of responses to external factors over a longer time scale than that available to modern ecologists, and it provides a temporal perspective on responses under different climatic conditions and in times of rapid climatic change. It thus helps to assess the results of 'experiments conducted by nature' (Deevey, 1969).

Methods of Assessing Stomatal Frequency

Stomata on modern and fossil leaves can be observed and counted either using epifluorescence or transmitted light microscope. By using an image analysis system connected to the microscope, additional morphometric measurements can be obtained, such as stomatal pore length, stomatal pore width and cell wall undulation (e.g. Kürschner, 1997). The stomatal frequency of angiosperm leaves is conventionally expressed in terms of SD and/or SI. Counting fields should be limited to interveinal areas because stomata are mostly absent on leaf veins. To obtain accurate estimates of the mean SF per leaf, it is necessary to analyse several counting fields per leaf and to limit the analysis to the central leaf lamina because, as shown on specimens of *Alnus glutinosa*, up to 2.5 fold variations of stomatal frequencies were observed (Poole et al., 1996; 2000). In addition, the separation of sun and shade leaves may be important as the position of leaves within the canopy influences the leaf growth, which in turn can affect SD (Poole et al., 1996; Kürschner, 1997). In conifers, the arrangement of stomata into rows occurring either in single file or grouped in bands parallel to the long needle axis prevents assessment of stomatal frequency based on small counting fields. Thus a series of techniques have been developed to estimate stomatal responses to CO_2 changes of conifer species, such as the stomatal number per length, the SD per length and the true SD per length, which is a measure of the number of stomata per unit needle length (see, for example, McElwain et al., 2002; Kouwenberg et al., 2003, Eide and Birks, 2006).

Models used to infer CO_2 based on SF have been developed using different approaches: (i) 'calibration in time' based on historical series as described above and (ii) 'calibration in space' using leaves collected along altitudinal transects. For calibration in space, responses are modelled using records along altitudinal transects because CO_2 partial pressure decreases with altitude. In several cases the two approaches have been combined to increase the gradient length of the modern datasets. Some training sets were developed including closely related species assuming that the species-specific responses to CO_2 changes are similar (e.g. Wagner et al., 1999; Jessen et al., 2005). Interestingly, the stomatal responses are not always equivalent between studies, as indicated by different results on stomatal responses of *Betula pubescens* (Wagner et al., 1999; Eide and Birks, 2006).

Ice Core and Stomatal Based Reconstructions of Atmospheric CO_2 Concentrations

Air bubbles in Antarctic ice cores offer the most direct measurements of past atmospheric CO_2 concentration (e.g. Siegenthaler et al., 2005b), providing a continuous record of CO_2 changes for the past 800,000 years (Lüthi et al., 2008). Older CO_2 concentrations have been estimated by various techniques, including stomatal frequencies (e.g. Beerling and Chaloner, 1992; Van der Burgh et al., 1993; Pagani et al., 1999; Kürschner et al., 2008). Stomatal-based inference models have, however, also been used to reconstruct CO_2 changes for time periods for which ice-core derived CO_2 records are available. The interest lies in the smoothed nature of ice-core gas measurements in comparison to the indirect and potentially less smoothed stomatal-based reconstructions from deciduous plants. In ice cores, air bubbles are not closed off from the atmosphere until they reach the base of the firn layer. The rate of this process varies in time and space. Nevertheless, analysis of air extracted from multiple suitable ice cores with different snow-accumulation rates are all consistent with pre-industrial variations of CO_2 during the past millennium. These variations are limited to changes of 12 ppmv at most (Siegenthaler et al., 2005a). This shows that ice-core derived CO_2 concentration measurements are highly comparable. Moreover, they are in good agreement with records of a large network of direct atmospheric measurement stations starting in AD 1958 on Mauna Loa and the South Pole (Keeling et al., 2005; 2008).

Further back in time, however, snow-accumulation rates were lower (e.g. during the late glacial-early Holocene transition). According to Neftel et al. (1988), CO_2 fluctuations with a duration of less than twice the bubble enclosure time (for example, up to 550 cal years in Dome Concordia (Monnin et al., 2001)) cannot be detected. By contrast, leaves of deciduous plants are formed each year and potentially provide an indirect record of CO_2 changes at annual resolution. While some stomatal-based reconstructions support the main trends in CO_2 changes as recorded by ice cores (e.g. Rundgren and Beerling, 1999), others suggest high-amplitude variations of past atmospheric CO_2 concentrations at centennial or even decadal timescales during the last deglaciation and the Holocene (Beerling et al., 1995; Wagner et al., 1999; McElwain et al., 2002; Wagner et al., 2004; Kouwenberg et al., 2005; Van Hoof et al., 2005; Jessen et al., 2007). For example, stomatal-inferred reconstructions and ice-core derived CO_2 measurements diverge during the Younger Dryas. Based on leaves of various species from records in the Northern Hemisphere (Norway, Beerling et al., 1995; Canada, McElwain et al., 2002) a CO_2 concentration drop of ca. 65 ppmv at the onset of the Younger Dryas and an increase of ca. 50 ppmv CO_2 at the transition to the Holocene (11,650 cal years BP) was reconstructed. Another reconstruction obtained with a composite record of three different plant species from Sweden suggested a drop of CO_2 concentrations of ca. 50 ppmv during 12,700–12,400 cal years BP followed by a gradual increase during the Younger Dryas stadial and stable CO_2 values during the transition to the Holocene (Rundgren and Björck, 2003). By contrast, there is no sign of a CO_2 reduction in Antarctic ice cores (Monnin et al., 2001; Lüthi et al., 2008). The causes for these discrepancies are strongly debated (see, e.g. Monnin et al., 2001; McElwain et al., 2002).

The difference between the stomatal-based CO_2 reconstructions and the ice-core derived CO_2 measurements is important for understanding the carbon cycle, and for modelling climate responses to changes in the atmospheric CO_2 concentration. For example, as noted by Joos and Spahni (2008), the twentieth-century increase in CO_2 and its radiative forcing occurred more than an order of magnitude faster than any sustained change during the past 22,000 years as recorded in ice cores. Because stomatal-based reconstructions may indicate that the past CO_2 concentration changed at rates similar to CO_2 concentration changes during the past 200 years, it is essential to evaluate uncertainties, potentials, and pitfalls of stomatal-based CO_2 reconstructions (Poole et al., 1996; Finsinger and Wagner-Cremer, 2009).

Signal or Noise: Validation of Stomatal-Based CO_2 Reconstructions?

The most powerful means to validate palaeo-proxy reconstructions is to compare proxy-based reconstructions to measured records. A methodical assessment of the CO_2-prediction quality of SF data with herbarium-training sets and measured-SF records for three subtropical species showed significant differences in prediction accuracy (Wagner et al., 2005). Although these results were promising, they emphasized the need to validate SF-inferred CO_2 reconstructions. More recently, using a well-dated stomatal record of subfossil *B. nana* leaves collected from a peat bog in Kiruna, northern Sweden, Finsinger and Wagner-Cremer (2009) were able to compare reconstructed vs. measured CO_2 concentration records for the past 140 years. They used two inference models developed with the same herbarium-training set of *B. nana* and *B. pubescens* (Fig. 5a, b) and reconstructed CO_2 changes based on the subfossil Kiruna leaf record. They could show that reconstruction uncertainties can be as large as ~70 ppmv, a span that corresponds to CO_2 changes that occurred during glacial/interglacial transitions (Petit et al., 1999). The large uncertainties were obtained by combining two different sources of uncertainties: (i) those of the inference models used for the reconstruction (method error) (Fig. 5a, b) and (ii) those derived from the difference between reconstructed and expected values (validation error) (Fig. 5c). These uncertainties are thought to result from the inherent variability of SF within and between leaves of individual plants. In addition, although it is recommended to limit the analysis to the central leaf lamina (Poole et al., 1996, 2000), it is often difficult to determine the location of the central leaf lamina if only leaf fragments are available for analysis.

On the other hand, as shown in Fig. 5d, decreasing residuals (reconstructed – measured) attests to the higher prediction ability of smoothed reconstructions. This is consistent with Van Hoof et al. (2005), who used stomatal-based CO_2 reconstructions between the tenth and sixteenth centuries and showed that the

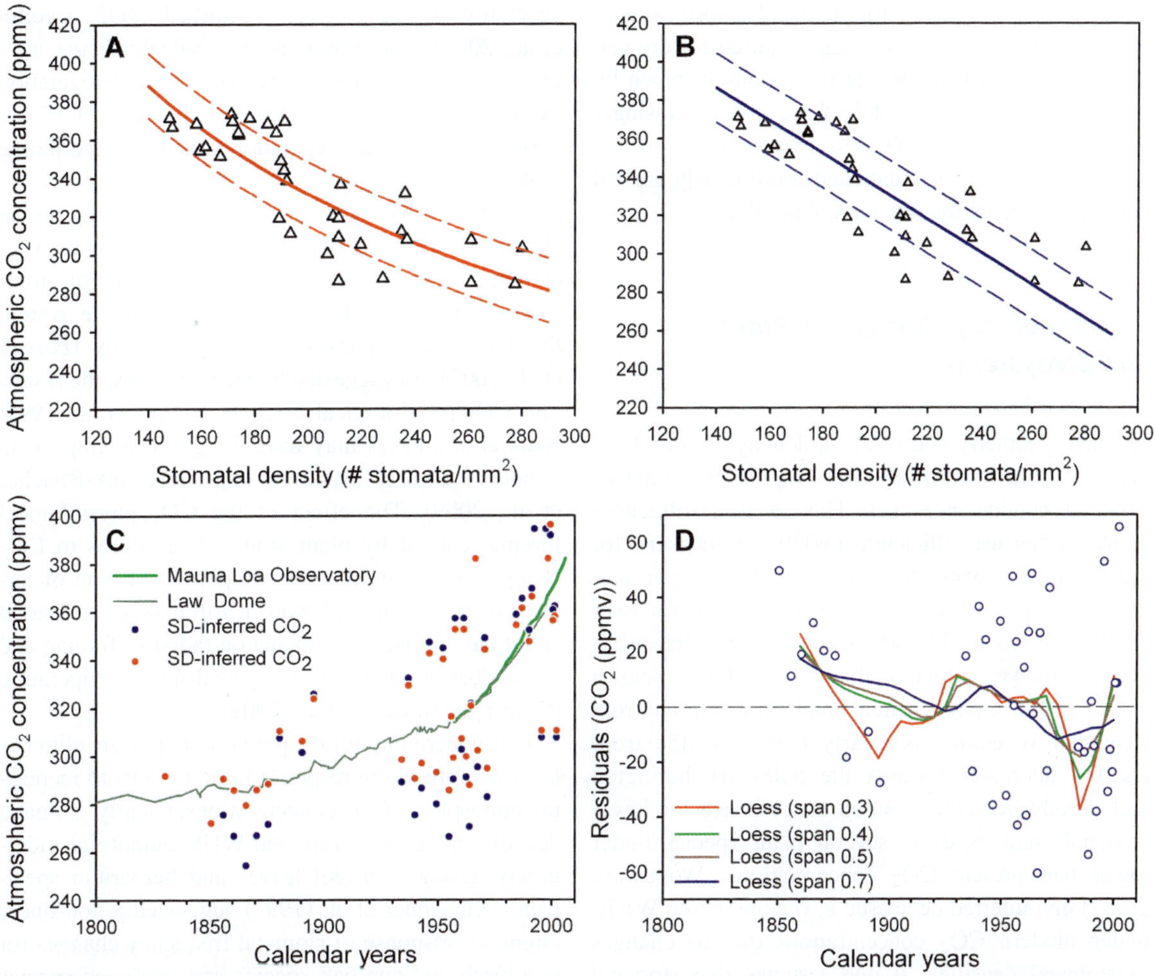

Fig. 5 **a** and **b** Stomatal density-based inference models (*thick lines*) for atmospheric CO_2 reconstruction based on herbarium leaves (*empty triangles*) (from Finsinger and Wagner-Cremer, 2009). *Thin lines*: ± 1 RMSE. **c** Stomatal density-inferred atmospheric CO_2 concentration reconstruction obtained with the inference models (**a** and **b**) and a stomatal density record from the Kiruna mire (Sweden) (not shown). Atmospheric CO_2 concentrations are from Mauna Loa (Keeling et al., 2008) and from Law Dome ice core (Antarctica, MacFarling Meure et al., 2006) (modified from Finsinger and Wagner-Cremer, 2009). **d** Residuals (expected–predicted) of stomatal density-inferred reconstructed values. *Empty blue circles*: unsmoothed reconstruction; *lines*: smoothed stomatal density-inferred CO_2 reconstruction. Smoothing was performed using locally weighted regression (Loess, Cleveland and Devlin, 1988) with different smoothing windows (spans)

discrepancies between the amplitudes estimated by the stomatal record and ice-core records diminished when the raw stomatal record was smoothed. However, these authors attributed the discrepancies to the smoothed nature of CO_2 records from ice cores. Finsinger and Wagner-Cremer (2009) could show that strong discrepancies occur between the stomata-inferred CO_2 concentration and the annual instrumental CO_2 record (Fig. 5d). This clearly indicates that the apparent disagreement between ice core and stomatal-based CO_2 reconstructions cannot be solely explained by the inherently smoothed nature of ice core records, as suggested by Blunier et al. (2005). It also indicates that it is essential to validate stomatal-based inference models to estimate uncertainties of reconstructed CO_2 concentrations. Clearly, a reconstruction cannot have a smaller uncertainty than the model uncertainty. Hence, stomatal-based CO_2 reconstructions should be limited to low-frequency changes after smoothing the records. It is therefore suggested that the degree of smoothing for subfossil records should be consistent with the smoothing required for the validation, instead of the degree of smoothing given by the firn densification model of ice cores (Finsinger and Wagner-Cremer, 2009). This may be feasible when subfossil leaves are abundant and continuously present in high-sedimentation rate natural archives.

Impacts of CO_2 Changes on Plants and Ecosystems

Stomatal geometry and stomatal density are the dominant variables controlling the complex parameter of stomatal conductance (g_s). This in turn influences plants' water-use efficiency (WUE), a measure for the amount of water loss at the leaf level per unit of carbon gain. Globally, annual precipitation over the land is about 110,000 km^3 and evapotranspiration accounts for a total of about 70,000 km^3 (Jackson et al., 2001), though annual total transpiration from terrestrial vegetation is clearly highest in the tropics and decreases towards the poles (Hetherington and Woodward, 2003). Based on evidence for lower stomatal frequencies in several plant species under lower-than present CO_2 concentrations, Woodward (1987) first inferred decreased g_s (i.e. increased WUE) under modern CO_2 concentrations due to changes in stomatal densities. If one assumes that stomatal geometry does not change after leaves are shed, then maximal stomatal conductance (g_{max}) can be inferred using SD, maximum pore length, and width (Franks and Beerling, 2009). Clearly, g_{max} measurements from subfossil leaves will have large uncertainties likely due to within-leaf, within-plant and between-plant variability of stomatal densities and geometry. However, the long-term trend of g_{max} changes may be filtered out by smoothing such records.

Present CO_2 concentrations already exceed the upper limit of the past ca. 800,000 years, and predictions for atmospheric CO_2 concentrations over the next two centuries suggest that CO_2 concentrations will likely increase further (Meehl et al., 2007). The continued rise in CO_2 concentrations will potentially further decrease g_s and increase WUEs, so plants will become more efficient at water use (Bunce, 2004; Saurer et al., 2004). However, species differ significantly in their CO_2 responses, which thus influences vegetation dynamics (e.g. Woodward, 2002; Zweifel et al., 2009). These aspects are gradually being integrated into models used to reconstruct past vegetation dynamics (e.g. Wu et al., 2007; Prentice and Harrison, 2009) and to predict vegetation changes at the biome scale.

In addition to exerting (i) a radiative forcing on the climate system and (ii) having an influence on vegetation dynamics, increasing concentrations of atmospheric CO_2 may also (iii) alter the climate system through direct effects on plant physiology (Forster et al., 2007). As suggested by recent studies, the resulting 'CO_2 physiological forcing' (Sellers et al., 1996; Betts et al., 1997) may have a significant impact on a model's climate sensitivity (e.g. Doutriaux-Boucher et al., 2009). The effect of the CO_2 physiological forcing, caused by plant stomatal responses to CO_2 changes that may alter the evapotranspiration of terrestrial ecosystems is, however, still poorly understood and it has an unknown impact on radiative forcing and on feedbacks dependent on global-mean temperature (Doutriaux-Boucher et al., 2009).

One uncertainty for the prediction of future climates lies therefore in the response rate of plants to increasing atmospheric CO_2 concentrations. Clearly, stomatal density can never be zero and WUE cannot rise indefinitely. Using subfossil leaves and herbarium specimens, Kürschner et al. (1997) suggested a non-linear (sigmoid) response of stomatal frequency changes for two birch and one oak species and predicted species specific response limits to CO_2 at ~350 and 400 ppmv, respectively. Accordingly, some species that did not alter their stomatal frequency in recent decades may be beyond their response limits (Woodward et al., 1991). The change in g_{max} of individual plants may be scaled up through to the community level assuming that plants do not tend to develop leafier canopies (Woodward et al., 1991). Consistent evidence for response limits comes from evidence of stable carbon isotope changes ($\delta^{13}C$) from tree rings, indicating that WUE in some trees across northern Europe is no longer rising (Waterhouse et al., 2004). Long-term records derived from subfossil leaves preserved in wetlands may thus be a valuable source to constrain the response limits of plants (e.g. in terms of g_{max}) to past atmospheric CO_2 changes. In combination with whole-plant WUE records (e.g. from $\delta^{13}C$ records) they may be used to better constrain the response limits of individual plant species to changing atmospheric CO_2 concentrations.

Climate Variability Reconstructed from Peat Bogs

Since the early part of the last century, peat bog stratigraphy has been used to reconstruct past climatic changes. Because the typical peat accumulation rate for the past ca. 2,000 years in Europe is 10–20 years/cm, finely sampled peat cores can provide data on a decadal time scale for this period (Blackford, 2000). Recent advances in dating methodologies allow for high-precision reconstructions of specific climatic events and their possible causes. However, a remaining difficulty is the interpretation of proxy data, since the exact climatic meaning of, for example, bog surface wetness (BSW) changes remains illusive. Increases in BSW are often described as being the results of 'wetter and/or cooler' conditions, which is not always helpful for climatic inferences (Charman, 2007). However, as described in the previous sections, the use of transfer functions and a multi-proxy approach could be the way forward to improve climatic reconstructions from peatlands (Blackford, 2000). Furthermore, studies with a high chronological precision can be compared from different localities, which allows for regional-scale climatic reconstructions with minimal chronological offsets (e.g. Swindles et al., 2007 and Mauquoy et al., 2008).

Environmental reconstructions from peatlands often register considerable changes that appear to be simultaneous over large regions (Hughes et al., 2000; Barber et al., 2003). Many studies find particularly large variability around 2,650 ^{14}C years BP (van Geel et al., 1996) and between 500 and 100 cal years BP (encompassing the so-called Little Ice Age, Mauquoy et al., 2002b; Lamentowicz et al., 2008a). These periods are characterized by rapid changes in bog surface wetness (BSW), as well as fluctuations in minerogenic influx (see Fig. 3). The causes for these BSW variations are debated (Plunkett and Swindles, 2008). Both mentioned time periods are characterized by large variations in solar activity, as reconstructed from ^{14}C variations. Therefore, a 'solar link' for these BSW variations is often implied (e.g. van Geel et al., 1996; Mauquoy et al., 2002a; 2008). Detailed ^{14}C wiggle-match dating of finely sampled peat records, mainly from the North Atlantic region, has provided strong arguments for a 'solar link' to these variations in BSW. For the time period ~2,650 ^{14}C years BP, findings from the Northern Hemisphere are supported by studies from Chile (van Geel et al., 2000) and southern Patagonia (Chambers et al., 2007a), which also show a rapid change in BSW at this time, pointing to a globally synchronous climatic response.

However, the response time of the recorded environmental changes to decreased solar irradiance appears to vary between sites (e.g. Swindles et al., 2007). This may in part be due to a natural 'smoothing' that occurs in peat bogs with low accumulation rates (Mauquoy et al., 2008). However, it may also point to a different process, such as oceanic circulation changes in response to solar forcing (Swindles et al., 2007; Plunkett and Swindles, 2008), which would in turn have a delayed impact on precipitation and evaporation rates in the oceanic climate areas of North-West Europe. More high-resolution, multi-proxy studies with very tight chronological precision from different regions are required to resolve the issue. In addition, comparison with other high-resolution records (e.g. lakes, speleothems) with independent dating would improve our understanding of the processes that caused rapid changes in BSW.

Impact of Climate Change on Peatlands and Potential Feedback Mechanisms on Climate

Palsa Mires in a Changing Climate

Marginal permafrost regions are very sensitive to environmental and climatic changes (Sollid and Sörbel, 1998; Fronzek et al., 2006; Parviainen and Luoto, 2007). As a result of projected climate change, the geographic position of the climatic envelope for discontinuous permafrost is likely to shift. 'Palsa mires' are wetland complexes with permanently frozen peat hummocks. Palsas and peat plateaus are characteristic permafrost features that are located in wetlands in the zone of discontinuous permafrost. The peat forms have permanently frozen cores elevated above the surrounding wetland. Palsas are isolated permafrost features generally less than 100 m in diameter with a height of 1–7 m. Peat plateaus are flat-topped, generally lower and more extensive landforms, which may, in some cases, cover areas over 1 km^2 (Thie, 1974).

In Canada palsas occur most frequently in the southern part of the discontinuous permafrost zone whereas peat plateaus occur predominantly in the north (Beilman et al., 2001). In Fennoscandia palsa mires are treeless bogs with dry surface vegetation communities dominated by bryophytes, dwarf shrubs and lichens. In contrast to this, in North America dense stands of black spruce are common on subarctic bogs containing permafrost (Zoltai and Tarnocai, 1971; Thie, 1974; Vitt et al., 1994). Palsas and peat plateaus have been intensively studied in Fennoscandia and Canada, whereas data from Russia are relatively scarce (Oksanen, 2005).

Both local and climatic factors are important for the presence or absence of permafrost in wetlands. Permafrost aggradation most often occurs in *Sphagnum* deposits where pore waters are significantly colder than in other types of wetland deposits (Vitt et al., 1995), but permafrost landforms also develop directly from sedge peat accumulated in fens (Brown, 1980). Optimal climatic conditions for the existence of palsas are long, cold winters with little precipitation or strong winds (Seppälä, 1986), and mean annual temperatures of –5 to –3°C (Luoto et al., 2004). However, palsas are found at higher mean annual temperatures; for example, in northern Fennoscandia, where the distribution of lowland permafrost is closely related to the precipitation shadow east of the Scandes Mountains (Luoto et al., 2004). Vegetation cover may also be decisive for the presence of permafrost. Whereas low vegetation or barren ground on palsas in Fennoscandia promote snowdrift and thereby exposure to low winter temperatures (Seppälä, 1986), the presence of a dense tree cover on palsas in subarctic North America reduces the amount of incoming solar radiation and snow to the ground surface (Zoltai and Tarnocai, 1971). Further, lichen cover is frequent on both wooded and treeless palsas and cause an increase in the albedo. In this way, both a sparse and dense vegetation cover on the palsa surface may in different ways contribute to a negative energy balance and thereby efficiently promote maintenance of permafrost belowground. Soil moisture is also an important factor for the formation and maintenance of permafrost. At the onset of freezing, a saturated peat layer increases the thermal conductivity and allows heat transfer from the ground to the atmosphere, whereas during summer, a dry peat layer insulates permafrost from high air temperatures (Seppälä, 1986).

Palsa features are cyclic in occurrence. They grow and when they mature they start to collapse at the edges by block erosion along surface cracks. The frozen core starts to thaw as soon as the peat layer covering it is lost. Such a development may result in internal lawns, collapse scars or thermokarst ponds, often the only remains of a palsa (e.g. Vitt et al., 1994; Payette et al., 2004). Permafrost degradation on treed palsas causes trees to tilt, and complete permafrost thaw may cause the trees to topple, become submerged and eventually overgrown by fen communities to create internal lawns. In the boreal Haut Fagnes Plateau in Belgium, so-called 'viviers' have been interpreted to represent remnants of past mineral-cored palsas (lithalsas) that aggraded permafrost during the Younger Dryas cold spell 11,500 years ago (Pissart, 2003).

Dating of Permafrost Phases in Palsa Mires

Various methods can be used to determine when permafrost formed in a peatland. Frost heave resulting from aggradation of permafrost promotes the development of a dry surface layer, suitable for the growth of certain shrubs and trees. The oldest trees on palsas may be used to indicate a minimum age of a permafrost phase. The frequent occurrence of wildfires in these ecosystems, however, affects the age distribution of trees on a palsa. The age distribution of black spruce on treed palsas may therefore reflect the occurrence of past wildfires (Zoltai, 1993), in turn potentially causing damage to local permafrost lenses in wetlands (Turetsky et al., 2002; Myers-Smith et al., 2008). Mortality dates of black spruce stems submerged in thermokarst ponds as determined by tree-ring analysis, can be used to assess the timing of permafrost decay (Vallée and Payette, 2007). Radiocarbon dating of the lowermost layers of older palsa deposit offers another possibility to date the timing of a permafrost phase (Oksanen, 2005; Seppälä, 2005). Peat stratigraphies and reported radiocarbon dates from palsa mires in northern continental Europe (including NW Russia) indicate the initiation of permafrost aggradation during the second half of the Holocene (reviewed by Oksanen, 2002, 2005). In Canada, early signs of permafrost aggradation have also been dated to the later part of the Holocene in response to regional cooling (e.g.

Zoltai, 1993; Kuhry and Turunen, 2006). Multiple permafrost aggradation and degradation phases within wetlands are frequently reported; the latest during the Little Ice Age after ca. 700 cal BP (e.g. Zuidhoff and Kolstrup, 2000; Kuhry, 2008; Kokfelt et al., 2010). Cautious sampling and knowledge of peat stratigraphy is necessary to interpret radiocarbon dates originating from dynamic permafrost environments, where highly humified peat layers as well as erosion surfaces may occur between strata. By combining detailed information on peatland stratigraphy, biogeochemical changes and a detailed chronology based on a combination of ^{210}Pb and ^{14}C dating, Kokfelt et al. (2010) demonstrated the existence of a low accumulation layer in a peat core from the Stordalen Mire in northern most Sweden. This layer spans ca. 1,500 years and has been interpreted to mirror an early period of permafrost dynamics in the wetland. In the same study, sediment sequences of adjacent lakes were used to support conclusions regarding permafrost dynamics of the peatland based on peat stratigraphy. Permafrost phases dated from peat deposits were found to correlate with periods of relatively acidic lake water conditions, resulting from the formation of poor fen and bog communities in the peatland in response to frost heave.

Ecosystem Protected Permafrost in Palsa Mires

Although the geographic position of the climatic envelope for discontinuous permafrost is likely to shift if global warming continues, the occurrence of palsas and peat plateaus does not necessarily follow this shift at the same pace. Peat thickness, vegetation cover and snow depth are also important factors for palsa development and preservation (Luoto et al., 2004, Seppälä, 2005; Fronzek et al., 2006). Permafrost is present in peatlands under climatic conditions that do not support permafrost in other types of soils because of the insulating properties of accumulated *Sphagnum* peat (Halsey et al., 1995; Yi et al., 2007). Such permafrost, so-called 'ecosystem protected permafrost' (Shur and Jorgensen, 2007), has formed under colder climates but can persist as patches in a warmer climate. 'Ecosystem protected permafrost' is typically found in climates where the mean annual air temperature is approximately 2 to –2°C (Shur and Jorgensen, 2007). Hence, small changes in precipitation and air temperature may cause erosion and degradation of permafrost in palsas and peat plateaus and thereby alter hydrology, ecology and biogeochemical cycling of these wetlands. In a warming climate, peat plateaus and palsas will be affected and are thus expected to decrease dramatically during the next 100 years (Fronzek et al., 2006).

Carbon Cycling in Relation to Degradation of Permafrost in Palsa Mires

Describing carbon cycling of a landscape includes a complex array of mechanisms and interactions between major carbon compartments: atmosphere, living vegetation, aquatic environments and sediment stores such as peat and lake sediments. Palsa mires in particular have distinct features and dynamics in relation to emissions of the greenhouse gas methane (Fig. 6); a very strong greenhouse gas with a radiative potential 25 times that of CO_2 over a time period of 100 years (IPCC, 2007). Intact palsas emit insignificant amounts of methane (Bubier et al., 1995). Permafrost degradation can, however, cause waterlogged conditions and changes in vegetation composition whereby the proportion of anaerobic to aerobic degradation of organic matter together with the transport of methane via sedges cause an increased production and flux of methane to the atmosphere (Christensen et al., 2004; Johansson et al., 2006; Ström and Christensen, 2007). At the Stordalen mire complex (northernmost Sweden) microtopography and hydrological changes as indicated by vegetation change alone were calculated to have caused an increase in the landscape-scale methane (CH_4) emissions ranging between 22 and 66% from AD 1970 to 2000 (Christensen et al., 2004). Ström and Christensen (2007) concluded that the hydrological changes promoted the expansion of plant species that stimulate the production and transport of methane to the atmosphere. Johansson et al. (2006) made a combined analysis including both CO_2 and CH_4 exchanges at the same mire, and showed an increased radiative forcing functioning of this ecosystem owing to the relative importance of increased CH_4 emissions. Similarly, Turetsky et al. (2002) showed that permafrost degradation features (such as internal lawns) emit 1.6–30 times more CO_2 and CH_4 relative to palsas.

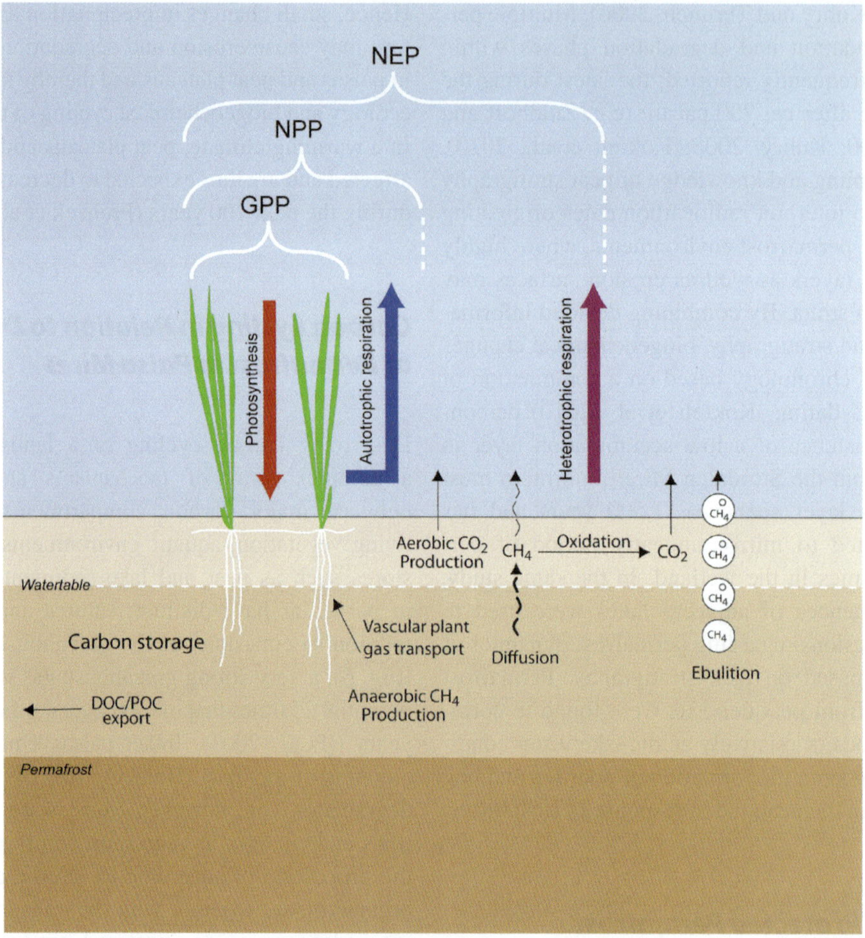

Fig. 6 Schematic diagram of the processes governing the net ecosystem carbon balance, including pathways of CO_2 and CH_4 exchange with the atmosphere, in a permafrost environment (Christensen et al., 2009)

Changes in the delivery of organic carbon to aquatic environments as a result of permafrost decay serve another important source of increased emissions of CO_2 and CH_4 to the atmosphere. In the aquatic environment organic carbon is efficiently respired by lake bacteria that cause lake water to become supersaturated with respect to CO_2, which is subsequently emitted to the atmosphere (Kling et al., 1991). Along this path, approximately 50% of the terrestrial derived carbon is emitted back to the atmosphere (Cole et al., 2007). Landforms representing degrading palsas, such as thermokarst ponds, internal lawns and collapse scars, all emit significantly higher amounts of methane than intact palsas and peat plateaus (Bubier et al., 1995; Christensen et al., 2004; Turetsky et al., 2007).

The sink or source function of a terrestrial ecosystem is typically measured by eddy covariance or chamber measurement techniques, as the balance between fluxes (photosynthesis versus respiration) to and from the atmosphere (e.g. Johansson et al., 2006). Measurements of this net ecosystem exchange (termed NEE) quantifies the carbon flux in and out of a system, where a positive value denotes a net source to the atmosphere and a negative value denotes a net sink from the atmosphere to the terrestrial system. However, NEE measurements typically overestimate the sink function of terrestrial ecosystems (Roulet et al., 2007), as carbon losses to aquatic environments are not accounted for. Such losses are important to incorporate in carbon budgets of catchments containing permafrost in peatlands, as the export of dissolved organic carbon (DOC) to rivers, streams and lakes may increase dramatically in response to thawing of permafrost (Frey and Smith, 2005).

Current and Future Development of Palsa Mires

Palsa mires are currently degrading at many sites throughout their distribution, probably due to regional climatic warming (Johansson et al., 2009). In Norway, Hofgaard (2006) reported on a decreasing peat plateau at Hagutjørnin, Dovre, which has been documented over a 31-year period (1974–2005). From the same area, Sollid and Sörbel (1998) reported that palsa plateaus that were almost intact in the 1960s had almost completely disappeared by the end of the 1990s. In sub-arctic Sweden, permafrost thawing has been reported from palsa mires and peat plateaus (Christensen et al., 2004; Johansson et al., 2006; Åkerman and Johansson, 2008; Sannel and Kuhry, 2009). In the Abisko area (68°20′E, 19°02′E) this has been associated with wetting of peat plateaus (Christensen et al., 2004; Johansson et al., 2006, Kokfelt et al., 2009) whereas in Tavvavuoma, north-east of Abisko, the opposite scenario was observed and thermokarst drainage has occurred during the last four decades (Sannel and Kuhry, 2009). Degrading permafrost in palsa mires has also been reported from Finland (Luoto and Seppälä, 2003), North America (Payette et al., 2004; Camill, 2005) and Western-Siberia (Kirpotin et al., 2007)

Active Layer and Permafrost Temperatures

The trend in active layer thickness is not uniform throughout the distribution range of palsas and peat plateaus. Wright et al. (2009) reported that the active layer thickness in a peat plateau in Canada at the end of the season had increased from 1999 to 2006, even though there was great inter-annual variability. In northernmost Sweden, active layer thickness had increased, ranging from 0.69 to 1.26 cm/year during the last three decades in nine peat mires in the Torneträsk catchment. In addition, permafrost had disappeared completely at one site and at 81% of all sampling points during the same time period (Fig. 7). The increased active layer thicknesses are correlated with increases in mean summer air temperature, Degree-Days of Thawing (DDT) and in five out of nine sites also with increases in snow depth (Åkerman and Johansson, 2008). In contrast Seppälä

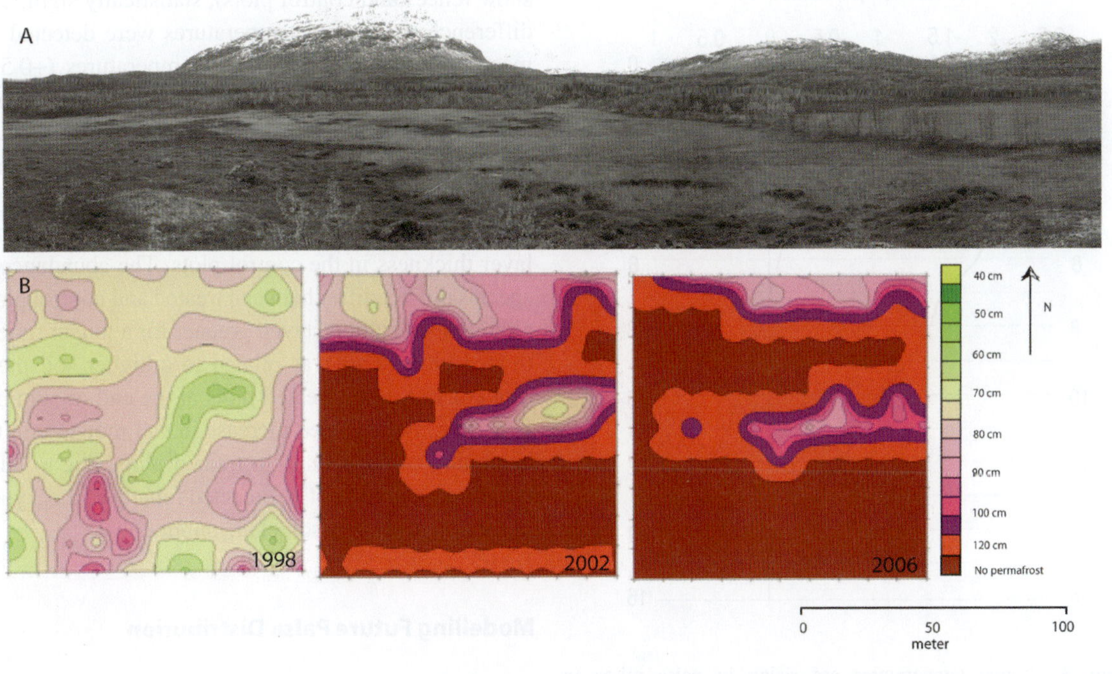

Fig. 7 The permafrost disappeared in 81% of a palsa mire in less than a decade (modified from Åkerman and Johansson, 2008). **a** Photo of the Katterjokk mire (Jonas Åkerman) and **b** active layer thickness measurements

(2003) reported that in Finland, no unusual development in the active layer on palsas had been detected and that new permafrost had formed in small peat hummocks.

Associated with the increases in air temperatures recorded in the Arctic during recent decades, ground temperatures in areas underlain by permafrost have increased (Walsh et al., 2005) by 0.5–2°C at the depth of zero annual amplitude (the distance from the ground surface downward to the level beneath which there is practically no annual fluctuation in ground temperature) (Brown and Romanovsky, 2008). Permafrost thawing occurs at the southern limits of the permafrost zone (Brown and Romanovsky, 2008) and this has contributed to a northward shift of the southern boundary of discontinuous permafrost (Lemke et al., 2007). Monitoring of permafrost temperatures in palsa mires in northernmost Sweden shows that lowland permafrost is currently degrading. Borehole recordings show increasing ground temperatures of 0.04–0.05°C per year in the upper 1 m, and 0.03–0.04°C per year in the lower 12–15 m between 1980 and 2002 (Fig. 8). However, no trend was detected in the middle of the profile. The changes in ground temperatures were correlated with increasing air temperature and increasing summer precipitation. At lower depths, the increases may be due to possible increased heating from a slightly warmer or more freely flowing ground water (Johansson et al., 2008).

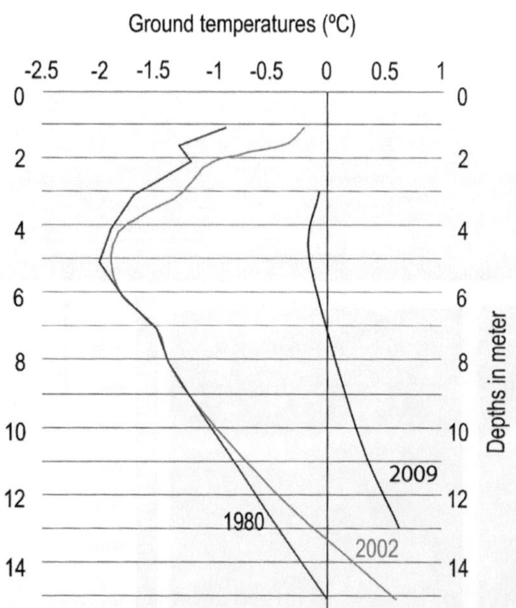

Fig. 8 Ground temperatures are rising in palsa mires in the Abisko area and permafrost is thawing (modified from Johansson et al., 2008)

Experimental Snow Depth Manipulation

A thin snow cover is required in winter to preserve the permafrost in palsa mires. In the Abisko area, northernmost Sweden, a century-long trend of increasing snow depth has been recorded during the last century (Kohler et al., 2006) and this is projected to continue until the end of the twenty-first century (Sælthun and Barkved, 2003). The observed change in snow cover has affected peat mires in the area as thawing of permafrost, increases in active layer thickness, and associated vegetation changes have been reported during the last decade (see above). To be able to predict future changes due to climate change, it is necessary to look at both vegetation and permafrost changes. An experimental manipulation was set up in autumn 2005 to simulate future snow cover on a peat mire, as was projected for the end of the century (Johansson et al., submitted). After 3 years of treatment (plots with snow fences and control plots), statistically significant differences in ground temperatures were detected for mean winter (October – May) temperatures (+0.5 to 1°C at the treated plots) as well as for minimum winter temperatures (+3 to 5°C at the treated plots). No significant difference was detected in the mean summer soil temperature and the maximum soil temperatures. A difference could also be detected between active layer thickness in the control plots. The abundance of species did not differ between treated and control plots, although vegetation stayed greener for a longer time in treated plots. This was related to a higher soil moisture content in the treated plots at the end of summer. However, the expected increase in soil moisture in the beginning of the season in treated plots, due to additional snow, could not be detected for any of the 3 years of treatment (Johansson et al., submitted).

Modelling Future Palsa Distribution

Climate envelopes of palsa mires are narrow and hence they can be expected to be extremely sensitive

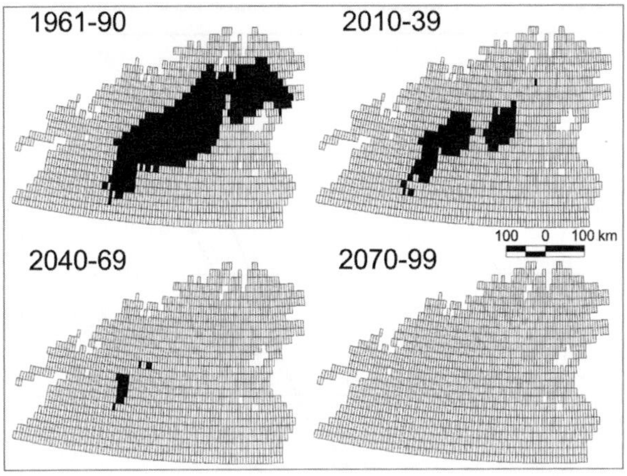

Fig. 9 Palsa mire distribution (*solid black grid cells*) according to simulations with present-day climate and for future time periods, 2010–2039, 2040–2069 and 2070–2099, using a scenario based on the HadCM3 General Circulation Model with the A2 emission scenario. Modified from Fronzek et al. (2006)

to changes in future climatic conditions (Parviainen and Luoto, 2007). Luoto et al. (2004) mapped the spatial distribution of palsas in northern Fennoscandia, north of the Arctic Circle on a regular 10′ × 10′ grid (Fig. 9). The distribution was then modelled using climate envelope techniques that determine a statistical relationship between climate variables and the spatial distribution of palsa mires. The models were studied with respect to their sensitivity to altered climate, and climate change scenarios were applied to assess possible impacts on the palsa distribution during the twenty-first century (Fronzek et al., 2006, 2010). These models achieved a good to very good agreement with the observed palsa distribution, which suggests a strong dependency on climate. Even small increases of temperature (1°C) and precipitation (10%) resulted in considerable losses of areas suitable for palsa development. The models predicted the total disappearance of suitable regions for palsa development with an increased mean annual temperature of 4°C. Under climate change scenarios based on seven Atmosphere-Ocean General Circulation Models (AOGCMs) the models indicated that the degradation of palsas might proceed very quickly (Fig. 9). All but one climate scenario resulted in the total disappearance of suitable regions for palsa development by the end of the twenty-first century (Fronzek et al., 2006).

Recent progress in estimating probabilities of future climate change from ensembles of model projections offers an opportunity to go beyond 'what if' types of scenario analysis to a quantified assessment of risks to natural or human systems. However, the application of a potentially high number of ensemble climate projections as inputs to impact models may prove impractical. An alternative method makes use of the probabilistic representation of future climate in combination with impact response surfaces (Fronzek et al., 2010). The palsa model was applied to construct impact response surfaces that show the change in palsa areas as a function of mean annual temperature and annual precipitation. These were combined with probabilistic climate change projections derived from an ensemble of 15 AOGCMs using a resampling method (Räisänen and Ruokolainen, 2006). These authors estimated it as very likely (>90% probability) that a loss of area suitable for palsa mires to less than half of the baseline distribution would occur by the 2030s and likely (>66%) that all suitable areas would disappear by the end of the twenty-first century under the A1B and A2 emissions scenarios. For the B1 scenario, it was more likely than not (>50%) that a small proportion of the current palsa mire distribution would remain until the end of the twenty-first century (Fig. 10).

Impact of Climate Change on Regenerating Cutover Bogs: The Future Does Not Look Bright

Northern peatlands have been heavily impacted by mining, drainage and forestry, causing the release of CO_2 to the atmosphere and the loss of their CO_2 sink function (Joosten and Clarke, 2002; Chapman et al., 2003). The restoration of mined peatlands, which in Western Europe cover more surface than natural peatlands, could reverse this trend (Moore, 2002) and there is hope that such regenerating peatlands could act as strong carbon sinks. It is, however, unclear if these peatlands could recover their C-sink function over the long-term (Waddington and Warner, 2001) and what effects climate change may have on the regeneration process and the carbon balance of secondary bogs.

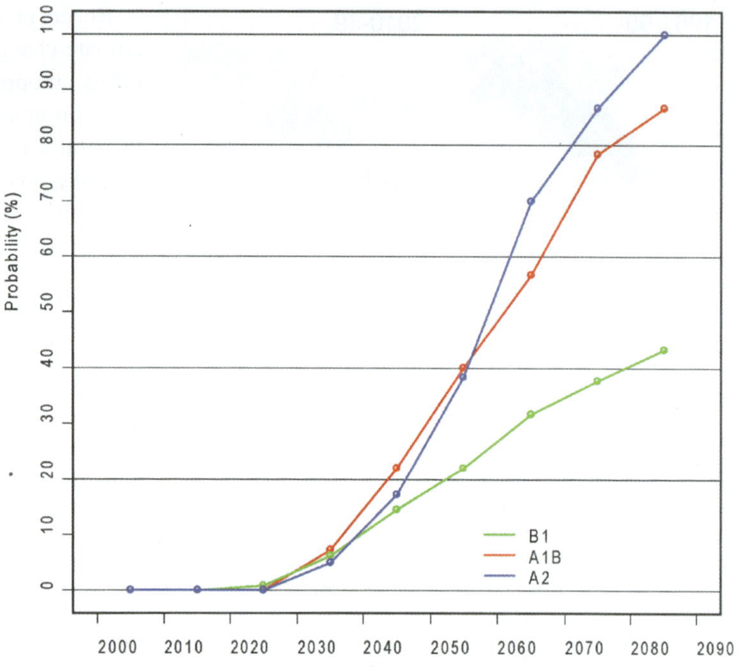

Fig. 10 Probability of the total loss permafrost in palsa mires during the twenty-first century based on probabilistic projections of climate change from a resampled ensemble of 15 GCM for three emission scenarios. Modified from Fronzek et al. (2010)

To assess how changes in temperature and water table fluctuations might affect the CO_2 balance and hence the long-term fate of the regeneration process of cutover bogs, we used models of photosynthesis (P_G) and respiration (R_{tot}) fluxes derived from gas flux measurements on three regeneration stages of a cutover peatland in the Swiss Jura Mountains, respectively, 29, 42 and 51 years after abandonment of exploitation (Samaritani et al., 2010). Three temperature increases (+0.5, +1.0 and +1.5°C – corresponding to observed increases in average annual temperature in the twentieth century in Switzerland (Rebetez and Reinhard, 2007)), and two increases in water table variability were used in the fitted models from the study site. We increased the hydrological variability by 10 and 20% in comparison to the median value of the water table depth. Temperature changes were applied to both air and soil temperatures. Net Ecosystem Exchange (NEE) for all different scenarios were also calculated on model parameters inferred from re-sampled datasets. To estimate the stability of each model, 500 new datasets of model parameters were generated by bootstrapping. R_{tot} and P_G were re-calculated for each of these pseudo-replicate datasets. The 1st and 3rd quartiles of NEE calculated on these pseudo-replicates are illustrated on the figures as error bars.

Modelling Net Ecosystem Exchange (NEE)

In spring and autumn all sites accumulated carbon, but in summer the youngest site was a net carbon source, the intermediate site alternated between sink and source, and the most advanced site acted as a sink. Further details are given elsewhere (Samaritani et al., 2010). Over the entire growing season, the youngest site was a carbon source, whereas the intermediate and advanced sites accumulated carbon (Fig. 11). This agrees with vegetation differences among the sites and especially the percentage cover of *Sphagnum* (a keystone genus for carbon accumulation) and changes from fen to bog vegetation (indicating differences in soil fertility and litter quality which affect net primary production and decomposition rates): (1) *Sphagnum* cover was higher in the intermediate and the most advanced sites than in the youngest site, and (2) the intermediate site, which had a poor fen vegetation with the highest *Eriophorum* cover, was more active (higher P_G and R_{tot}) than the most advanced site, which had a bog-like vegetation (Kummerow et al., 1988; Tuittila et al., 1999).

The responses to simulated climate change contrasted along the regeneration sequence (Fig. 11).

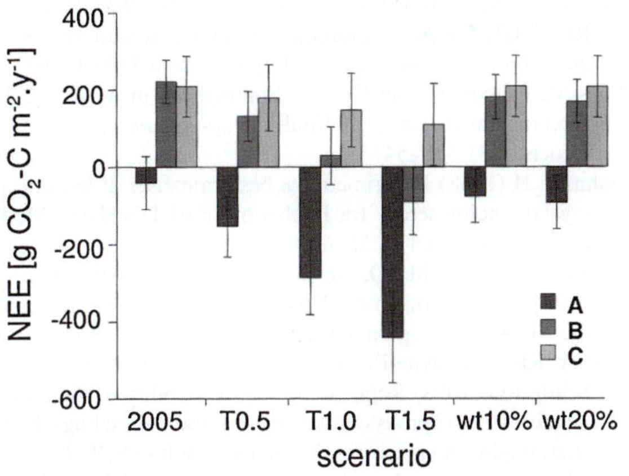

Fig. 11 Simulated effect of climate change on Net Ecosystem Exchange (NEE) of early to advanced bog regeneration stages. NEE estimated for the growing season 2005 in the three sites *A* (29 years after abandonment of peat extraction; *dark grey*), *B* (42 years; *intermediate grey*) and *C* (51 years; *light grey*), for three air and peat temperature increase scenarios and for two water table depth (wt) variation scenarios. Air temperature and peat temperature at 30 cm depth were increased by 0.5, 1.0 and 1.5°C. The water table variation around the median value was increased by 10 and 20%. *Error bars* indicate the 1st and 3rd quartiles values obtained by bootstrapping

In the youngest site, all scenarios resulted in strong increases in carbon release. In the intermediate site, NEE decreased both with increasing water table depth variations and temperature, becoming a source of C when warming exceeded 1.0°C. In the most advanced site the NEE was little affected by the different simulated climate changes.

The decreasing temperature effect on the carbon balance along the regeneration gradient suggests a higher resistance to climate change in the more advanced site. After 30 years of regeneration, not only does the younger site still release CO_2, but our models suggest that such early regeneration stages are also highly sensitive to climate change. Both a warmer climate and a more variable precipitation would cause a large loss of carbon, making it less likely for these abandoned mined peatlands to return to the structure and carbon-sequestering function that is characteristic of natural bogs. In the present climatic conditions, about 40 years of regeneration are needed for minerotrophic plant species to be replaced by typical bog species. However, the NEE models suggest that the carbon-sequestrating function is still fragile, and increases in temperature can easily convert these regenerating peatlands into a source of atmospheric CO_2. After five decades, secondary bogs are more similar to natural pristine peatlands both in terms of vegetation and carbon dynamics.

This study has important implications for conservation and management of mined bogs and the associated potential carbon sequestration. There might be a tipping point in the secondary succession beyond which regenerating peatlands become more resistant to climate fluctuations. However, the risk is that rather than evolving towards bogs and having a cooling effect on global climate, regenerating peatlands may revert to fens and thereby contribute to warming. Any action that speeds-up the regeneration of mined peatlands, such as the blocking of drainage ditches which decreases water table fluctuations and thus increases the thermal inertia of soils, will therefore increase the likelihood of mined peatlands to recover fully and durably and contribute to mitigating increases in atmospheric CO_2 concentrations.

Concluding Remarks

The scope of topics covered in this chapter clearly illustrates the breadth of research that is concerned with peatlands. Peatland research includes the reconstruction of past climatic and environmental conditions, which can range thousands of years back in time. Peatland areas are an important archive for terrestrial palaeo-environmental and climatic reconstructions. For these reconstructions a good chronological control is crucial. Recent advances in dating methods and age-depth modelling have strongly contributed to the development of highly resolved reconstructions with minimal chronological errors. The research field has also gained strongly from the introduction of novel proxies and the development of large training sets, which allow for a quantitative approach to reconstructions. Well-dated, multi-proxy studies have provided comprehensive reconstructions that can be used for the interpretation of climatic changes and their causes.

Peatland research also focusses on the functioning of plants and ecosystems, ecosystem sensitivity and responses to environmental changes. Peatland areas that are situated near environmental thresholds, such as palsas and high-alpine peatlands, were shown to be particularly sensitive to change. Changes in past and modern palsa characteristics can be used for environmental reconstructions as well as for future projections of palsa distributions. Furthermore, peatland experiments and monitoring studies provide important clues to processes controlling greenhouse gas storage and release, potential climatic feedbacks and peatland responses to environmental changes. Research activity in this field is increasing.

Peatland research thus has the potential to provide detailed information on past conditions, as well as recent and continuing responses to climatic and environmental changes. Thus far, research activity has been focussed predominantly on (ombrotrophic) bogs in the northern Hemisphere. With new techniques becoming available, the scope and resolution of such studies can be increased to include detailed reconstructions of different climatic factors, which can be compared more easily with other records. In addition, since peatland areas are found worldwide, the potential in this field of research is vast.

Acknowledgments WF is thankful to M. Gagen, D. McCarroll, F. Wagner-Cremer and H.H. Birks for discussions, to the Millennium Project (EU-FP6 contract 017008) and the AXA Research Fund for financial support. Funding to ML by the Foundation for Polish Science (FNP; Outgoing Fellowship KOLUMB) is acknowledged. Work by EADM and ES was funded by EU project RECIPE (n° EVK2-2002-00269). The Swiss part of RECIPE was funded by the State Secretariat for Education and Research.

References

Aaby B, Tauber H (1975) Rates of peat formation in relation to degree of humification and local environment, as shown by studies of a raised bog in Denmark. Boreas 4:1–14.

Aaby B (1976) Cyclic variations in climate over the past 5,500 years reflected in raised bogs. Nature 263:281–284.

Ainsworth EA, Rogers A (2007) The response of photosynthesis and stomatal conductance to rising [CO_2]: Mechanisms and environmental interactions. Plant Cell Environ 30:258–270.

Åkerman HJ, Johansson M (2008) Thawing permafrost and thicker active layers in sub-arctic Sweden. Permafr Periglac Process 19(3):279–292.

Alexandersson H, Schmith T, Iden K, Tuomenvirta H (1998) Long-term variations of the storm climate over NW Europe. Glob Atmos Ocean Syst 6:97–120.

Aubert D, Le Roux G, Krachler M, Cheburkin A, Kober B, Shotyk W, Stille P (2006) Fate of atmospheric REE entering an ombrotrophic peat bog in Black Forest (SW Germany): Evidence from snow, lichens and mosses. Geochim Cosmochim Acta 70:2815–2826.

Avsejs LA, Nott CJ, Xie S, Maddy D, Chambers FM, Evershed RP (2002) 5-n-Alkylresorcinols as biomarkers of sedges in an ombrotrophic peat section. Org Geochem 33:861–867.

Baas M, Pancost R, van Geel B, Sinninghe Damsté JS (2000) A comparative study of lipids in *Sphagnum* species. Org Geochem 31:535–541.

Bahnson H (1968) Kolorimetriske bestermmelser af humificeringstal i højmosetørv fra Fuglsø mose på Djursland. Medd Dansk Geol Foren 18:55–63.

Bakke J, Lie Ø, Dahl SO, Nesje A, Bjune AE (2008) Strength and spatial patterns of the Holocene wintertime westerlies in the NE Atlantic region. Glob Planet Change 60:28–41.

Barber KE, Dumayne-Peaty L, Hughes PDM, Mauquoy D, Scaife RG (1998) Replicability and variability of the recent macrofossil and proxy-climate record from raised bogs: Field stratigraphy and macrofossil data from Bolton Fell Moss and Walton Moss, Cumbria, England. J Quat Sci 13:515–528.

Barber KE, Chambers FM, Maddy D (2003) Holocene palaeoclimates from peat stratigraphy: Macrofossil proxy climate records from three oceanic raised bogs in England and Ireland. Quat Sci Rev 22:521–539.

Barber KE, Chambers FM, Maddy D (2004) Late Holocene climatic history of northern Germany and Denmark: Peat macrofossil investigations at Dosenmoor, Schleswig-Holstein, and Svanemose, Jutland. Boreas 33:132–144.

Beer J, van Geel B (2008) Holocene climate change and the evidence for solar and other forcings. In RW Battarbee, HA Binney (Eds), Natural Climate Variability and Global Warming: A Holocene perspective (pp. 138–162). Oxford: Wiley-Blackwell.

Beerling DJ, Chaloner WG (1992) Stomatal density as an indicator of atmospheric CO_2 concentration. Holocene 2:71–78.

Beerling DJ, Birks HH, Woodward FI (1995) Rapid late-glacial atmospheric CO_2 changes reconstructed from the stomatal density record of fossil leaves. J Quat Sci 10:379–384.

Beilman DB, Vitt DH, Halsey LA (2001) Localized permafrost peatlands in Western Canada: Definition, distributions and degradation. Arct Antarct Alp Res 33(1):70–77.

Beilman DW, Vitt DH, Bhatti JS, Forest S (2008) Peat carbon stocks in the southern Mackenzie River Basin: Uncertainties revealed in a high-resolution case study. Glob Chang Biol 14:1–12.

Belyea LR, Clymo RS (2001) Feedback control of the rate of peat formation. Proc R Soc Lond Biol Sci 268:1315–1321.

Belyea LR, Baird AJ (2006) Beyond the "limits to peat bog growth": Cross-scale feedback in peatland development. *Ecol Monogr* 76:299–322.

Bennett KD (1994) Confidence intervals for age estimates and deposition times in late-quaternary sediment sequences. Holocene 4:337–348.

Bennett KD, Fuller JL (2002) Determining the age of the mid-Holocene Tsuga canadensis (hemlock) decline, eastern North America. Holocene 12:421–429.

Berglund BE, Malmer N, Persson T (1991) Landscape-ecological aspects of long-term changes in the Ystad area. In BE Berglund (Ed), The Cultural Landscape During 6000 Years in Southern Sweden – The Ystad Project (vol. 41). Ecological Bulletins, Stockholm.

Betts RA, Cox PM, Lee SE, et al. (1997) Contrasting physiological and structural vegetation feedbacks in climate change simulations. Nature 387:796–799.

Birks HH (2007) Plant macrofossil introduction. In SA Elias (Ed), Encyclopedia of Quaternary Science (vol. 3, pp. 2266–2288). Amsterdam: Elsevier.

Björck S, Clemmensen LB (2004) Aeolian sediment in raised bog deposits, Halland, SW Sweden: A new proxy record for winter storminess variation in southern Scandinavia? Holocene 14:677–688.

Blaauw M, Heuvelink GBM, Mauquoy D, van der Plicht J, van Geel B (2003) A numerical approach to 14C wiggle-match dating of organic deposits: Best fits and confidence intervals. Quat Sci Rev 22:1485–1500.

Blaauw M, Christen JA (2005) Radiocarbon peat chronologies and environmental change. Appl Statist 54:805–816.

Blackford JJ, Chambers FM (1991) Blanket peat humification: Evidence for a Dark Age (1400 BP) climatic deterioration in the British Isles. Holocene 1:63–67.

Blackford JJ, Chambers FM (1993) Determining the degree of peat decomposition for peat-based palaeoclimatic studies. Int Peat J 5:7–24.

Blackford J (2000) Palaeoclimatic records from peat bogs. Trends Ecol Evol 15:193–198.

Blockley SPE, Blaauw M, Bronk Ramsey C, van der Plicht J (2007) Building and testing age models for radiocarbon dates in Lateglacial and Early Holocene sediments. Quat Sci Rev 26:1915–1926.

Blundell A, Charman DJand, Barber KE (2007) Multiproxy late Holocene peat records from Ireland: Towards a regional palaeoclimate curve. J Quat Sci 23:59–71.

Blunier T, Monnin E, Barnola J-M (2005) Atmospheric CO_2 data from ice cores: Four climatic cycles. In JR Ehleringer, TE Cerling MD Dearing (Eds), A History of Atmospheric CO_2 and Its Effects on Plants, Animals, and Ecosystems (vol. 177). Ecological Studies. New York, NY: Springer.

Blytt A (1876) Essay on the Immigration of the Norwegian Flora During Alternating Rainy and Dry Periods (89p). Christiania: Cammermeye.

Booth RK, Jackson ST (2003) A high resolution record of late Holocene moisture variability from a Michigan raised bog. Holocene 13:865–878.

Booth RK (2008) Testate amoebae as proxies for mean annual waiter-table depth in *Sphagnum*-dominated peatlands of North America. J Quat Sci 23:43–57.

Borgmark A (2005) Holocene climate variability and periodicities in south-central Sweden, as interpreted from peat humification analysis. Holocene 15:387–395.

Bragazza L (2008) A climatic threshold triggers the die-off ofpeat mosses during an extreme heat wave. Glob Chang Biol 14:2688–2695.

Bronk Ramsey C (2007) Deposition models for chronological records. Quat Sci Rev 27.42–60.

Brown G (1980) Palsas and other permafrost features in a lower rock creek valley, west-central Alberta. Arctic Alpine Res 12:31–40.

Brown J, Romanovsky VE (2008) Report from the International Permafrost Association: State of permafrost in the first decade of the 21st century. Permafr Periglac Process 19: 255–260.

Bubier JL, Moore TR, Juggins S (1995) Predicting methane emissions from bryophyte distribution in northern Canadian peatlands. Ecology 76(3):677–693.

Bunce JA (2004) Carbon dioxide effects on stomatal responses to the environment and water use by crops under field conditions. Oecologia 140:1–10.

Buttler A, Warner BG, Grosvernier P, Matthey Y (1996) Vertical patterns of testate amoebae (Protozoa: Rhizopoda) and peat forming vegetation on cutover bogs in the Jura, Switzerland. New Phytol 134:371–382.

Bücher A, Lucas C (1984) Sédimentation éolienne intercontinentale, poussières sahariennes et géologie. Bulletin du Centre de recherche, exploration et production Elf-Aquitaine 8:151–165.

Camill P (2005) Permafrost thaw accelerates in boreal peatlands during late-20th century climate warming. Clim Change 68:135–152.

Caseldine C, Baker A, Charman D, Hendon D (2000) A comparative study of optical properties of NaOH peat extracts: Implications for humification studies. Holocene 10:649–658.

Caseldine CJ, Thompson G, Langdon C, Hendon D (2005) 'Evidence for an extreme climatic event on Achill Island, Co. Mayo Ireland around 5200–5100 cal yr BP'. J Quat Sci 20:169–178.

Chambers FM (1984) Studies on the initiation, growth-rate and humification of 'blanket peats' in South Wales, Occasional Paper No. 9. Keele University: Department of Geography.

Chambers FM (1988) Peat humification: Proxy climatic record or indicator of land-use history? In S Limbrey (Ed), Soils and Human Settlement (pp. 27–44). Bangor: Welsh Soils Discussion Group, Report 26.

Chambers FM, Barber KE, Maddy D, Brew J (1997) A 5500-year proxy-climate and vegetational record from blanket mire at Talla Moss, Peebleshire, Scotland. Holocene 7:391–399.

Chambers FM, Mauquoy D, Todd PA (1999) Recent rise to dominance of Molinia caerulea in environmentally sensitive areas: New perspectives from palaeoecological data. J Appl Ecol 36:719–733.

Chambers FM, Charman DJ (2004) Holocene environmental change: Contributions from the peatland archive. Holocene 14(1):1–6.

Chambers FM, Mauquoy D, Brain SA, Blaauw M, Daniell JRG (2007a) Globally synchronous climate change 2800 years ago: Proxy data from peat in South America. Earth Planet Sci Lett 253:439–444.

Chambers FM, Mauquoy D, Gent A, Pearson F, Daniell JRG, Jones PS (2007b) Palaeoecology of degraded blanket mire in South Wales: Data to inform conservation management. Biol Conserv 137:197–209.

Chapman S, Buttler A, Francez A-J, Laggoun-Défarge F, Vasander HMS, Combe J, Grosvernier P, Harms H, Epron D, Gilbert D, Mitchell EAD (2003) Commercial exploitation of peatlands and maintenance of biodiversity – A conflict between economy and ecology. Front Ecol Environ 1:525–532.

Charman DJ (1997) Modelling hydrological relationships of testate amoebae (Protozoa: Rhizopoda) on New Zealand peatlands. J R Soc N Z 27:465–483.

Charman DJ, Warner B (1997) The ecology of testate amoebae (Protozoa: Rhizopoda) in oceanic peatlands in Newfoundland, Canada: Modelling hydrological relationships for paleoenvironmental reconstruction. Ecoscience 4:555–562.

Charman DJ, Hendon D, Woodland WA (2000) The Identification of Testate Amoebae (Protozoa: Rhizopoda) in Peats. Technical Guide No. 9. London: Quaternary Research Association.

Charman DJ (2001) Biostratigraphic and palaeoenvironmental applications of testate amoebae. Quat Sci Rev 20:1753–1764.

Charman DJ (2002) Peatlands and Environmental Change (301p). Chichester: Wiley.

Charman DJ, Blundell A, Chiverrell RC, Hendon D, Langdon PG (2006) Compilation of non-annually resolved Holocene proxy climate records: Stacked Holocene peatland palaeo-water table reconstructions from northern Britain. Quat Sci Rev 25:336–350.

Charman DJ (2007) Summer water deficit variability controls on peatland water-table changes: Implications for Holocene palaeoclimate reconstructions. Holocene 17:217–227.

Charman DJ, Blundell A, ACCROTELM members (2007) A new European testate amoebae transfer function for palaeohydrological reconstruction on ombrotrophic peatlands. J Quat Sci 22:209–221.

Charman DJ, Barber KE, Blaauw M, Langdon PG, Mauquoy D, Daley T, Hughes PDM, Karofeld E (2009) Climate drivers for peatland palaeoclimate records. Quat Sci Rev 28:1811–1819.

Christen JA, Clymo D, Litton CD (1995) A Bayesian approach to the use of 14C dates in the estimation of the age of peat. Radiocarbon 37:431–442.

Christensen TR, Johansson T, Åkerman HJ, Mastepanov M, Malmer N, Friborg T, Crill P, Svensson BH (2004) Thawing sub-arctic permafrost: Effects on vegetation and methane emissions. Geophys Res Lett 31:L04501.

Christensen TR, Johansson T, Olsrud M, Ström L, Lindroth A, Mastepanov M, Malmer N, Friborg T, Crill PM, Callaghan TV (2009) A catchment scale process study of carbon and greenhous gas exchange in a subarctic landscape. In S Fronzek, M Johansson, TR Christensen, TR Carter, T Friborg, M Luoto (Eds), Proceedings of the PALSALARM Symposium, Abisko, Sweden, 28–30 October 2008 (vol. 3:pp. 41–43). Reports of Finnish Environment Institute.

Clark I, Fritz P (1997) Environmental Isotopes in Hydrogeology. Boca Raton, FL: CRC Press.

Clemmensen LB, Murray A, Heinemeier J, De Jong R (2009) The evolution of Holocene coastal dunefields, Jutland, Denmark: A record of climate change over the past 5000 years. Geomorphology 105:303–313.

Cleveland WS, Devlin SJ (1988) Locally weighted regression: An approach to regression analysis by local fitting. J Am Statist Assoc 83:596–610.

Clymo RS (1984) The limits to peat bog growth. Philos Trans R Soc Lond B 303:605–654.

Cole JJ, Prairie YT, Caraco NF (2007) Plumbing the global carbon cycle: Integrating inland waters into the terrestrial carbon budget. Ecosystems 10:171–184.

Davis SR, Wilkinson DM (2004) The conservation management value of testate amoebae as 'restoration' indicators: Speculations based on two damaged raised mires in northwest England. Holocene 14:135–143.

De Jong R, Björck S, Björkman L, Clemmensen LB (2006) Storminess variation during the last 6500 years as reconstructed from an ombrotrophic peat bog in Halland, southwest Sweden. J Quat Sci 21:905–919.

De Jong R, Hammarlund D, Nesje A (2009) Late Holocene effective precipitation variations in the maritime regions of south-west Scandinavia. Quat Sci Rev 28:54–64.

de Vleeschouwer F, Cheburkin A, Le Roux G, Piotrowska N, Sikorski J, Lamentowicz M, Fagel N, Mauquoy M (2009) Multiproxy evidences of Little Ice Age palaeoenvironmental changes in a peat bog from northern Poland. Holocene 19:625–637.

Deevey E (1969) Coaxing history to conduct experiments. BioScience 19:40–43.

Doutriaux-Boucher M, Webb MJ, Gregory JM, et al. (2009) Carbon dioxide induced stomatal closure increases radiative forcing via a rapid reduction in low cloud. Geophys Res Lett 36:L02703. doi: 10.1029/2008GL036273.

Eide W, Birks HH (2006) Stomatal frequency of *Betula pubescens* and *Pinus sylvestris* shows no proportional relationship with atmospheric CO_2 concentration. Nordic J Bot 24:327–339.

Finsinger W, Wagner-Cremer F (2009) Stomatal-based inference models for reconstruction of atmospheric CO_2 concentration: A method assessment using a calibration and validation approach. Holocene 19:757–764.

Fletcher M-S, Thomas I (2007) Holocene vegetation and climate change from near Lake Pedder, south-west Tasmania, Australia. J Biogeogr 34:665–677.

Forster P, Ramaswamy V, Artaxo P, et al. (2007) Changes in atmospheric constituents and in radiative forcing. In S Solomon, D Qin, M Manning, Z Chen, M Marquis, KB Averyt, M Tignor, HL Miller (Eds), Climate Change 2007: The Physical Science Basis. Contribution of Working Group I to the fourth assessment report of the Intergovernmental Panel on Climate Change. Cambridge, UK; New York, NY: Cambridge University Press.

Franks PJ, Beerling DJ (2009) Maximum leaf conductance driven by CO_2 effects on stomatal size and density over geologic time. Proc Natl Acad Sci USA 106: 10343–10347.

Frey KE, Smith LC (2005) Amplified carbon release from vast West Siberian peatlands by 2100. Geophys Res Lett 32:L09401.

Fronzek S, Luoto M, Carter TR (2006) Potential effect of climate change on the distribution of palsa mires in subarctic Fennoscandia. Clim Res 32:1–12.

Fronzek S, Carter TR, Räisänen J, Ruokolainen L, Luoto M (2010) Applying probabilistic projections of climate change with impact models: A case study for sub-arctic palsa mires in Fennoscandia. Clim Change 99:515–534.

Görres M, Frenzel B (1993) The Pb, Br, and Ti content in peat bogs as indicator for recent and past depositions. Naturwissenschaften 80:333–335.

Goudie AS, Middleton NJ (2006) Desert Dust in the Global System (287 pp). Heidelberg: Springer.

Gray JE, Holroyd GH, Van der Lee FM, et al. (2000) The HIC signalling pathway links CO_2 perception to stomatal development. Nature 408:713–716.

Hajdas I, Lowe DJ, Newnham RM, Bonani G (2006) Timing of the late-glacial climate reversal in the Southern Hemisphere using high-resolution radiocarbon chronology for Kaipo bog, New Zealand. Quat Res 65(2):340–345.

Hájek M, Horsák M, Hájkova P, Dítě D (2006) Habitat diversity of central European fens in relation to environmental gradients and an effort to standardise fen terminology in ecological studies. Perspect Plant Ecol Evol Syst 8:97–114.

Halsey LA, Vitt DA, Zoltai SC (1995) Disequilibrium response of permafrost in boreal continental western Canada to climate change. Clim Change 30:57–73.

Heal WO (1961) The distribution of testate amoebae (Rhizopoda: Testacea) in some fens and bogs in northern England. J Linnean Soc Zool 44:369–382.

Heal OW (1964) Observations on the seasonal and spatial-distribution of Testacea (Protozoa, Rhizopoda) in *Sphagnum*. J Anim Ecol 33:395–412.

Heger TJ, Mitchell EAD, Todorov M, Golemansky V, Lara E, Pawlowski J (2010) Molecular phylogeny of euglyphid testate amoebae (Cercozoa: Euglyphida) suggest transitions between marine supralittoral and freshwater/terrestrial environments are infrequent. Mol Phylogenet Evol 55:113–122.

Heijmans MMPD, Mauquoy D, van Geel B, Berendse F (2008) Long-term effects of climate change on vegetation and carbon dynamics in peat bogs. J Veg Sci 19:307–320.

Hendon D, Charman D, Kent M (2001) Palaeohydrological records derived from testate amoebae analysis from peatlands in northern England: Within-site variability, between-site comparability and palaeoclimatic implications. Holocene 11:127–148.

Hetherington AM, Woodward FI (2003) The role of stomata in sensing and driving environmental change. Nature 424:901–908.

Hofgaard A (2006) Overvåkning av palsmyr. Forstegangsundersökelse i Dovre 2005: Haukskardmyrin og Haugtjörnin. NINA Rapport 154.

Hölzer A, Hölzer A (1998) Silicon and titanium in peat profiles as indicators of human impacts. Holocene 8:685–696.

Hughes PDM, Mauquoy D, Barber KE, Langdon P (2000) Mire-development pathways and palaeoclimatic records from a full Holocene peat archive at Walton Moss, Cumbria, England. Holocene 10:465–479.

Hughes PDM, Barber KE (2004) Contrasting pathways to ombrotrophy in three raised bogs from Ireland and Cumbria, England. Holocene 14:65–77.

Hughes PDM, Blundell A, Charman DJ, Bartlett S, Daniell JRG, Wojatschke A, Chambers FM (2006) A 9000 cal year multi-proxy climate record from a bog in eastern Newfoundland: Contributions of meltwater discharge and solar forcing. Quat Sci Rev 25:1208–1227.

IPCC (2007) Climate change 2007: The physical science basis. In S Solomon, D Qin, M Manning, Z Chen, M Marquis, KB Averyt, M Tignor, HL Miller (Eds), Contribution of Working Group I to the Fourth Assessment Report of the Intergovernmental Panel on Climate Change (996 pp). Cambridge, UK; New York, NY : Cambridge University Press.

Jackson RB, Carpenter SR, Dahm CN, et al. (2001) Water in a changing world. Ecol Appl 11:1027–1045.

Jessen CA, Rundgren M, Björck S, et al. (2005) Abrupt climatic changes and an unstable transition into a late Holocene Thermal Decline: A multiproxy lacustrine record from southern Sweden. J Quat Sci 20:349–362.

Jessen CA, Rundgren M, Björck S, et al. (2007) Climate forced atmospheric CO_2 variability in the early Holocene: A stomatal frequency reconstruction. Glob Planet Change 57:247–260.

Johansson M, Callaghan TV, Åkerman J, Jackowicz-Korczynski M, Christensen TR (submitted) Rapid response of active layer thickness and vegetation greenness in sub-arctic Sweden to experimentally increased snow cover. Arct Antarct Alp Res.

Johansson T, Malmer N, Crill PM, Friborg T, Akerman JH, Mastepanov M, Christensen TR (2006) Decadal vegetation changes in a northern peatland, greenhouse gas fluxes and net radiative forcing. Glob Chang Biol 12:2352–2369.

Johansson M, Akerman HJ, Jonasson C, Christensen TR, Callaghan TV (2008) Increasing permafrost temperatures in subarctic Sweden. In DL Kane, KM Hinkel (Eds), Ninth International Conference on Permafrost (pp. 851–856). University of Alaska Fairbanks (I): Institute of Northern Engineering.

Johansson M, Fronzek S, Christensen TR, Luoto M, Carter TR (2009) Risk of disappearing sub-arctic palsa mires in Europe. In J Settele, L Penev, T Georgiev, R Grabaum, V Grobelnik, V Hammer, S Klotz, I Kühn (Eds), Atlas of Biodiversity Risks – From Europe to the Globe, from Stories to Maps. Sofia and Moscow: Pensoft (in press).

Johnson LC, Damman AWH (1991) Species controlled *Sphagnum* decay on a south Swedish raised bog. Oikos 61:234–242.

Jones MC, Peteet DM, Kurdyla D, Guilderson T (2009) Climate and vegetation history from a 14,000-year peatland record, Kenai Peninsula, Alaska. Quat Res 72:207–217.

Joos F, Spahni R (2008) Rates of change in natural and anthropogenic radiative forcing over the past 20,000 years. Proc Natl Acad Sci 105:1425–1430.

Joosten H, Clarke D (2002) Wise Use of Mires and Peatlands. Background and Principles Including a Framework for Decision-Making. Finland: International Mire Conservation Group and International Peat Society.

Kamenov GD, Brenner M, Tucker JL (2009) Anthropogenic versus natural control on trace element and Sr-Nd-Pb isotope stratigraphy in peat sediments of southeast Florida (USA), ~1500 AD to present. Geochim Cosmochim Acta 73(12):3549–3567. ISSN 0016-7037, DOI: 10.1016/j.gca.2009.03.017.

Keeling CD, Piper SC, Bacastow RB, et al. (2005) Atmospheric CO_2 and $^{13}CO_2$ exchange with the terrestrial biosphere and oceans from 1978 to 2000: Observations and carbon cycle implications. In JR Ehleringer, TE Cerling, MD Dearing (Eds), A History of Atmospheric CO_2 and Its Effects on Plants, Animals, and Ecosystems (pp. 83–113). New York, NY: Springer. http://scrippsco2.ucsd.edu/data/atmospheric_co2.html

Keeling RF, Piper SC, Bollenbacher AF, et al. (2008) Atmospheric CO_2 records from sites in the SIO air sampling network. In Trends: A Compendium of Data on Global Change, Carbon Dioxide Information Analysis Center, Oak Ridge National Laboratory, U.S. Department of Energy, Oak Ridge, Tenn.

Kilian MR, van der Plicht J, van Geel B (1995) Dating raised bogs: New aspects of AMS 14C wiggle matching, a reservoir effect and climatic change. Quat Sci Rev 14: 959–966.

Kilian MR, van Geel B, van der Plicht J (2000) ^{14}C AMS wiggle matching of raised bog deposits and models of peat accumulation. Quat Sci Rev 19:1011–1033.

Kirpotin SN, Naumov AV, Vorobiov SN, Mironycheva-Tokareva NP, Kosych NP, Lapshina ED, Marquand J, Kulizhski SP, Bleuten W (2007) Western-siberian peatlands: Indicators of climate change and their role in global carbon balance. In R Lal, M Suleimenov, BA Stewart, DO Hansen, P Doraiswamy (Eds), Chapter 33 in Climate Change and Terrestrial Carbon Sequestration in Central Asia (pp. 453–472). Amsterdam, Holland: Taylor and Francis.

Kling GW, Kipphut GW, Miller MC (1991) Arctic lakes and streams as gas-conduits to the atmosphere: Implications for tundra carbon budgets. Science 251:298–301.

Kohler J, Brandt O, Johansson M, Callaghan TV (2006) A long-term Arctic snow depth record from Abisko, northern Sweden, 1913–2004. Polar Res 25:91–113.

Kokfelt U, Reuss N, Struyf E, Sonesson M, Rundgren M, Skog G, Rosén P, Hammarlund D (2010) Wetland development, permafrost history and nutrient cycling inferred from Holocene peat and lake sediment records in subarctic Sweden. J Paleolimnol 44:327–342.

Kokfelt U, Rosén P, Schoning K, Christensen TR, Förster J, Karlsson J, Reuss N, Rundgren M, Callaghan TV, Jonasson C, Hammarlund D (2009) Ecosystem responses to increased precipitation and permafrost decay in sub-arctic Sweden inferred from peat and lake sediments. Glob Chang Biol 15(7):1652–1663.

Kouwenberg LLR, McElwain JC, Kürschner WM, et al. (2003) Stomatal frequency adjustment of four conifer species to historical changes in atmospheric CO_2. Am J Bot 90:610–619.

Kouwenberg L, Wagner R, Kürschner W, et al. (2005) Atmospheric CO_2 fluctuations during the last millennium reconstructed by stomatal frequency analysis of Tsuga heterophylla needles. Geology 33:33–36.

Kuhry P, Turunen J (2006) The postglacial development of boreal and subarctic peatlands. In RK Wieder, DH Vitt (Eds), Boreal Peatland Ecosystems (pp. 25–46). Berlin, Heidelberg: Springer.

Kuhry P (2008) Palsa and peat plateau development in the Hudson Bay Lowlands, Canada: Timing, pathways and causes. Boreas 37:316–327.

Kummerow J, Mills JM, Ellis BA, Kummerow A (1988) Growth dynamics of cotton-grass (*Eriophorum vaginatum*). Can J Bot Rev Can Bot 66:253–256.

Kylander ME, Weiss DJ, Cortizas AM (2005) Refining the pre-industrial atmospheric Pb isotope evolution curve in Europe using an 8000 year old peat core from NW Spain. Earth Planet Sci Lett 240:467–485. ISSN: 0012–821X.

Kylander ME, Muller J, Wust RAJ (2007) Rare earth element and Pb isotope variations in a 52 kyr peat core from Lynch's Crater (NE Queensland, Australia): Proxy development and application to paleoclimate in the Southern Hemisphere. Geochim Cosmochim Acta 71:942–960. ISSN: 0016–7037.

Körner C (2006) Plant CO_2 responses: An issue of definition, time and resource supply. New Phytol 172:393–411.

Kürschner WM (1997) The anatomical diversity of recent and fossil leaves of the durmast oak (Quercus petraea Lieblein/Q. pseudocastanea Goeppert) – Implications for their use as biosensors of palaeoatmospheric CO_2 levels. Rev Palaeobot Palynol 96:1–30.

Kürschner WM, Wagner F, Visscher EH, et al. (1997) Predicting the response of leaf stomatal frequency to a future CO_2-enriched atmosphere: Constraints from historical observations. Geologische Rundschau 86:512–517.

Kürschner WM, Kvacek Z, Dilcher DL (2008) The impact of Miocene atmospheric carbon dioxide fluctuations on climate and the evolution of terrestrial ecosystems. Proc Natl Acad Sci 105:449–453.

Lake JA, Quick WP, Beerling DJ, et al. (2001) Signals from mature to new leaves. Nature 411:154.

Lake JA, Woodward FI (2008) Response of stomatal numbers to CO_2 and humidity: Control by transpiration rate and abscisic acid. New Phytol 179:397–404.

Lamentowicz Ł, Lamentowicz M, Gąbka M (2008a) Testate amoebae ecology and a local transfer function from a peatland in western Poland. Wetlands 28:164–175.

Lamentowicz M, Obremska M, Mitchell EAD (2008b) Autogenic succession, land-use change, and climatic influences on the Holocene development of a kettle hole mire in Northern Poland. Rev Palaeobot Palynol 151: 21–40.

Lamentowicz M, Milecka K, Gałka M, Cedro A, Pawlyta J, Piotrowska N, Lamentowicz Ł, van der Knaap WO (2009a) Climate- and human-induced hydrological change since AD 800 in an ombrotrophic mire in Pomerania (N Poland) tracked by testate amoebae, macro-fossils, pollen, and tree-rings of pine. Boreas 38:214–229.

Lamentowicz M, Van der Knaap P, Lamentowicz Ł, Van Leeuwen JFN, Mitchell EAD, Goslar T, Kamenik C (2009b) A near-annual palaeohydrological study based on testate amoebae from an Alpine mire: Surface wetness and the role of climate during the instrumental period. J Quat Sci 24:190–202.

Lamentowicz M, Lamentowicz L, van der Knaap WO, Gąbka M, Mitchell EAD (2010) Contrasting species-environment relationships in communities of testate amoebae, bryophytes, and vascular plants along the fen-bog gradient. Microb Ecol 59:499–510. doi: 10.1007/s00248-009-9617-6.

Lara E, Heger TJ, Mitchell EAD, Meisterfeld R, Ekelund F (2007) SSU rRNA reveals a sequential increase in shell complexity among the Euglyphid testate amoebae (Rhizaria: Euglyphida). Protist 158:229–237.

Lara E, Heger TJ, Ekelund F, Lamentowicz M, Mitchell EAD (2008) Ribosomal RNA genes challenge the monophyly of the Hyalospheniidae (Amoebozoa: Arcellinida). Protist 159:165–176.

Le Roux G, Shotyk W (2006) Alteration of minerals in peat bogs. In P Martini, A Martinez-Cortizas (Eds), Peatlands: Evolution and Records of Environmental and Climatic Changes (vol. 9, 606p). Developments in Earth Surface Processes, Amsterdam: Elsevier.

Le Roux G, Laverret E, Shotyk W (2006) Fate of minerals in ombrotrophic peat bogs. J Geol Soc Lond 163:641–646.

Lemke P, Ren J, Alley RB, Allison I, Carrasco J, Flato G, Fujii Y, Kaser G, Mote P, Thomas RH, Zhang T (2007) Observations: Changes in snow, ice and frozen ground. In S Solomon, D Qin, M Manning, Z Chen, M Marquis, KB Averyt, M Tignor, HL Miller (Eds), Climate Change 2007: The Physical Science Basis (pp. 337–384). Contribution of Working group I to the Fourth Assessment Report of the Intergovernmental Panel on Climate Change. Cambridge, UK; New York, NY: Cambridge University Press.

Luoto M, Seppälä M (2003) Thermokarst ponds indicating former distribution of palsas in Finnish Lapland. Permafr Periglac Process 14:19–27.

Luoto M, Fronzek S, Zuidhoff FS (2004) Spatial modelling of palsa mires in relation to climate in Northern Europe. Earth Surf Process Landforms 29:1373–1387.

Lüthi D, Le Floch M, Bereiter B, et al. (2008) High-resolution carbon dioxide concentration record 650,000–800,000 years before present. Nature 453:379–382.

MacFarling Meure C, Etheridge D, Trudinger C, et al. (2006) Law Dome CO_2, CH_4 and N_2O ice core records extended to 2000 years BP. Geophys Res Lett 33:L14810. doi: 10.1029/2006GL026152.

Marx SK, McGowan HA, Kamber BS (2009) Long-range dust transport from eastern Australia: A proxy for Holocene aridity and ENSO-type climate variability. Earth Planet Sci Lett 282:167–177.

Mattson S, Koutler-Andersson E (1954) Geochemistry of a raised bog. Kungl Lantbrushögskolans Annaler 21: 321–366.

Mauquoy D, Engelkes T, Groot MHM, Markesteijn F, Oudejans MG, van der Plicht J, van Geel B (2002a) High-resolution records of late Holocene climate change and carbon accumulation in two north-west European ombrotrophic peat bogs. Palaeogeogr Palaeoclimatol Palaeoecol 186: 275–310.

Mauquoy D, van Geel B, Blaauw M, van der Plicht J (2002b) Evidence from North-west European bogs shows 'Little Ice Age' climatic changes driven by changes in solar activity. Holocene 12:1–6.

Mauquoy D, van Geel B, Blaauw M, Speranza A, van der Plicht J (2004) Changes in solar activity and Holocene climate shifts derived from 14C wiggle-match dated peat deposits. Holocene 14:45–52.

Mauquoy D, van Geel B (2007) Mire and peat macros. In SA Elias (Ed), Encyclopedia of Quaternary Science (vol. 3, pp. 2315–2336). Amsterdam: Elsevier.

Mauquoy D, Yeloff D, van Geel B, Charman D, Blundell A (2008) Two decadally resolved records from north-west European peat bogs show rapid climate changes associated with solar variability during the mid-late Holocene. J Quat Sci 23:745–763.

Mazei YA, Tsyganov AN, Bubnova OA (2009) The species composition and community structure of testate amoebae in sphagnum bogs of northern Karelia (The White Sea Lowland). Zool J 88:1–12.

McCarroll D, Loader NJ (2004) Stable isotopes in tree rings. Quat Sci Revi 23:771–801.

McElwain JC, Mayle FE, Beerling DJ (2002) Stomatal evidence for a decline in atmospheric CO_2 concentration during the Younger Dryas stadial: A comparison with Antarctic ice core records. J Quat Sci 17:21–29.

McLennan T (2001) Relationships between the trace element composition of sedimentary rocks and upper continental crust. Geochem Geophys Geosyst 2:2000GC000109.

McMullen JA, Barber KE, Johnson B (2004) A palaeoecological perspective of vegetation succession on raised bog microforms. Ecol Monogr 74:45–77.

Meehl GA, Stocker TF, Collins WD, et al. (2007) Global climate projections. In S Solomon, D Qin, M Manning, Z Chen, M Marquis, KB Averyt, M Tignor, HL Miller (Eds), Climate Change 2007: The Physical Science Basis. Contribution of Working Group I to the Fourth Assessment Report of the Intergovernmental Panel on Climate Change. Cambridge, UK; New York, NY: Cambridge University Press.

Meeker LD, Mayewski PA (2002) A 1400-year high-resolution record of atmospheric circulation over the North Atlantic and Asia. Holocene 12:257–266.

Meisterfeld R (2001a) Order Arcellinida, Kent 1880. In JJ Lee, GF Leedale, P Bradbury (Eds), The Illustrated Guide to the Protozoa (pp. 827–860). Lawrence: Allen Press.

Meisterfeld R (2001b) Testate amoebae with filopoda. In JJ Lee, GF Leedale, P Bradbury (Eds), The Illustrated Guide to the Protozoa (pp. 1054–1084). Lawrence: Allen Press.

Mitchell EAD, Buttler A, Warner BG, Gobat JM (1999) Ecology of testate amoebae (Protozoa: Rhizopoda) in *Sphagnum* peatlands in the Jura mountains, Switzerland and France. Ecoscience 6:565–576.

Mitchell EAD, Borcard D, Buttler AJ, Grosvernier P, Gilbert D, Gobat JM (2000) Horizontal distribution patterns of Testate Amoebae (Protozoa) in *Sphagnum magellanicum* Carpet. Microb Ecol 39:290–300.

Mitchell EAD (2004) Response of testate amoebae (Protozoa) to N and P fertilization in an Arctic wet sedge tundra. Arct Antarct Alp Res 36:77–82.

Mitchell E, Charman D, Warner B (2008a) Testate amoebae analysis in ecological and paleoecological studies of wetlands: Past, present and future. Biodivers Conserv 17:2115–2137.

Mitchell EAD, Payne RJ, Lamentowicz M (2008b) Potential implications of differential preservation of testate amoebae shells for paleoenvironmental reconstruction in peatland. J Paleolimnol 40:603–618.

Monnin E, Indermuhle A, Dallenbach A, et al. (2001) Atmospheric CO_2 concentrations over the Last Glacial termination 10.1126/science.291.5501.112. Science 291:112–114.

Moore PD (2002) The future of cool temperate bogs. Environ Conserv 29:3–20.

Morgan TJ, Herod AA, Brain SA, Chambers FM, Kandiyoti R (2005) Examination of soil contaminated by coal-liquids by size-exclusion chromatography in 1-methyl-2-pyrrolidinone solution to evaluate interference from humic and fulvic acids and extracta from peat. J Chromatogr 1095:81–88.

Muller J, Kylander M, Martinez-Cortizas A, Wuest RA:J, Weiss D, Blake K, Coles B, Garcia-Sanches R (2008) The use of principle component analyses in characterising trace and major elemental distribution in a 55 kyr peat deposit in tropical Australia: Implications to paleoclimate. Geochim Cosmochim Acta 72:449–463.

Muller S, Bobrov AA, Schirrmeister L, Andreev AA, Tarasov PE (2009) Testate amoebae record from the Laptev Sea coast and its implication for the reconstruction of Late Pleistocene and Holocene environments in the Arctic Siberia. Palaeogeogr Palaeoclimatol Palaeoecol 271:301–315.

Myers-Smith IH, Harden JW, Wilmking M, Fuller CC, McGuire AD, Chapin III FS (2008) Wetland succession in a permafrost collapse: Interactions between fire and thermokarst. Biogeosciences 5:1273–1286.

Neff JC, Ballantyne AP, Famer GL, Mahowald NM, Conroy JL, Landry CC, Overpeck JT, Painter TH, Lawrence CR,

Reynolds. RL (2008) Recent increase in eolian dust deposition related to human activity in the Western United States. Nat Geosci 1:189–195.

Neftel A, Oeschger H, Staffelbach T, et al. (1988) CO_2 record in the Byrd ice core 50,000–5000 years BP. Nature 331:609–611.

Nichols JE, Booth RK, Jackson ST, Pendall EG, Huang Y (2006) Paleohydrologic reconstruction based on n-alkane distributions in ombrotrophic peat. Org Geochem 37: 1505–1513.

Nichols JE, Huang Y (2007) C23–C31 n-alkan-2-ones are biomarkers for the genus *Sphagnum* in freshwater peatlands. Org Geochem 38:1972–1976.

Nichols JE, Walcott M, Bradley R, Pilcher J, Huang Y (2009) Quantitative assessment of precipitation seasonality and summer surface wetness using ombrotrophic sediments from an Arctic Norwegian peatland. Quat Res doi: 10.1016/j.yqres.2009.07.007.

Nilssen D, Vorren KD (1991) Peat humification and climate history. Norsk Geologisk Tiddskrift 71:215–217.

Nott CJ, Xie S, Avsejs LA, Maddy D, Chambers FM, Evershed RP (2000) n-Alkane distribution in ombrotrophic mires as indicators of vegetation change related to climatic variation. Org Geochem 31:231–235.

Oksanen PO (2002) Holocene permafrost dynamics in palsa and peat plateau mires of continental Europe. Licentiate thesis, University of Oulu, Finland.

Oksanen PO (2005) Development of palsa mires on the northern European continent in relation to Holocene climatic and environmental changes. Doctoral thesis, University of Oulu, Finland

Oldfield F, Thompson R, Crooks PRJ, Gedye SJ, Hall VA, Harkness DD, Housley RA, McCormac FG, Newton AJ, Pilcher JR, Renberg I, Richardson N (1997) Radiocarbon dating of a recent high-latitude peat profile: Stor Åmyrân, northern Sweden. Holocene 7:283–290.

Pagani M, Arthur MA, Freeman KH (1999) Miocene evolution of atmospheric carbon dioxide. Paleoceanography 14:273–292.

Painter TH, Barrett AP, Landry CC, Neff JC, Cassidy MP, Lawrence CR, Thatcher KP, Farmer. L (2007) Impact of disturbed desert soils on duration of mountain snowcover. Geophys Res Lett 34(12):L12502. 10.1029/2007GL030208.

Pancost R, Baas M, van Geel B, Sinninghe Damste JS (2002) Biomarkers as proxies for plant inputs to peats: An example from a sub-boreal ombrotrophic bog. Org Geochem 33:675–690.

Parviainen M, Luoto M (2007) Climate envelopes of mire complex types in Fennoscandia. Geogr Ann Ser A Phys Geogr 89A:137–151.

Payette S, Delwaide A, Caccianiga M, Beauchemin M (2004) Accelerated thawing of sub-arctic peatland permafrost over the last 50 years. Geophys Res Lett 31:L18208.

Payne RJ, Kishaba K, Blackford JJ, Mitchell EAD (2006) Ecology of testate amoebae (Protista) in south-central Alaska peatlands: Building transfer-function models for palaeoenvironmental studies. Holocene 16:403–414.

Payne R, Mitchell EAD (2007) Ecology of testate amoebae from Mires in the Central Rhodope Mountains, Greece and development of a transfer function for palaeohydrological reconstruction. Protist 158:159–171.

Payne RJ, Charman DJ, Matthews S, Eastwood WJ (2008) Testate amoebae as palaeohydrological proxies in Surmene Agacbasi Yaylasi peatland (Northeast Turkey). Wetlands 28:311–323.

Payne RJ, Mitchell EAD (2009) How many is enough? Determining optimal count totals for ecological and palaeoecological studies of testate amoebae. J Paleolimnol 42: 483–495.

Pellerin S, Lavoie C (2003) Reconstructing the recent dynamics of mires using a multitechnique approach. J Ecol 91:1008–1021.

Petit JR, Jouzel J, Raynaud D, et al. (1999) Climate and atmospheric history of the past 420,000 years from the Vostok ice core, Antarctica. Nature 399:429–436.

Pilcher JR, Hall VA, McCormac FG (1995) Dates of Holocene Icelandic volcanic eruptions from tephra layers in Irish peats. Holocene 5:103–110.

Pissart A (2003) The remnants of Younger Dryas lithalsas in the Hautes Fagnes plateau in Belgium and elsewhere in the world. Geomorphology 52:5–38.

Plunkett G, Swindles GT (2008) Determining the Sun's influence on Lateglacial and Holocene climates: A focus on climate response to centennial-scale solar forcing at 2800 cal BP. Quat Sci Rev 27:175–184.

Poole I, Weyers JDB, Lawson T, et al. (1996) Variations in stomatal density and index: Implications for palaeoclimatic reconstructions. Plant Cell Environ 19:705–712.

Poole I, Lawson T, Weyers JDB, et al. (2000) Effect of elevated CO_2 on the stomatal distribution and leaf physiology of *Alnus glutinosa*. New Phytol 145:511–521.

Prentice IC, Harrison SP (2009) Ecosystem effects of CO_2 concentration: Evidence from past climates. Clim Past 5:297–307.

Raghoebarsing A, Smolders A, Schmid M (2005) Methanotrophic symbionts provide carbon for photosynthesis in peat bogs. Nature 436:1153–1156.

Räisänen J, Ruokolainen L (2006) Probabilistic forecasts of near-term climate change based on a resampling ensemble technique. Tellus 58A:461–472.

Rebetez M, Reinhard M (2007) Monthly air temperature trends in Switzerland 1901–2000 and 1975–2004. Theor Appl Climatol 91:27–34.

Reimer PJ, Baillie MGL, Bard E, Bayliss A, Beck JW, Bertrand CJH, Blackwell PG, Buck CE, Burr GS, Cutler KB, Damon PE, Edwards RL, Fairbanks RG, Friedrich M, Guilderson TP, Hogg AG, Hughen KA, Kromer B, McCormac G, Manning S, Bronk Ramsey C, Reimer RW, Remmele S, Southon JR, Stuiver M, Talamo S, Taylor FW, van der Plicht J, Weyhenmeyer CE (2004) IntCal04 terrestrial radiocarbon age calibration, 0–26 cal kyr BP. Radiocarbon 46:1029–1058.

Roulet NT, Lafleur PM, Richard PJH, Moore TR, Humphreys ER, Bubier J (2007) Contemporary carbon balance and late Holocene carbon accumulation in a northern peatland. Glob Chang Biol 13:397–411.

Royer DL (2001) Stomatal density and stomatal index as indicators of paleoatmospheric CO_2 concentration. Rev Palaeobot Palynol 114:1–28.

Rundgren M, Beerling D (1999) A Holocene CO_2 record from the stomatal index of subfossil Salix herbacea L. leaves from northern Sweden. Holocene 9:509–513.

Rundgren M, Björck S (2003) Late-glacial and early Holocene variations in atmospheric CO_2 concentration indicated by high-resolution stomatal index data. Earth Planet Sci Lett 213:191–204.

Saez AG, Probert I, Geisen M, Quinn P, Young JR, Medlin LK (2003) Pseudo-cryptic speciation in coccolithophores. PNAS 100:7163–7168.

Samaritani E, Siegenthaler A, Yli-Petäys M, Buttler A, Christin P-A, Mitchell EAD (2010) Seasonal net ecosystem carbon exchange of a regenerating cutover bog in the Swiss Jura Mountains. Restoration Ecol doi: 10.1111/j.1526-100X.2010.00662.X.

Sannel ABK, Kuhry P (2009) Monitoring permafrost and thermokarst dynamics in a subarctic peat plateau complex in northern Sweden. In S Fronzek, et al. (Eds), Proceeding of the PALSALARM Symposium, Abisko, Sweden, 28–30 October 2008 (pp. 59–60).

Sapkota A (2006) Mineralogical, Chemical, and Isotopic (Sr, Pb) Composition of Atmospheric Mineral Dusts in an Ombrotrophic Peat Bog, Southern South America (162p). Doctoral Thesis, University of Heidelberg.

Saurer M, Siegwolf RTW, Schweingruber FH (2004) Carbon isotope discrimination indicates improving water-use efficiency of trees in northern Eurasia over the last 100 years. Glob Chang Biol 10:2109–2120.

Schröter D, Wolters V, De Ruiter PC (2003) C and N mineralisation in the decomposer food webs of a European forest transect. Oikos 102:294–308.

Scott EM (2007) Radiocarbon dating: Sources of error. In SA Elias (Ed), Encyclopedia of Quaternary Science (pp. 2918–2923). Oxford: Elsevier.

Sellers PJ, Bounoua L, Collatz GJ, et al. (1996) Comparison of radiative and physiological effects of doubled atmospheric CO_2 on climate. Science 271:1402–1406.

Seppä H, Hammarlund D, Antonsson K (2005) Low-frequency and highfrequency changes in temperature and effective humidity during the Holocene in south-central Sweden: Implications for atmospheric and oceanic forcings of climate. Clim Dyn 25:285–297.

Seppälä M (1986) The origin of palsas. Geogr Ann Ser A Phys Geogr 68:141–147.

Seppälä M (2003) Surface abrasion of palsas by wind action in Finnish Lapland. Geomorphology 52:141–148.

Seppälä M (2005) Dating of palsas. Geol Surv Finl Spec Paper 40:79–84.

Sernander R (1908) On the evidence of postglacial changes of climate furnished by the peat-mosses of northern Europe. Geol Fören I Stockh Förh 30:467–478.

Shotyk W, Weiss D, Appleby PG, Cheburkin AK, Frei R, Gloor M, Kramers JD, Reese S, van Der Knaap WO (1998) History of atmospheric lead deposition since 12,370 ^{14}C yr BP from a peat bog, Jura mountains, Switzerland. Science 281:1635–1640.

Shur YL, Jorgensen MT (2007) Patterns of permafrost formation and degradation in relation to climate and ecosystems. Permafr Periglac Process 18:7–19.

Siegenthaler U, Monnin E, Kawamura K, et al. (2005a) Supporting evidence from the EPICA Dronning Maud Land ice core for atmospheric CO_2 changes during the past millennium. Tellus B 57:51–57.

Siegenthaler U, Stocker TF, Monnin E, et al. (2005b) Stable carbon cycle-climate relationship during the late Pleistocene. Science 310:1313–1317.

Sjögren P (2009) Sand mass accumulation rate as a proxy for wind regimes in the SW Barents Sea during the past 3 ka. Holocene 19(4):591–598.

Smith AG, Pilcher JR (1973) Radiocarbon dates and vegetational history of the British Isles. New Phytol 72:903–914.

Sollid JL, Sörbel L (1998) Palsa bogs as a climate indicator – Examples deom Dovrefjell, Southern Norway. Ambio 27(4):287–291.

Speranza A, van der Plicht J, van Geel B (2000) Improving the time control of the Subboreal/Subatlantic transition in a Czech peat sequence by ^{14}C wiggle-matching. Quat Sci Rev 19:1589–1604.

Speranza A, van Geel B, van der Plicht J (2002) Evidence for solar forcing of climate change at ca. 850 cal BC from a Czech peat sequence. Glob Planet Change 35:51–65.

Ström L, Christensen TR (2007) Below ground carbon turnover and greenhouse gas exchanges in a sub-arctic wetland. Soil Biol Biochem 39:1689–1698.

Swindles GT, Blundell A, Roe HM, Hall VA (2010) A 4500-year proxy climate record from peatlands in the North of Ireland: The identification of widespread summer 'drought phases'? Quat Sci Rev 29:1577–1589.

Swindles GT, Plunkett G, Roe HM (2007) A delayed climatic response to solar forcing at 2800 cal BP: Multiproxy evidence from three Irish peatlands. Holocene 17:177–182.

Sælthun NR, Barkved L (2003) Climate Change Scenarios for the SCANNET Region (74 pp). NIVA Report SNO 4663-2003.

Telford RJ, Heegaard E, Birks HJB (2004) All age-depth models are wrong: But how badly? Quat Sci Rev 23:1–5.

Thie J (1974) Properties of a wooded palsa in Northern Manitoba. Arct Alp Res 3(2):115–129.

Todorov M, Golemanski V, Mitchell EAD, Heger TJ (2009) Morphology, biometry, and taxonomy of freshwater and marine interstitial cyphoderia (Cercozoa: Euglyphida). J Eukaryot Microbiol 56:279–289.

Tolonen K, Warner BG, Vasander H (1994a) Ecology of Testaceans (Protozoa, Rhizopoda) in mires in Southern Finland: 2. Multivariate-analysis. Arch Protistenkunde 144:97–112.

Tolonen K, Warner BG, Vasander H (1994b) Ecology of Testaceans (Protozoa: Rhizopoda) in mires in Southern Finland: II. Multivariate analysis. Arch Protistenkunde 144:97–112.

Tricker P, Trewin H, Kull O, et al. (2005) Stomatal conductance and not stomatal density determines the long-term reduction in leaf transpiration of poplar in elevated CO_2. Oecologia 143:652–660.

Troels-Smith J (1955) Karakterisering af løse jordater. Characterisation of unconsolidated sediments. Danm Geol Unders Ser IV 3(10):73.

Tuittila E-S, Komulainen VM, Vasander H, Laine J (1999) Restored cut-away peatland as a sink for atmospheric CO_2. Oecologia 120:563–574.

Turetsky MR, Kelman Wieder R, Vitt DH (2002) Boreal peatland fluxes under varying permafrost regimes. Soil Biol Biochem 34:907–912.

Turetsky MR, Manning S, Wieder. RK (2004) Dating recent peat deposits. Wetlands 24:324–356.

Turetsky MR, Wieder RK, Vitt DH, Evans RJ, Scott KD (2007) The disappearance of relict permafrost in boreal north America: Effects on peatland carbon storage and fluxes. Glob Chang Biol 13:1922–1934.

Väliranta M, Korhola A, Seppä H, Tuittila E-S, Sarmaja-Korjonen K, Laine J, Alm J (2007) High-resolution reconstruction of wetness dynamics in a southern boreal raised bog, Finland, during the late Holocene: A quantitative approach. Holocene 17:1093–1107.

Vallée S, Payette S (2007) Collapse of permafrost mounds along a subarctic river over the last 100 years (northern Québec). Geomorphology 90:162–170.

Van der Burgh J, Visscher H, Dilcher DL, et al. (1993) Paleoatmospheric signatures in Neogene fossil leaves. Science 260:1788–1790.

Van der Linden M, Vickery E, Charman DJ, van Geel B (2008) Effects of human impact and climate change during the last 350 years recorded in a Swedish raised bog deposit. Palaeogeogr Palaeoclimatol Palaeoecol 262:1–31.

Van der Putten N, Stieperaere H, Vervruggen C, Ochyra R (2004) Holocene palaeoecology and climate history of South Georgia (sub-Antarctica) based on a macrofossil record of bryophytes and seeds. Holocene 14:382–392.

Van Geel B, Mook WG (1989) High-resolution ^{14}C dating of organic deposits using natural atmospheric ^{14}C variations. Radiocarbon 31:151–155.

Van Geel B, Buurman J, Waterbolk HT (1996) Archaeological and palaeoecological indications of an abrupt climate change in The Netherlands, and evidence for climatic teleconnetctions around 2650 BP. J Quat Sci 11:451–460.

Van Geel B, Heusser CJ, Renssen H, Schuurmans CJE (2000) Climatic change in Chile at around 2700 BP and global evidence for solar forcing: A hypothesis. Holocene 10: 659–664.

Van Hoof TB, Kaspers KA, Wagner F, et al. (2005) Atmospheric CO_2 during the 13th century AD: Reconciliation of data from ice core measurements and stomatal frequency analysis. Tellus Ser B Chem Phys Meteorol 57:351–355.

Vitt DH, Halsey LA, Zoltai SC (1994) The bog landforms of continental western Canada in relation to climate and permafrost patterns. Arct Alp Res 26:1–13.

Vitt DH, Bayley SE, Jin T-L (1995) Seasonal variation in water chemistry over a bog-rich fen gradient in continental western Canada. Can J Fish Aquat Sci 52:587–606.

Von Post L, Sernander R (1910) Pflanzenphysiognomische Studien aus Torfmooren in Närke (48p). XI International Geological Congress, Excursion Guide No. 14, Stockholm.

Von Post L, Granlund E (1926) Södra Sveriges torvtillganger. I. Sveriges. Geol Undersökn C335(19):1–128.

Von Post L (1946) The prospect for pollen analysis in the study of the Earth's climatic history. New Phytol 45:193–217.

Waddington JM, Warner KD (2001) Atmospheric CO_2 sequestration in restored mined peatlands. Ecoscience 8:359–368.

Wagner F, Below R, De Klerk P, et al. (1996) A natural experiment on plant acclimation: Lifetime stomatal frequency response of an individual tree to annual atmospheric CO_2 increase. Proc Natl Acad Sci USA 93:11705–11708.

Wagner F, Bohncke SJP, Dilcher DL, et al. (1999) Century-scale shifts in early Holocene atmospheric CO_2 concentration. Science 284:1971–1973.

Wagner F, Kouwenberg LLR, Van Hoof TB, et al. (2004) Reproducibility of Holocene atmospheric CO_2 records based on stomatal frequency. Quat Sci Rev 23:1947–1954.

Wagner F, Dilcher DL, Visscher H (2005) Stomatal frequency responses in a hardwood-swamp vegetation from Florida during a 60 years continuous CO_2 increase. Am J Bot 92:690–695.

Wall AAJ, Gilbert D, Magny M, Mitchell EAD (2010) Testate amoeba analysis of lake sediments: Impact of filter size and total count on estimates of density, diversity and community structure. J Paleolimnol 43:689–704.

Walsh JE, Anisimov OA, Hagen JOM, Jakobsson T, Oerlemans J, Prowse TD, Romanovsky V, Savelieva N, Serreze M, Shiklomanov A, Shikomanov I, Solomon I, Arendt A, Atkinson D, Dermuth MN, Dowdeswell J, Dyurgerov M, Glazovsky A, Koerner RM, Meier M, Reeh N, Sigurdsson O, Steffen K, Truffer M (2005) Cryosphere and hydrology. In Arctic Climate Impact Assessment (pp. 183–242). Cambridge: Cambridge University Press.

Warner BG (1987) Abundance and diversity of testate amoebae (Rhizopoda: Testacea) in *Sphagnum* peatlands in southwestern Ontario, Canada. Arch Protistenkd 133:173–189.

Warner B, Asada T, Quinn N (2007) Seasonal influences on the ecology of testate amoebae (Protozoa) in a small/*Sphagnum*/peatland in Southern Ontario, Canada. Microb Ecol 54:91–100.

Waterhouse JS, Switsur VR, Barker AC, et al. (2004) Northern European trees show a progressively diminishing response to increasing atmospheric carbon dioxide concentrations. Quat Sci Rev 23:803–810.

Weiss D, Shotyk W, Rieley J, Page S, Gloor M, Reese S, Martinez-Cortizas A (2002) The geochemistry of major and selected trace elements in a forested peat bog, Kalimantan, SE Asia, and its implications for past atmospheric dust deposition. Geochim Cosmochim Acta 66:2307–2323. ISSN: 0016–7037.

Wilkinson DM (2008) Testate amoebae and nutrient cycling: Peering into the black box of soil ecology. Trends Ecol Evol 23:596–599.

Williams T, Flannagan L (1996) Effect of changes in water content on photosynthesis, transpiration and discrimination agains $^{13}CO_2$ and $C^{18}O^{16}O$ in *Pleurozium* and *Sphagnum*. Oecologia 108:38–46.

Woodland WA, Charman DJ, Sims PC (1998) Quantitative estimates of water tables and soil moisture in Holocene peatlands from testate amoebae. Holocene 8:261–273.

Woodward FI (1987) Stomatal numbers are sensitive to increases in CO_2 from pre-industrial levels. Nature 327:617–618.

Woodward FI, Thompson GB, McKee IF (1991) How plants respond to climate change: Migration rates, individualism and the consequences for plant communities. Ann Bot 67:23–38.

Woodward FI (2002) Potential impacts of global elevated CO_2 concentrations on plants. Curr Opin Plant Biol 5:207–211.

Wright N, Hayashi M, Quinton WL (2009) Spatial and temporal variations in active layer thawing and their implications on runoff generation in peat-covered permafrost terrain. Water Resour Res 45:W05414. doi: 10.1029/2008WR006880.

Wu H, Guiot J, Brewer S, et al. (2007) Dominant factors controlling glacial and interglacial variations in the treeline elevation in tropical Africa. Proc Natl Acad Sci 104: 9720–9724.

Xie S, Nott C, Avsejs L, Maddy D, Chambers F (2004) Molecular and isotopic stratigraphy in an ombrotrophic mire for paleoclimate reconstruction. Geochim Cosmochim Acta 68:2849–2862.

Yeloff D, Mauquoy D (2006) The influence of vegetation composition on peat humification: Implications for palaeoclimatic studies. Boreas 35:662–673.

Yeloff D, Bennett KD, Blaauw M, Mauquoy D, Sillasoo Ü, van der Plicht J, van Geel B (2006) High precision C-14 dating of Holocene peat deposits: A comparison of Bayesian calibration and wiggle-matching approaches. Quat Geochronol 1:222–235.

Yeloff D, Charman D, van Geel B, Mauquoy D (2007) Reconstruction of hydrology, vegetation and past climate change in bogs using fungal microfossils. Rev Palaeobot Palynol 146:102–145.

Yi S, Woo M-K, Arain MA (2007) Impacts of peat and vegetation on permafrost degradation under climate warming. Geophys Res Lett 34:L16504.

Young JJ, Mehta S, Israelsson M, et al. (2006) CO_2 signaling in guard cells: Calcium sensitivity response modulation, a Ca^{2+}-independent phase, and CO_2 insensitivity of the gca2 mutant. Proc Natl Acad Sci USA 103:7506–7511.

Yu Z, Campbell ID, Vitt DH, Apps MJ (2001) Modelling long-term peatland dynamics. I. Concepts, review, and proposed design. Ecol Model 145:197–210.

Zoltai SC, Tarnocai C (1971) Properties of a wooded palsa in Northern Manitoba. Arct Alp Res 3(2):115–129.

Zoltai SC (1993) Cyclic development of permafrost in the peatlands of northwestern Alberta, Canada. Arct Alp Res 25(3):240–246.

Zona D, Oechel WC, Kochendorfer J, Paw UKT, Salyuk AN, Olivas PC, Oberbauer SF, Lipson DA (2009) Methane fluxes during the initiation of a large-scale water table manipulation experiment in the Alaskan Arctic tundra. Glob Biogeochem Cycles 23. doi: 10.1029/2009GB003487.

Zuidhoff F, Kolstrup E (2000) Changes in palsa distribution in relation to climate change in Laivadalen, northern Sweden, especially 1960–1997. Permafr Periglac Process 11:55–59.

Zweifel R, Rigling A, Dobbertin M (2009) Species-specific stomatal response of trees to drought – A link to vegetation dynamics? J Veg Sci 20:442–454.

Climate and Lacustrine Ecosystems

Isabelle Larocque-Tobler, Isabelle Laurion, Robert Moschen, and Monique Stewart

Abstract Terrestrial aquatic environments cover about 2% of the Earth's surface (Wetzel, 1983, Limnology (767p). Philadelphia, PA: Saunders College Publishing House). However, this small percentage of surface coverage act as important sources of drinking water and sustains biodiversity. The UN estimates that by 2050 there will be an additional 3.5 billion people with most of the population growth in developing countries which already suffer water stress. Climate change could have significant impacts on water resources around the world because of the close connections between the climate and hydrologic cycle. Studying the possible impact of climate on these water resources is thus of primary importance, and on-going research is being made at multiple levels: neo-ecology (studying the actual functioning of these ecosystems), palaeoecology (studying how these ecosystems have functioned in the past) and modelling (predicting their future functioning under climate and anthropogenic influences). Although this chapter is mainly concerned with the past functioning of these ecosystems, a synergy between these three research levels is necessary to better comprehend and estimate the impact of climate change on water resources. An effort is thus made to include some aspects of neo-ecology and modelling of past, present and future terrestrial aquatic environments (Fig. 1). Lakes can act in two ways to help better understand climate change and its effect on terrestrial aquatic ecosystems: (1) they contain multiple indicators which can be used to quantitatively reconstruct climate (e.g. temperature, precipitation, atmospheric circulation). Although the global temperature seems to have increased in the last century, the regional and even local climate can vary, and might be more important for the management of water resources. Only by looking at the climate variation through time at regional and local scales can we better understand and predict the future climate and its impact on water availability and biodiversity. (2) when linked with independent records of climate change, multi-proxy analysis in lakes can provide insight on the possible impacts of climate on water resources. For example, by linking the historical reconstruction of total productivity of various biological assemblages preserved in lake sediments and the meteorological data of the last 200 years, Korhola et al. (2002) have shown that these changes were due to climate warming. This chapter will look at how climate can be reconstructed using proxies preserved in lakes and second how climate has impacted these ecosystems in the past and how it might impact terrestrial aquatic ecosystems in the future.

Keywords Quantitative climate reconstructions · Biodiversity · Eutrophication · Late Glacial to present

I. Larocque-Tobler (✉)
Institute of Geography, University of Bern, 3012 Bern, Switzerland; Institut National de Recherche Scientifique, Centre Eau, Terre et Environnement, Québec, Québec, Canada G1K 9A9
e-mail: larocque@giub.unibe.ch; isabelle_larocque@ete.inrs.ca

Introduction

Lakes are being used around the world as water, power and recreational (e.g. water sports, fishing) resources. Although it is expected that climate might

J. Dodson (ed.), *Changing Climates, Earth Systems and Society*, International Year of Planet Earth,
DOI 10.1007/978-90-481-8716-4_6, © Springer Science+Business Media B.V. 2010

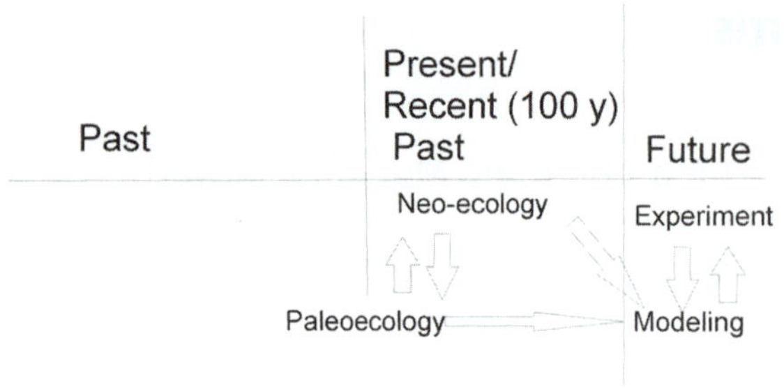

Fig. 1 Desired synergy between different research levels. The arrows indicate the direction in which conceptual development from one level can help in setting research goals at a different level

affect these resources in various ways (e.g. increasing eutrophication, decreasing lake levels), the lack of long-term and spatial coverage of climatic and lacustrine ecosystem data restrain the predictions of the effect of climate changes on lakes. However, the lakes themselves can be used to provide the data needed to better predict these effects. Lake sediments have been described as "paleoarchives" as they contain multi proxies, either biotic (e.g. chironomids, diatoms, chrysophytes, pollen) or abiotic (e.g. quartz/plagioclaste ratios, grain size, percentage of organic matter, mineralogy), which can be used to qualitatively and quantitatively reconstruct past environmental and climatic parameters (Fig. 2).

For climate reconstruction, the most appropriate study lakes are either closed basins where external loadings are rectricted (Fig. 2) or, for example, a lake connected to a glacier where the external loadings are climatically driven (and thus increase runoff to the lake when the glacier melts). It should also be located at ecotonal boundaries where climate has the most impact on biological and sedimentological proxies. In such lakes, the impact of human activities is limited thus it might be easier to decipher the climate signal from anthropogenic factors. However, if palaeoclimatology in lakes was restricted to these two types of lakes, the geographical coverage of studied lakes would be extremely restricted.

When the goal of the study is also to determine the impact of climate on water resources, lakes used for drinking, feeding or recreational activities are more interesting. In these lakes, the biological and sedimentological parameters can be used to study the impact of climate, however ways of disentangling the climate from the anthropogenic signals should be found. Various textbooks cover methods to reconstruct climate and environmental changes using multi proxies in lake sediments (e.g. Bradley, 1999; Cohen, 2003; Smol et al., 2001).

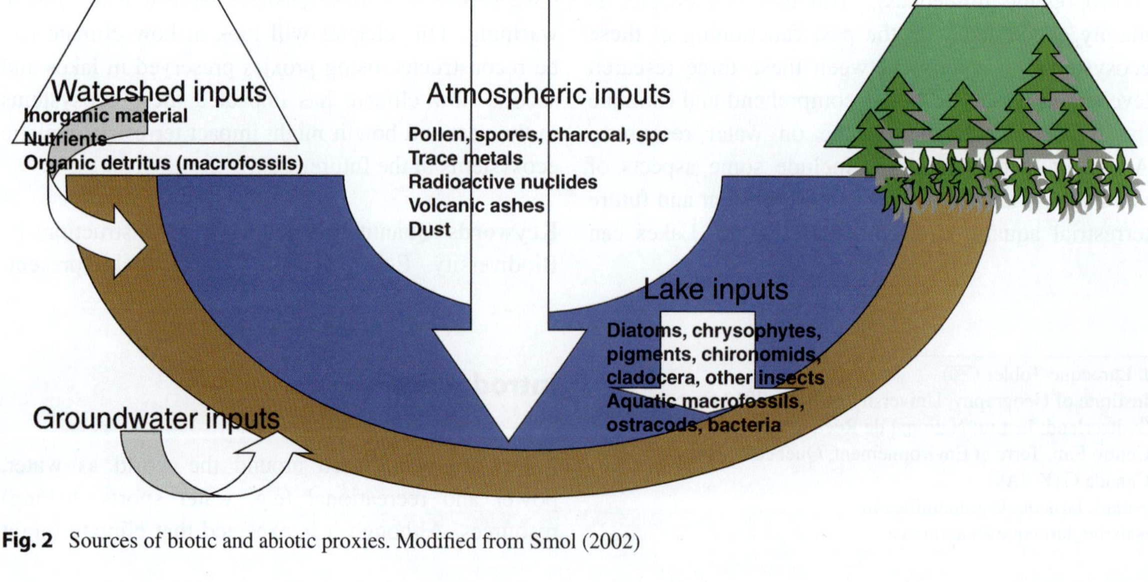

Fig. 2 Sources of biotic and abiotic proxies. Modified from Smol (2002)

Future climate change will possibly impact the trophic status of lakes (Whitehead et al., 2009), induce changes in biodiversity (Folke et al., 2004), increase the primary production of Cyanobacteria (Paerl and Huisman, 2009a), affect oxygen concentration in lakes (Whitehead et al., 2009), alter fish production (Massol et al., 2007) and might change lake levels (Stow, 2008). Some of these topics will be discussed below.

Quantitative Temperature Reconstructions

This section will consider the use of biotic and abiotic indicators to quantitatively reconstruct temperature during the Late Glacial, the Holocene, the last millennium and recent decades. In the early 1990s, powerful tools (e.g. transfer functions) were developed to reconstruct temperature using proxies and some validation of this technique has recently emerged (e.g. Bigler and Hall, 2003; Blass et al., 2007; Larocque and Hall, 2003; Larocque et al., 2009; Trachsel et al., 2008). Temperature is rarely the sole factor explaining the variations of any of the proxies used for reconstruction and we need to keep in mind that most proxies respond to variations in different parameters that might act synchronously with climate change. Finding ways to disentangle these various parameters is still a major challenge for palaeoclimatology. However, when multi-proxy reconstructions are compared at local, regional and hemispheric scale, a better insight of climate change can be obtained.

Specific guides have also recently been published on image analysis, sediments and palaeoenvironments (Francus, 2004) and chironomid taxonomy (Brooks et al., 2007; Larocque and Rolland, 2006). Others are available online (for pollen: www.ncdc.noaa.gov/paleo/gpd.html; for diatoms: craticula.ncl.ac.uk/Eddi/jsp/; gsc.nrcan.gc.ca/paleo/diatoms/index_e.php;chysophytes: silicasecchidisk.conncoll.edu/Center_Chrysophyte_Home_Page.html). A reader interested in sediment processing and taxonomy should consult these references.

The main reason why proxies can be used to reconstruct climate is because they are directly or indirectly linked to temperature. Chironomids, for example, have four larval stages and their development is partly influenced by water temperature, while the flying adult stage is influenced by air temperature for its distribution, growth and egg laying (Fig. 3).

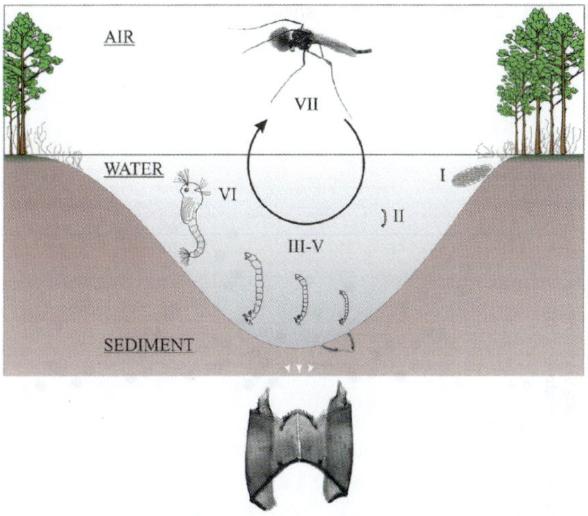

Fig. 3 Life cycle of chironomids. The four developmental stages of the larvae are influenced by water temperature; however, the adult is influenced by air temperature (figure from Larocque and Rolland, 2006)

Strong debates are still going on the use of proxies (specifically biological) to reconstruct climate (e.g. Anderson et al., 2008). Here, we will focus on cases where climate reconstruction was successful. However, the reader should keep in mind that this is not always the case.

Various statistical methods can be used to reconstruct climate from proxies: calibration in space by sampling many lakes across large temperature gradients (e.g. Larocque et al., 2006), calibration in time where a temporal sequence from the sediment is associated with instrumental data (e.g. Blass et al., 2007b), modern analogue techniques where fossil assemblages are associated to the closest (lowest dissimilarity) modern assemblages (e.g. Overpeck et al., 1985), Bayesian techniques (Toivonen et al., 2001) and artificial neural networks (Racca et al., 2007). Each technique has its pros and cons; however, most will produce an error of prediction around 1°C. This range of errors has been previously described as a pitfall for temperature reconstruction using proxies preserved in lake sediment (Battarbee, 2000; Broecker, 2001). However, when the inferences obtained by chironomid analysis in varved Lake Silvaplana were compared to instrumental data from a meteorological station located on

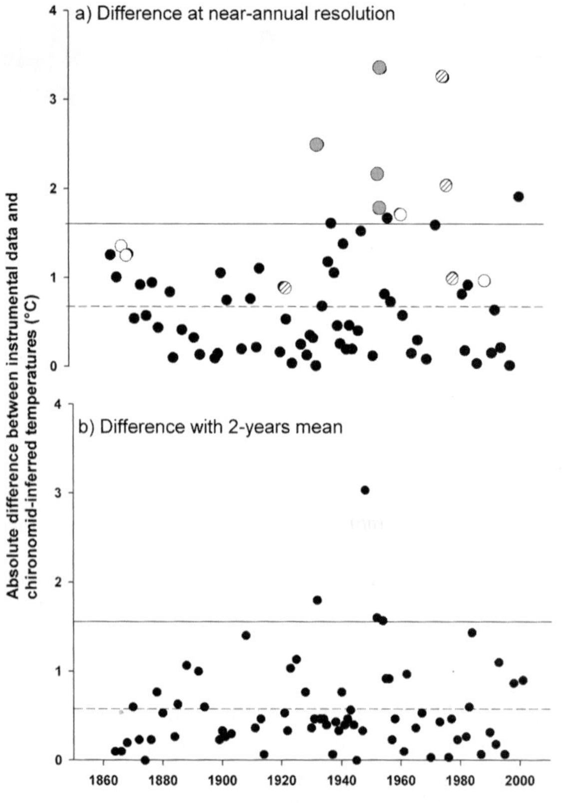

Fig. 4 Differences between instrumental data and chironomid-inferred temperatures in varved Lake Silvaplana, Switzerland. Most of the inferences have differences with the instrumental data below the root-mean-square-error of prediction given by the transfer function and 59% of all inferences have deviations with the instrumental data below 0.5°C (figure modified from Larocque et al., 2009a)

the shore of the lake, it was shown that the error of estimates were generally much lower than the prediction error given by the transfer function (Larocque et al., 2009b; Fig. 4). These results suggest that changes of small amplitude could be reconstructed using proxies preserved in lake sediments.

Although the relationship between temperature and proxies seems to be adequate during the last century for quantitative temperature estimates, it does not mean that this relationship is stable through time. Validation on longer temporal scales can be obtained by comparing the reconstructions with other archives. Such a validation was attempted for chironomid inferences in a varved lake in Switzerland (Fig. 5).

The chironomid-inferred temperature record compared well with early instrumental and documentary proxy evidence at local and regional scale, suggesting that quantitative temperatures estimated by chironomids are mostly accurate until ca. 1580 AD (Larocque et al., 2009b). Between ca. 1740 and 1783 AD, there was no significant relationship between the chironomid and any other record used for comparison. This is probably due to increased precipitation at that site, which overrode the effect of temperature. This is an example of the importance of disentangling the effects of various parameters on biotic (and probably abiotic) proxies.

Validation of other proxies has also been obtained. For example, Quartz/Mica ratios were used to infer summer temperatures in Lake Silvaplana, and the temperature reconstruction was compared with early instrumental data (Auer et al., 2007) showing that the inferences were highly correlated with measured data until ca. 1780 AD (Trachsel et al., 2008, Fig. 6).

Recent development in the indicators of climate, primary production and nutrient content has brought stable carbon and oxygen isotopes of biological materials into focus. The next section discusses the use of stable carbon and oxygen isotopes in lake sediments.

Stable Carbon and Oxygen Isotopes of Biological Materials from Lacustrine Archives as Proxies for Phytoplankton Primary Production, Nutrient Conditions and Water Temperature

Stable isotopes of biological materials from lacustrine archives have become common proxies to trace the course of past environmental and climatological fluctuations. Frequently, robust linkages exist between external forcing parameters and lake internal reactions making lakes excellent systems to study past climate and/or ecological changes (Hostetler, 1995). If a lake is sensitive to specific external forcing and if continuously shifting conditions prevail over longer time intervals, the physical and biotic structure of the lake will change, resulting in direct (e.g. temperature) and indirect (e.g. nutrient availability, pH value) modifications of the lacustrine ecosystem (Schleser et al., 1999). Changes of annually developing properties are frequently recorded by the biogenic fraction of the sedimentary matter, thus, reflecting the lake's environmental and/or climate history. The observation of

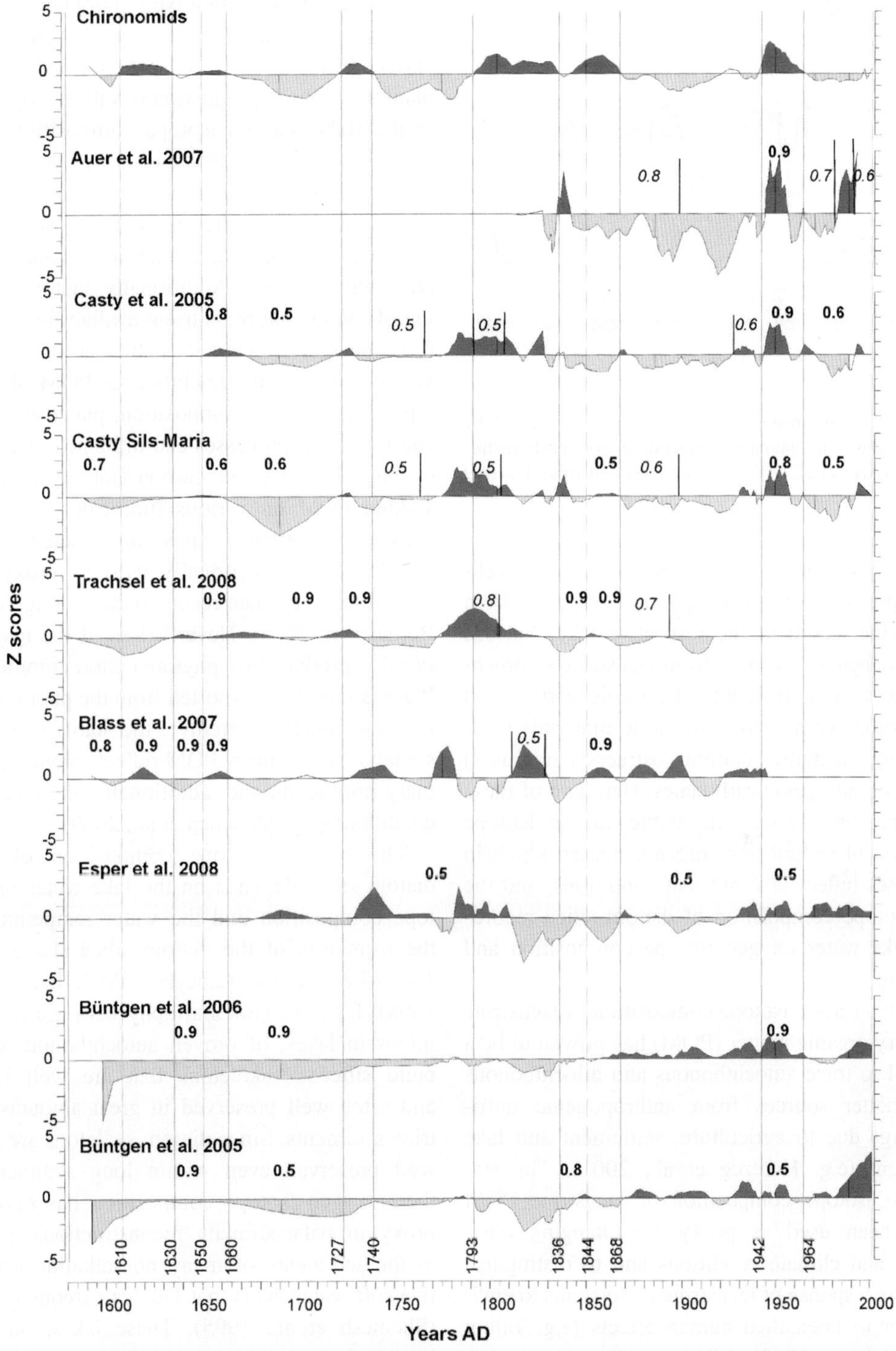

Fig. 5 *Z*-scores comparison between the chironomid record, early instrumental record (Auer et al., 2007), early instrumental and documentary proxy evidence at regional (Casty et al., 2005) and local (Casty Sils-Maria) scales, quartz/mica ratios reconstructing summer temperature in Lake Silvaplana (Trachsel et al., 2008), mass accumulation rate in Lake Silvaplana (Blass et al., 2007), and dendrochronology at regional (Büntgen et al., 2005, 2006) and local (Esper et al., 2008) scales. (Figure modified from Larocque et al., 2009b; see references in Larocque et al., 2009b)

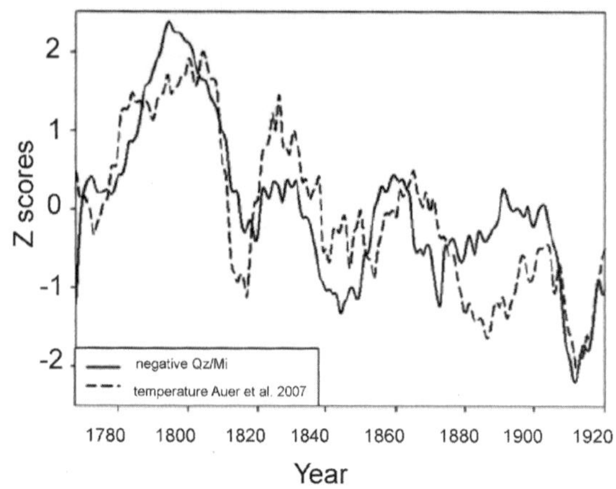

Fig. 6 *Z*-scores of mean summer temperature inferred by Quartz/Mica ratios and summer temperature from early instrumental data (Auer et al., 2007). Graph modified from Trachsel et al. (2008)

such annual variations is a prerequisite for a reliable interpretation of isotope proxies derivable from the lacustrine record of the respective lake. Various isotope geochemical proxies from lacustrine sedimentary archives exist, including the stable carbon and oxygen isotope composition of biotic materials (sedimentary organic matter, diatoms, ostracods, molluscs) as well as of authigenic carbonates. Here two of these proxies are summarized: the stable carbon isotope composition of sedimentary organic matter which in certain cases reflect lake nutrient conditions, and the oxygen isotope composition of diatom silica recording the lake water oxygen isotope composition and temperature.

The stable carbon isotope composition of lacustrine sedimentary organic matter (POM) has proven to be a useful tool to trace autochthonous and allochthonous organic matter sources from anthropogenic nutrient loadings due to agriculture, settlement and lake management (e.g. Herczeg et al., 2001). The stable carbon isotope composition of lacustrine POM has also been used as proxy for changing environmental and climate conditions and to distinguish between the response of lacustrine ecosystems to environment or to intensified human effects (e.g. Talbot and Johannessen, 1992). Like vascular plants, phytoplankton discriminates against $^{13}CO_2$ during CO_2 uptake. The stable carbon isotope composition of sedimentary organic matter from phytoplankton primary production which is exported from the photic zone by sedimentation thus preferentially removes ^{12}C. Thereby, the dissolved inorganic carbon pool (DIC pool) in the photic zone becomes gradually enriched in ^{13}C. As a consequence, increased phytoplankton primary production is reflected by an increase in the stable carbon isotope composition of sinking POM (e.g. Hodell and Schelske, 1998). POM buried in lacustrine sediments however integrates material originating from different sources, such as terrestrial plants, aquatic macrophytes, plankton and heterotrophic sources. Additionally factors like changing pH, temperature, nutrient availability, atmospheric CO_2 concentration and its stable carbon isotope value, varying amounts of allochthonous POM, shifts in phytoplankton species composition, plankton growth rate and microbial processes can influence internal carbon cycling and the stable carbon isotope composition of sinking POM on various timescales. Thus, the stable carbon isotope composition of sedimentary POM could not be categorically used as proxy for environmental or climate reconstructions (e.g.Teranes and Bernasconi, 2005). Nevertheless, if the major part of POM is produced by phytoplankton communities and POM is rapidly transported from the photic zone to the lake's sedimentary archive, the stable carbon isotope signal of sedimentary POM reflects phytoplankton primary production and, additionally, the lake's nutrient conditions (e.g. Moschen et al., 2009).

The oxygen isotope composition of lacustrine diatom silica depends on the lake water oxygen isotope composition and the water temperature during the formation of the diatom silica (Labeyrie, 1974; Juillet-Leclerc and Labeyrie, 1987; Brandriss et al., 1998). Diatoms (Bacillariophyceae) are almost ubiquitous in lakes, of proven autochthonous origin and build siliceous skeletons that are well identifiable and often well preserved in great amounts in lacustrine sediments. Since diatom skeletons are frequently well preserved even within long sediment records their oxygen isotope composition has been used as proxy in palaeoclimate reconstructions, particularly in the sediments of many non-alkaline and productive soft-water lakes that lack calcareous microfossils (Shemesh et al., 1995). These lakes, on the other hand, often provide the continuous, high-resolution sedimentary records that are necessary for consistent regional climate reconstructions (Moschen et al., 2006). Despite the potential applicability of diatom frustules in terms of palaeo-thermometry, to date most studies using oxygen isotope composition of lacustrine diatom silica have been conducted in areas where the oxygen isotope composition of diatom skeletons records changes in the oxygen isotope composition of the lake water (Leng and Barker, 2006). In the sole Holocene lacustrine record available from the Southern hemisphere (e.g. from an island of South Georgia), the oxygen isotope composition of diatom skeletons, however, was interpreted largely in terms of temperature change (Rosquist et al., 1999). Leng and Barker (2006) give an excellent review on the application of this proxy for palaeoenvironmental and palaeoclimate reconstruction.

Lake Holzmaar is a small mesotrophic and dimictic lake located in a maar crater situated in the Westeifel Volcanic Field, Germany. Due to its small water body, a high depth to surface ratio, and a relatively short water residence time of under 1 year, the lake responds quickly to environmental affects and, therefore, serves as an ideal "natural laboratory" to study the controls on the seasonal and interannual dynamics of the lake's biotic fraction and on the generation of isotope-geochemical proxy signals. A recurrent annual oscillation of the stable carbon isotope composition of POM with low $\delta^{13}C_{POM}$ values during winter and more positive values during summer is a result of phytoplankton primary production. Also a close $\delta^{13}C_{POM}$-POM flux relationship is observed, testifying to the primary dependence of $\delta^{13}C_{POM}$ on lacustrine primary production (Fig. 7).

In Lake Holzmaar the stable carbon isotope composition of POM ($\delta^{13}C_{POM}$), is a very valuable parameter for gaining insight into variations of net primary production in depth and time (Moschen et al., 2009). Due

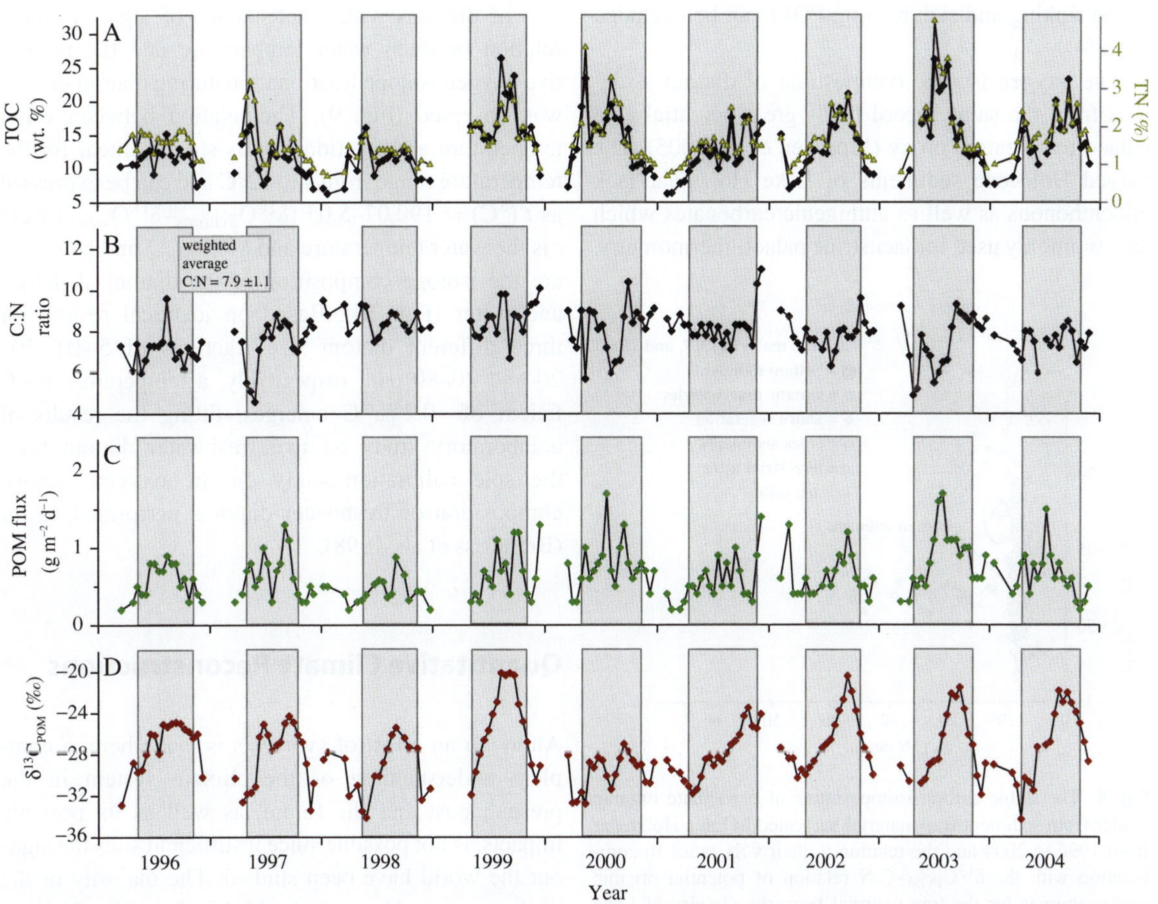

Fig. 7 Geochemical parameters of bulk sediment-trap material collected biweekly at a depth of 7 m in the centre of Lake Holzmaar determined for the period from 1996 to 2004. *Grey areas* denote the stratified period of the individual years: **a** total organic carbon content (TOC, black diamonds) and total nitrogen content (TN, yellow triangles), **b** C:N ratios, **c** daily POM flux and **d** stable carbon isotope composition of the sediment-trap material

to the close relationship between the stable carbon isotope composition of POM and the phytoplankton primary production, the $\delta^{13}C_{POM}$ of sedimentary POM from the varved Holocene sediments of Lake Holzmaar has been successfully used as proxy for the lake's palaeo-productivity and nutrient conditions (Lücke et al., 2003). This proxy parameter, however, has its limitations that should be carefully considered when applied to reconstructions of primary production in comparable lacustrine ecosystems. The emergence of the relation between the stable carbon isotopic composition of POM and phytoplankton primary production seems to be bound to mesotrophic nutrient conditions and is valid only as long as phytoplankton primary production is much larger than the sum of other organic matter sources. This is clearly the case in Lake Holzmaar, where a major contribution of particulate organic matter from sources other than plankton to the sinking and sedimenting POM can be excluded (Fig. 8).

The oxygen isotope composition of diatom skeletons from the same record holds great potential as a palaeo-temperature proxy (Moschen et al., 2005). The varved Holocene sediments of Lake Holzmaar lack allochthonous as well as authigenic carbonates which are commonly used for lacustrine palaeo-thermometry.

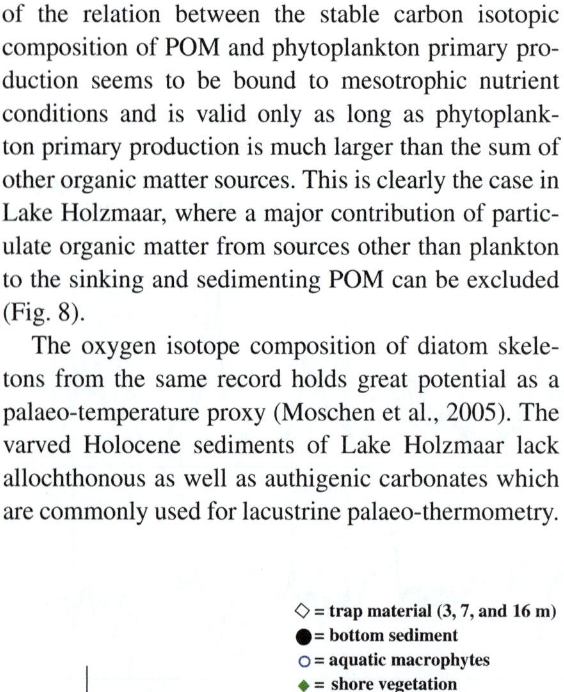

Fig. 8 The stable carbon isotope ratios of particulate organic matter from sediment-trap material harvested in Lake Holzmaar from 1996 to 2004 and the relation to their C:N ratios in combination with the $\delta^{13}C_{POM}$–C:N relation of potential organic matter sources for the trap material from the vicinity of Lake Holzmaar. The relatively high C:N ratios of the encircled data point to a contribution of aquatic macrophytes. Their $\delta^{13}C_{POM}$ values are much higher than those of all potential allochthonous organic matter sources

The record, however, contains great amounts of well preserved diatom skeletons, which can be used as an alternative to carbonates in stable isotope studies. However, a reliable quantitative application of the oxygen isotope composition of diatom skeletons from lacustrine records is limited by the unknown isotopic composition of ambient water in the past but also by uncertainties on the nature and the magnitude of the temperature influence on the oxygen isotope fractionation during formation of the diatom silica (e.g. Jones et al., 2004). Therefore a substantial sediment-trap calibration study dedicated to the investigation of the temperature-dependent oxygen isotope fractionation between diatom silica and water directly in a lacustrine ecosystem has been carried out (Moschen et al., 2005). In Lake Holzmaar, irrespective of variations in the seasonal abundance of larger and smaller taxa, of nutrient supply or of the degree of competition of diatoms with other classes of algae, a linear relation between water temperature and the respective oxygen isotope fractionation during diatom growth was observed (Fig. 9). The relation between water temperature and fractionation is strictly linear for the temperature range from 4 to 22°C and can be expressed as $t\ (°C) = 190.07–5.05\ (\delta^{18}O_{diatoms}–\delta^{18}O_{water})$ were t is the water temperature and $\delta^{18}O_{diatom}$ and $\delta^{18}O_{water}$ are the isotope composition of the diatom skeletons and water (Fig. 10). Based on identical results for three different diatom size fractions of 5–10, 10–20 and 20–80 μm, respectively, a temperature coefficient of –0.2‰/°C emerges, fitting the results of a laboratory study on two freshwater diatom taxa, the sole calibration study on the oxygen isotope composition of freshwater diatoms performed so far (Brandriss et al., 1998).

Quantitative Climate Reconstructions

Although an effort of synthesis is made here, a complete understanding of the climate system in the present, past and the future, as well as its possible impacts, is not possible since insufficient sites throughout the world have been studied. The majority of the studies reported here are in the Northern Hemisphere, which represents the actual distribution of studied sites. Furthermore, sites in the Southern Hemisphere, in Africa and in part of China, are mostly used to

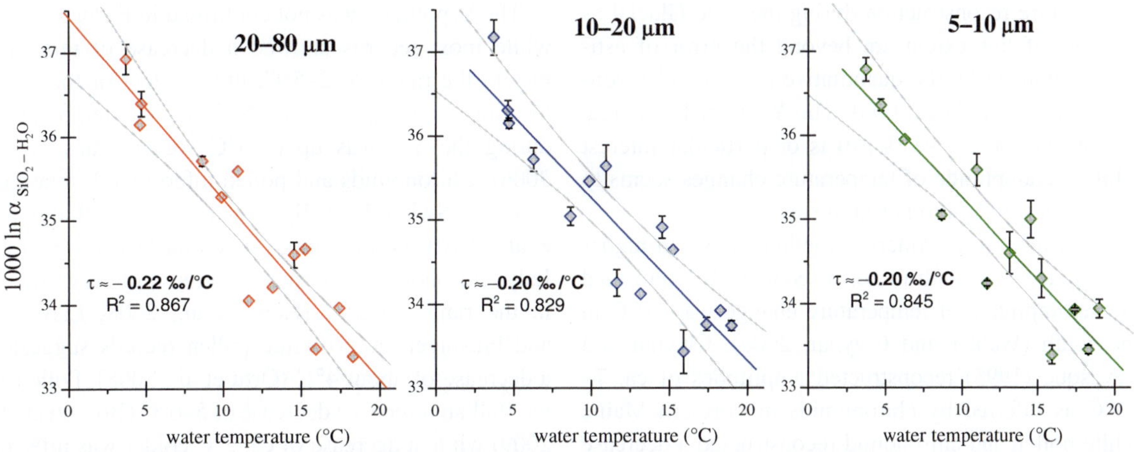

Fig. 9 The dependency of the oxygen isotope fractionation between diatom silica and water on the water temperature for three different size classes of freshwater diatoms harvested with sediment traps in the centre of Lake Holzmaar. Each symbol represents the average of up to four measurements (error bars represent 1σ). Centre lines are regressions; neighbouring curves express 95% confidence intervals. The values of these regression coefficients are identical at $p < 0.05$

reconstruct precipitations and/or lake level changes. Lake level changes will be presented below.

Only the newest studies (i.e. papers published after 2005) are presented here, to complete the existing literature on Quaternary climate (e.g. Elias, 2006a, b). The quantitative temperature reconstructions have been separated into four time periods (Late glacial, Holocene, last millennium and past decades) and specific questions within these time periods are evaluated.

Late Glacial

The Late Glacial period experienced large amplitude temperature changes (i.e. 2–10°C during the Younger Dryas), as previously reconstructed by proxies preserved in other types of archives (e.g. ice cores, speleotherms). These large amplitudes justify the use of proxies preserved in lake sediments for

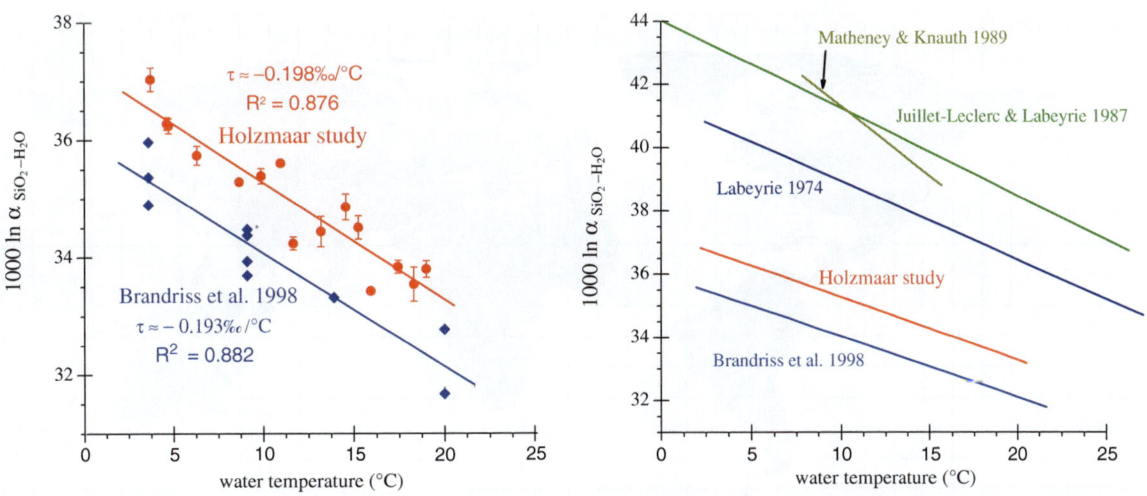

Fig. 10 Comparison of the Lake Holzmaar investigations on the temperature dependency of the oxygen isotope fractionation between diatom silica and water with the results of all earlier studies performed so far. **a** Temperature dependence of fractionation for the weighted average of all diatom size classes of the Lake Holzmaar study in comparison with results of a laboratory study on freshwater diatoms by Brandriss et al. (1998). **b** All calibration studies performed so far. Only the Lake Holzmaar study and the Brandriss et al. (1998) investigation compare well with each other

temperature reconstruction during the Late Glacial as changes of this extent are beyond the error of estimates obtained by the quantitative methods of reconstructions (Battarbee, 2000). The Younger Dryas (ca. 12,500–11,700 cal years BP) is or particular interest while the amplitude of temperature changes seems to vary from southern to northern sites.

In eastern North America, the hypothesis of higher amplitude from south to north seems to be confirmed as the amplitude of temperature change was 10°C in the south (Walker and Cwynar, 2006). Cwynar and Levesque (1995) reconstructed amplitudes of ca. 7–13°C as inferred by chironomids in northern Maine while pollen and chironomid reconstructed a decrease of 5–6°C during the Younger Dryas in Nova Scotia (Dieffenbacher-Krall et al., 2009; Whitney et al., 2005). However, the rest of North America seems to have experience a Younger Dryas with a decrease of temperaure in the range of 4–6°C as reconstructed by pollen records throughout Canada (Viau et al., 2006), pollen, chironomids and ostracods in Yukon (Bunbury and Gajewski, 2009) and chironomids in the Siera Nevada (MacDonald et al., 2008). In Alaska, the temperature reconstructed using chironomids did not show any climate deterioration during the Younger Dryas (Kurek et al., 2009).

The hypothesis was not confirmed in Europe either, while most records showed a decrease of temperature in the range of 2–5°C, independent of the site's location (see Fig. 11). In Norway, the temperature during the YD was up to 3°C colder (Paus et al., 2006). Chironomids and pollen inferred a decrease of 3–4°C in Switzerland (Ilyashuk et al., 2009; Larocque et al., 2010) and in France (Heiri and Millet, 2005). In Italy, chironomids inferred a decrease of temperature in the range of 2°C (Heiri et al., 2008; Larocque and Finsinger, 2008) while pollen records suggested a decrease of about 4°C (Ortu et al., 2008). Pollen in the Balkan inferred a decrease of 5–6°C (Bordon et al., 2009) while a decrease of ca. 2°C colder was inferred from pollen in Portugal (Feurdean et al., 2008). In Russia, pollen records suggested a decrease of 3–4°C (Seppä et al., 2008; Tarasov et al., 2009).

The Younger Dryas was possibly caused by a significant reduction or shutdown of the North Atlantic thermohaline circulation in response to a sudden influx of fresh water from Lake Agassiz and deglaciation in North America (Broecker, 2006; Coleman, 2007). However, other alternate hypotheses have started to arise, as the possibility of Earth's collision with a rare swarm of comets at the onset of the Younger Dryas in North America (Kennett et al., 2009). This alternate

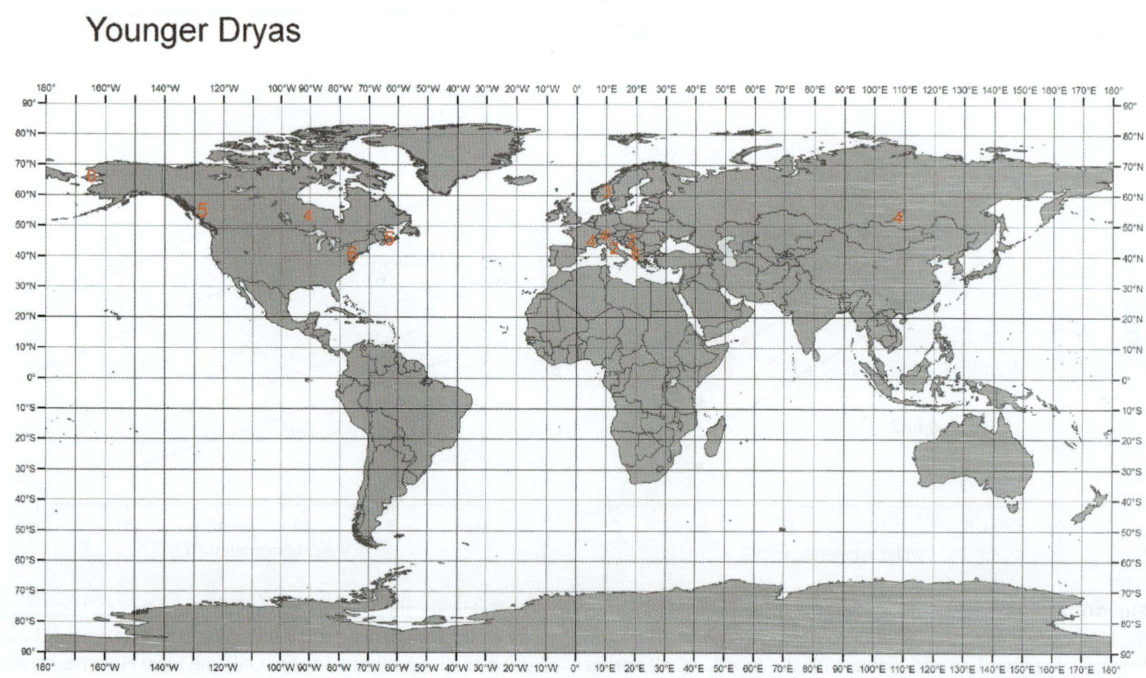

Fig. 11 Amplitude of temperature changes inferred by different proxies during the Younger Dryas. See text for references

hypothesis brings a lot of debate in the scientific world, but will probably trigger more research showing the change in thermohaline circulation.

The Holocene

In 2000, Battarbee suggested that Holocene quantitative reconstructions using proxies preserved in lake sediments needed to be tested to determine if indicators could be used to infer small amplitudes in climate change as experienced during the last 10,000 years BP. Validations comparing diatoms (Bigler and Hall, 2003) and chironomids (Larocque and Hall, 2003; Larocque et al., 2009a) with instrumental data have shown that biological proxies can infer accurately changes in amplitude of ca. 0.3°C or less. Thus, biotic proxies should be able to reconstruct changes during the Holocene, which were previously estimated at ca. 2°C. Two questions are of particular interest considering climate change in the Holocene: (a) when was the thermal maximum? and (b) was a 8,200 cal years BP event widespread?

The Holocene thermal maximum (HTM) was between deglaciation until ca. 5,000 years BP in many studied sites in Northern America (e.g. Victoria Island, Canada (Fortin and Gajewski, 2010); northern Québec, Canada (Fallu et al., 2005; Saulnier-Talbot et al., submitted); British Columbia, Canada (Chase et al., 2008)) and Europe (Finland (Ojala et al., 2008b) and Portugal (Fletcher et al., 2007)). In Spain, the decrease of temperature started at ca. 4,000 years BP (Pla and Catalan, 2005). On Southampton Island (Canada) (Rolland et al., 2008), in Alaska (McKay et al., 2008) and most of Canada and the USA (Viau et al., 2006) the temperature started to decrease after ca. 3,000 years BP. A similar timing was observed in Finnmark (Allen et al., 2007). In Nova Scotia (Canada) the temperatures started to decrease already after 8,000 years BP (Lennox et al., 2010) and around 6,000 years BP in Venezuela (Rull et al., 2008) and Iceland (Caseldine et al., 2006).

A consensus on the timing of the HTM was not obtained since it varied depending on the proxy used. In general, aquatic proxies indicated earlier thermal maximum than pollen. However, the climate after deglaciation was generally warmer than in the last 1,000 years (Fig. 12). This warmer climate in the early Holocene might have been associated with Milankovtich cycles. Around 9,000 cal years BP, the

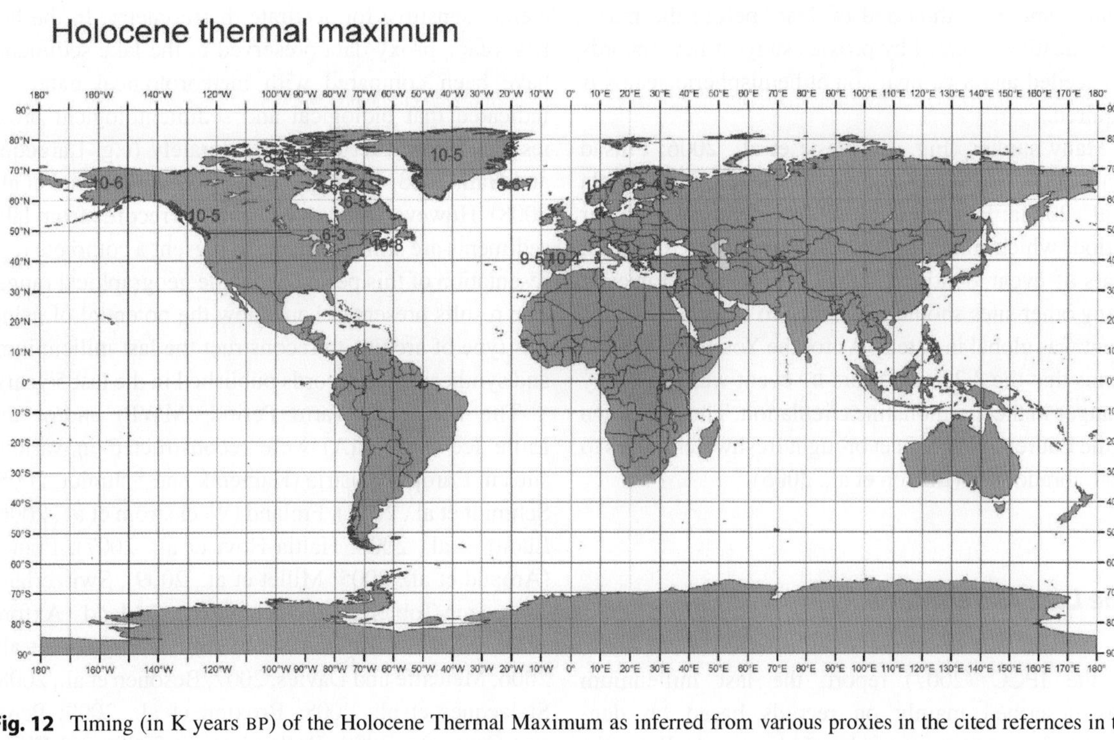

Fig. 12 Timing (in K years BP) of the Holocene Thermal Maximum as inferred from various proxies in the cited refernces in the text above

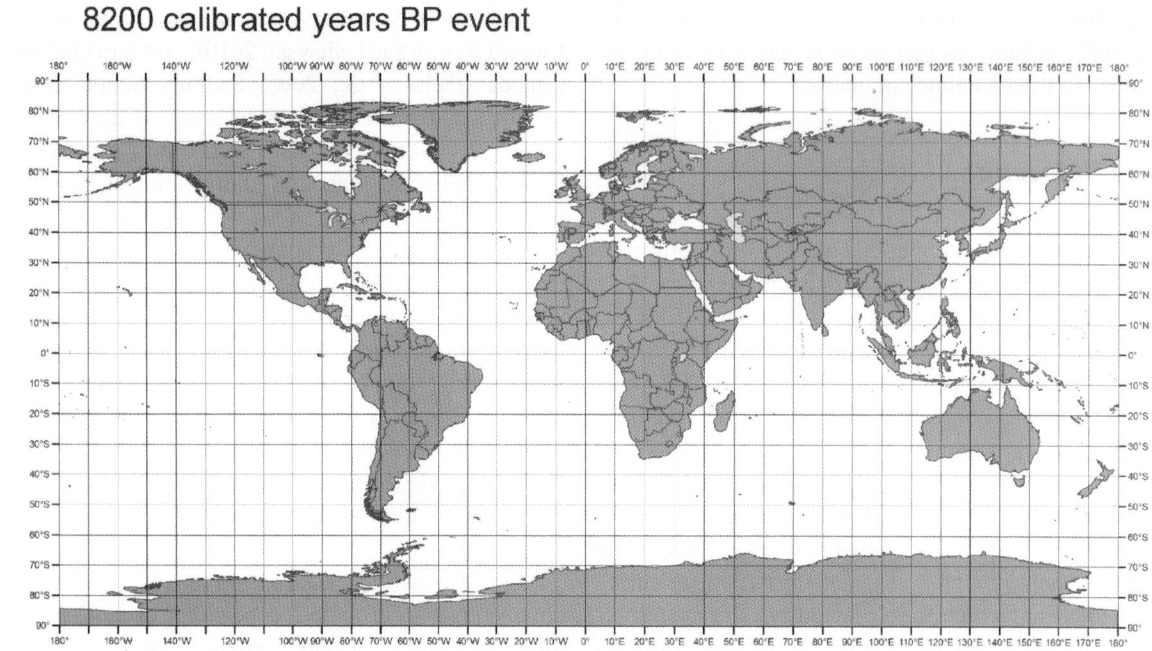

Fig. 13 Sites where the 8,200 calibrated years BP event was recorded by one or many proxies. The studies represented here are only the ones cited in the text above

earth was closer to the sun leading to maximum heating during summer. However, the timing of the maximum being sometimes thousand of years before the maximum heating inferred by proxies suggest new records are needed and a comparison at hemispheric scales is required.

Many studies (Fig. 13; Nesje et al., 2006; Prasad et al., 2006; Sarmaja-Korjonen and Seppä, 2007; Ojala et al., 2008a; Panizzo et al., 2008) identified a colder period, which might be associated with the 8,200 cal years BP event identified in the GRIP record. However, many other sites should be studied to determine if this event was global in extent. As for the Younger Dryas, it seems that the 8,200 cal years BP event was caused by changes in the thermohaline circulation. The meltdown of the Laurentide ice sheet brought freshwater pulses to the Labrador sea (Ellison et al., 2006).

The Last Millennium

In the IPCC (2007) report, the last millennium was described mainly on records based on dendrochronology, isotope and borehole records. Broecker (2001) mentioned that the accuracy of the temperature estimates based on remains from lake sediments is likely no better than ±1.3°C and hence not sufficiently sensitive for accurate thermometry. In the last few years, proxy data preserved in the lake sediments have been compared with meteorological data and indicated that biological and sedimentological proxies can reconstruct climate accurately (e.g. Larocque and Hall, 2003; Larocque et al., 2009a; Trachsel et al., 2008). However, the last millennium records from lake sediments are still too sparse to present a complete representation of this period at a wide geographical range. The results presented here show the potential of using this type of archive to reconstruct the last millennium, and synthesize the records published in the last 5 years.

The Medieval Warm Period (MWP) and/or the Little Ice Age (LIA) were reconstructed in various sites in Europe (Austria (Kamenik and Schmidt, 2005; Schmidt et al., 2007), Finland (Weckström et al., 2006; Luoto et al., 2008; Haltia-Hovi et al., 2007), France (Arnaud et al., 2005; Millet et al., 2009), Switzerland (Larocque-Tobler et al., submitted), Iceland (Axford et al., 2009), North and South Americas (Loso et al., 2006; Metcalfe and Davies, 2007; Besonen et al., 2008; St-Jacques et al., 2008; Broxton et al., 2009; Peros and Gajewski, 2009; Rolland et al., 2009) and China (Holmes et al., 2009)). See Figs. 14 and 15.

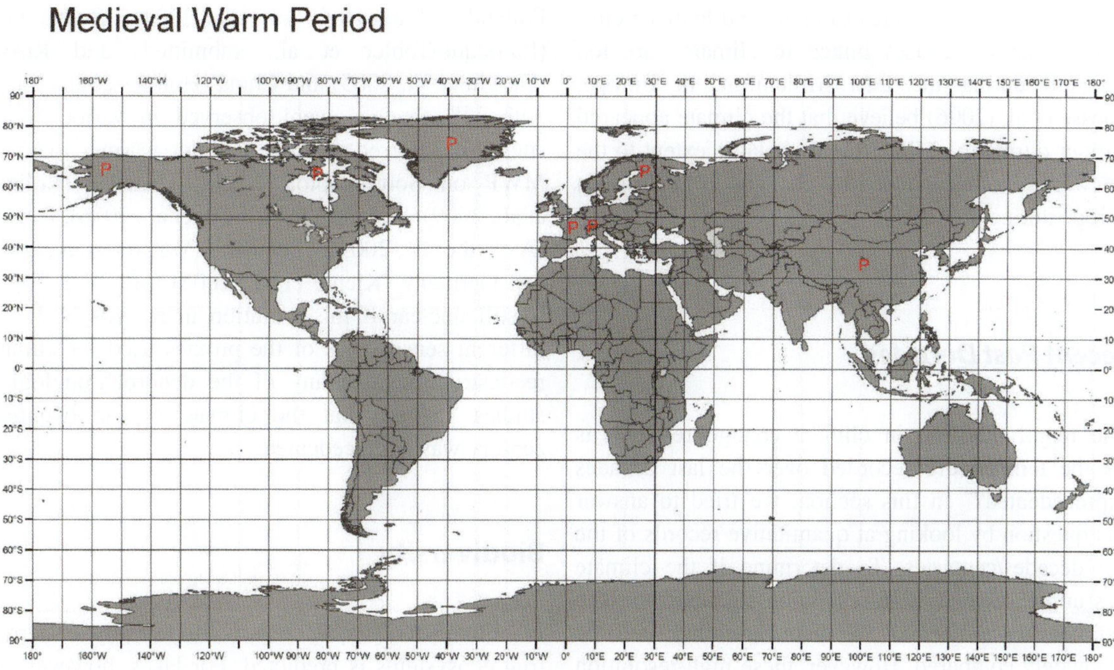

Fig. 14 Extent of the "Medieval Warm Period." P = warmer that today period reconstructed

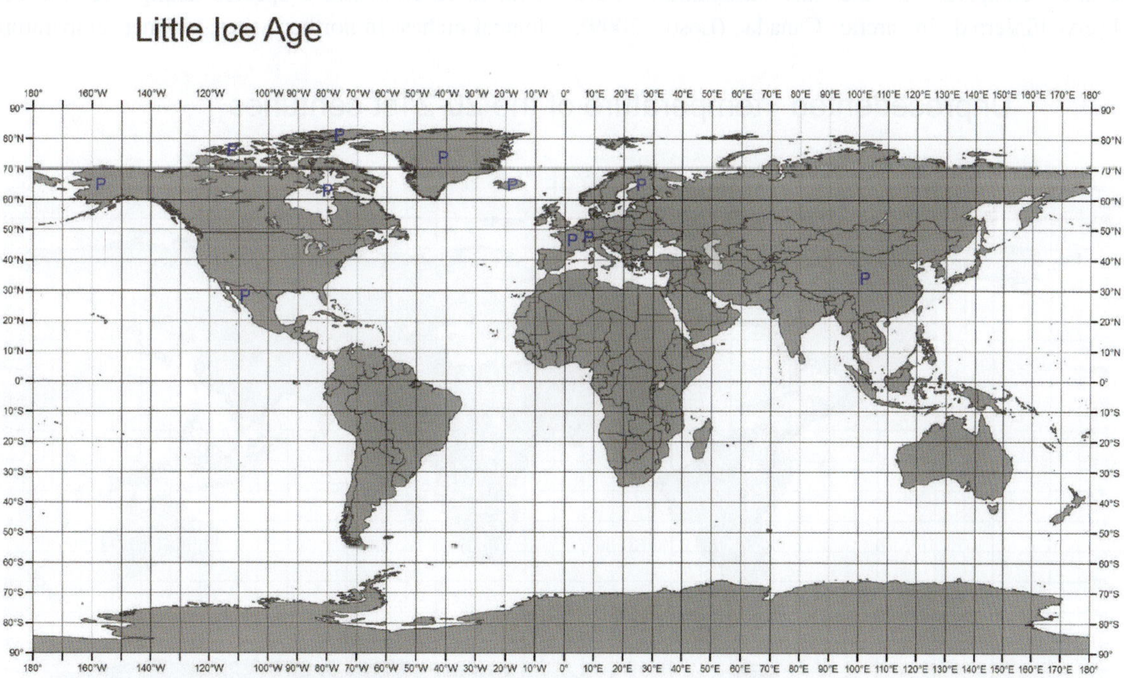

Fig. 15 Presence of a cold "Little Ice Age" period

Solar forcing, volcanic eruptions and human activities are among causes linked to climate variation observed during the Last Millenium. In Europe, Goosse et al. (2006) believe that the climate appeared warmer during the MWP due to "a large extent to the long term cooling induced by changes in land-use in Europe during the LIA.

Recent Past Decades

One major question in climate change research is "Is the temperature recorded over the last decades unprecedented?" In this section, we tried to answer this question by looking at quantitative records of the last decades/centuries. To determine if the climate was unprecedented, high-resolution studies from lake sediments are needed so that variability can be adequately compared. However, these high-resolution studies over the last millennium/centuries are still sparse.

Unprecedented temperatures in the past decades compared to the last millennium were observed/inferred in arctic Canada (Loso, 2009), Finland (Haltia-Hovi et al., 2007), Switzerland (Larocque-Tobler et al., submitted) and Russia (Daryin et al., 2005) and China (Holmes et al., 2009). Although changes were observed in various other sites, the inferred climate was not warmer than the MWP on Southampton Island (Canada) (Rolland et al., 2009), Finland (Weckström et al., 2006), Iceland (Axford et al., 2009), Greenland (Wagner et al., 2008) and Germany (Kienel et al., 2005) (See. Fig. 16). It is still unclear if this variation in records is due to different sensitivities of the proxies used for climate reconstruction as many of the dendrochronological studies suggest that the climate of the twentieth century was unprecedented.

Biodiversity

In climate models, decreased biodiversity of terrestrial ecosystems is predicted. For lakes, however, the changes might be linked to the latitude of studied sites and different indicators. In southern sites, warmer temperatures might lead to a decrease in biodiversity with more competitive species taking over the ecological niches. In northern sites, warmer temperatures

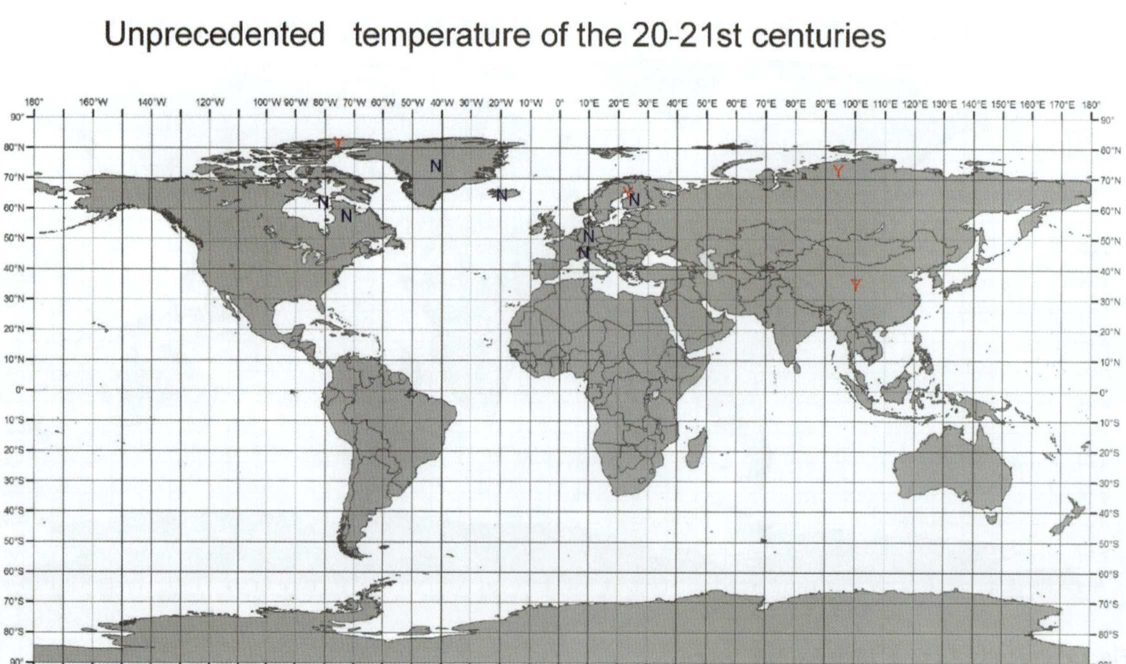

Fig. 16 Fig. 16 Was the climate of the twentieth to twenty-first centuries unprecedented? Y is for yes, N is for no

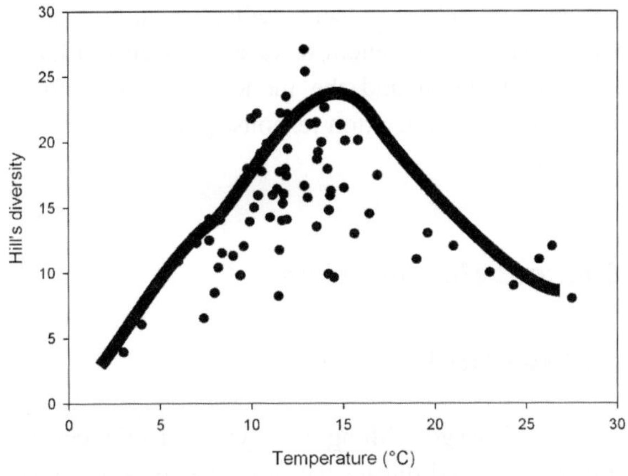

Fig. 17 Relationship between Hill's diversity numbers of chironomid assemblages in 80 lakes of the eastern Canada training sets and mean August air temperature. Data from Larocque (2008)

favour the establishment of a more diverse fauna. This hypothesis is shown in Fig. 17 where the Hill's diversity index of chironomid assemblages has been plotted against mean August air temperature. Until the lakes reach a mean temperature of ca. 15°C in August, an increase in the diversity of chironomid communities will occur while in lakes with warmer mean August air temperature, the diversity will decrease.

Organisms in lakes might also respond differently to warmer temperatures depending on their tolerance. Cold water fish, such as salmon and arctic char, are expected to reach the limit of their temperature tolerances by 2100 AD and might disappear (EPA, 1999). Here we will concentrate our review on other biological organisms in lakes. Studying the past variations associated with changes in temperature might give us insight on which taxa are at risk.

Quantifying biodiversity by indexes is controversial and it is difficult to determine which of these indexes should be used. Not all organisms are preserved in lake sediments, thus the reconstructed diversity is a partial representation of biodiversity in lakes. Here, the variations of diversity are described for taxonomical groups.

Chironomid diversity has been shown to increase with warmer climate in the Canadian Arctic (Gajewski et al., 2005; Quinlan et al., 2005; Rolland et al., 2008, 2009). Cladocera species diversity also increased in periods of warmer climate in the past in the Czeck Republic (Pražáková et al., 2006), in warmer lakes in the Italian and Swiss Alps (Guilizonni et al., 2006) and with the most recent warming in a remote Scottish mountain loch (Kattel et al., 2008). Higher diatom diversity were observed during warmer periods in northeast Siberia (Cherapanova et al., 2007) while less diverse diatom communities were linked to colder conditions in France (Ampel et al., 2008). Diatom diversities have increased in recent decades in arctic Canada and Greenland, but while the changes were unprecedented in arctic Canada (Quinlan et al., 2005; Finkelstein and Gajewski, 2008; Lim et al., 2008) they were similar to those observed during the Medieval Warm Period in Greenland (Perren et al., 2009). In Lake Khubsugul (Mongolia), the maximum ostracod diversity followed the temperature optima over the last 230,000 years (Poberezhnaya et al., 2006). Present, past and future climate (in modelling) have been linked with increases of aquatic plants (Lacoul and Freedman, 2006; Väliranta, 2006; Abrahams, 2008).

Although long-term studies and present conditions suggest that most indicators in lakes (chironomids, diatoms, ostracods, aquatic plants) had increased diversity with periods of warmer climate, it is still impossible to determine to which extent these will continue to increase while most taxa will not have reached their higher limit of tolerance by 2100 AD.

The Future

For future changes, it might be more important to determine which species will disappear and the most sensitive are probably those restricted to cold lakes. Transfer functions for different biological indicators might be a way to assess these disappearances in the future. For example, in eastern Canada, if mean July temperature of cold lakes increase to more than 12°C, five of the chironomid taxa which have optima and tolerances below that level (Larocque et al., 2006) (Fig. 18), three in the western part of Canada (Barley et al., 2006) and five in central Arctic Canada (Porinchu et al., 2009) might disappear. In Northern Sweden, 18 diatom species would be at risk (Rosèn et al., 2000) and four chironomid taxa (Larocque et al., 2001). In Ugandan lakes, eight chironomid taxa could disappear if cold lakes had temperature rise above 12°C (Eggermont et al., 2010). However, based on the diversity diagram in Fig. 17, the diversity in these lakes

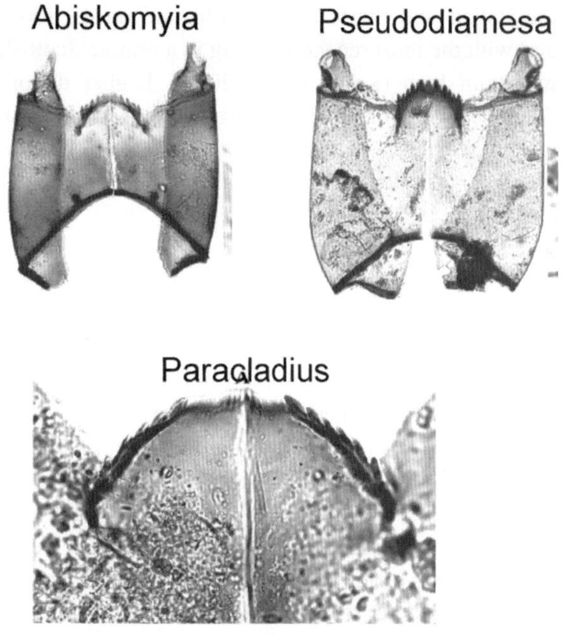

Fig. 18 Some of the chironomid taxa with temperature optima and tolerances below 12°C in Arctic Canada lakes. These taxa might disappear with climate warming although the chironomid diversity might increase (photos by Nicolas Rolland)

would probably increase with warmer temperatures. It could be important to look at diversity, and disappearances and replacements of taxa as well.

Cyanobacteria

The long evolutionary history of cyanobacteria has resulted in survival strategies and persistence during climatic changes that have taken place on Earth during the past 3.5 billion years. Today, some cyanobacterial species form massive surface blooms that produce toxins, cause oxygen depletion and alter food webs, posing a major threat to drinking and irrigation water supplies, fishing and recreational use of surface waters worldwide (Paerl and Huisman, 2009a; Fig. 19). Here we look at the link between cyanobacteria and climate in the past. In lake sediment, cyanobacteria are not preserved as complete algae, but can be deduced using pigment analysis. For example, echinenone, zeaxanthin, canthaxanthin, myxoxanthophyll and scytonemin are used to determine cyanobacteria preserved in lake sediment (see Leavitt and Hodgson, 2001). Second, we will present the most recent studies revealing that climatic change may benefit to various species of harmful cyanobacteria and the models predicting future changes in phytoplankton responses to climate change scenarios.

Changes Through Time

The Last Glacial

In Lake Hovsgol, Mongolia, cyanobacteria concentration increased during the post-glacial period when increased precipitation and runoff to the lake resulted in higher nutrient supply (Nara et al., 2005). The concentrations decreased during the Younger Dryas, and attained their maximum at the start of the Holocene through recent times. A similar pattern of temporal changes in cyanobacteria concentration was observed in Lake Baikal (Soma et al., 2007). In the period between ca. 16,100 and 8,470 years BP, lake levels were possibly low at Potrok Aike (Argentina) explaining the low concentration of cyanobacteria, which increased after 7,400 BP when the lake level increased (Mayr et al., 2009). Cyanobacteria in the Arctic have been shown to survive elevated UVR exposure by increasing extra cellular concentration of photoprotective compounds (Castenholz et al., 2000). In Lake Reid, east Antarctica, the mean exposure of the benthic cyanobacteria during the last glacial was more than three times higher than during the Holocene. This threefold increase was possibly linked to the reduction of snow cover on the lake ice, consistent with cooler conditions and leading to an increase of UV penetration (Hodgson et al., 2005).

The Holocene

In pristine lakes of Greenland, climate and lake age were the dominant factors explaining the variations in cyanobacteria through time. N_2-fixing and colonial cyanobacteria were more abundant in the early stage of lake development and decreased within 1,000 years. The conditions favouring their establishment at that time might be a combination of the water's alkaline state and low DOC, increasing UV penetration (McGowan et al., 2008).

Fig. 19 Examples of algal blooms. From Paerl and Huisman (2009a, b)

The Last Millennium

Over the last 200 years, cyanobacteria decreased during periods of low water level and then increased during high water level of Lagunillo del Tejo, Spain. These level changes were linked to climate change (Romero-Viana et al., 2009). In Lake Mattamuskeet, North Carolina, USA, cyanobacteria increased during the Little Ice Age, between ca. 1584 and 1860 AD when precipitation favoured wetland expansion and the connectivity between the lake and the watershed. However, the concentration further increased following human settlement in this area (Waters et al., 2009).

Recent Decades

According to Paerl and Huisman (2009a) climate change is a potent catalyst for further expansion of cyanobacterial blooms in waters already over-enriched by urban, agricultural and industrial developments. This is likely to happen for several reasons. First, cyanobacteria are more competitive in warmer waters because their optimal growth temperature is higher (often above 25°C) than for diatoms and green algae (e.g. Jöhnk et al., 2008). Second, higher water temperature promotes the establishment of stronger thermal stratification and its longer persistence (earlier stratification in spring and later destratification in autumn), conditions that favour buoyant and UV resistant cyanobacteria (e.g. Huisman et al., 2004). Cyanobacterial blooms can locally enhance surface water temperature or pH, which favours their dominance over eukaryotic phytoplankton (Hense, 2007). Changes in hydrological cycle (patterns of precipitation and droughts or damming, related to lake residence time) are also likely to affect the competition and promote cyanobacterial blooms under certain conditions (e.g. Uwins et al., 2007).

In a comparison of 2 years of contrasting weather conditions, Dupuis and Hann (2009) showed that the filamentous cyanobacteria biomass increased under the warmer spring-summer in three shallow polymictic lakes (which dominate the world's freshwater systems, Downing et al., 2006), along with shifts in zooplankton composition. The degree-day metric for epilimnetic water was the controlling variable, while N, P, N:P or water column stability did not change significantly.

An invasive toxin-producing cyanobacterium originating from the tropics, *Cylindrospermopsis raciborskii*, was reported for the first time in Lake Constance, Ontario, Canada. Between 1998 and 2001 AD, the highest biomass of this species was linked to warmer temperature rather than nutrients (Hamilton et al., 2005). The same species was present in two lakes in Germany, and long-term studies (1993–2005) showed that an earlier rise in water temperature has promoted the spread of *C. raciborskii* to temperate zones over the last few decades (Wiedner et al., 2007). Warming earlier in the season permits early germination, thereby exposing the cells to increased irradiance, which advances growth and the population establishment. This was also suggested by the results of Briand et al. (2004) showing the large tolerance of *C. raciborskii* to varying climatic conditions with optimal growth at elevated temperature (30°C).

Jacquet et al. (2005) studied *Planktothrix rubescens* in Lac Bourget, France, which started to bloom in 1996 even though a restoration program reducing nutrient loads and pollution in the lakes had occurred during the 1970s and 1980s. Among the factors influencing these blooms were warmer than average winter/spring periods permitting earlier stratification, which favoured *P. rubescens*.

The application of modelling experiments by De Senerpont Domis et al. (2007) indicated that whereas the successional sequence, from diatoms to green algae to cyanobacteria, was not affected by the different climate warming scenarios (cold, average and warm spring), cyanobacteria showed a stronger response than diatoms or green algae, with higher growth rates and peak abundances in the average and warm spring scenarios. Shatwell et al. (2008) showed that in a warm year, the cyanobacteria peak was four times more important than during a cold year (Fig. 20).

The summer of 2003 was the hottest of the past 500 years in Europe, and harmful cyanobacterium blooms of *Microcystis* were observed at Lake Nieuwe Meer located in the city of Amsterdam in the Netherlands. Jöhnk et al. (2008) developed a model to explain these blooms and found that high temperatures benifit to cyanobacteria directly through increased growth rate and by increasing the stability of the water column favouring the buoyant taxa.

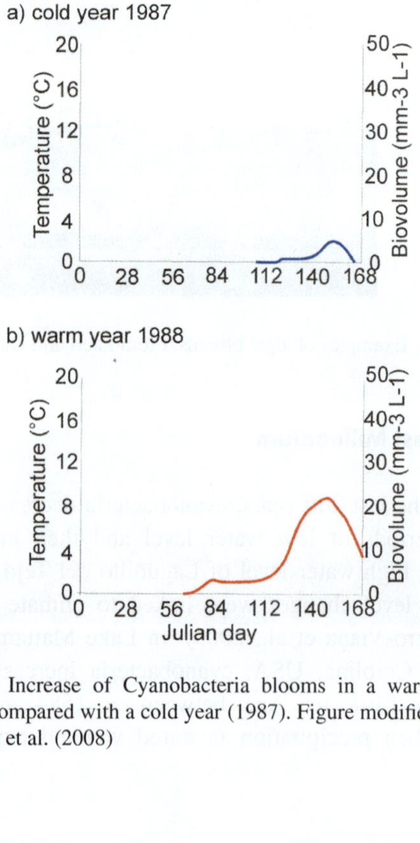

Fig. 20 Increase of Cyanobacteria blooms in a warm year (1988) compared with a cold year (1987). Figure modified from Shatwell et al. (2008)

In the wetlands of the Everglades, Florida, USA, the cyanobacterium *Gleotrichia*-type increased in the upper 50 cm of the studied core, probably associated with increased light availability (Chmura et al., 2006). The length of the spring-summer period and light availability has also been shown to affect the competition between toxic and non-toxic strains in *Microcyctis*, which has major consequences on the overall bloom toxicity (Kardinaal et al., 2007). According to a recent study by Davis et al. (2009) on the dynamics of toxic and non-toxic strains of *Microcystis* during cyanobacteria blooms, enhanced temperatures (+4°C) significantly increased growth rates of toxic *Microcystis* in 83% of experiments conducted, while only 33% of experiments showed increased growth of non-toxic cells. The concurrent increases in temperature and phosphorus concentrations yielded the highest growth rates of toxic *Microcystis* cells in most experiments, suggesting that future eutrophication and climatic warming may lead to blooms with higher microcystin contents.

The cyanobacterium *Microcyctis aeruginosa* has been shown to tolerate increasing salinities in freshwater ecosystems (specific growth rate, microcystin cell quota and microcystin production remained unaffected by salinity levels up to 10 g/L, Tonk et al., 2007). Increasing salinities can result from agricultural practices and specific water management strategies, but are also associated to droughts or rising seawater levels. *Microcystis* spp. Blooms have been observed in brackish waters (e.g. Robson and Hamilton, 2003). This tolerance might provide another competitive advantage to cyanobacteria under a warming climate, relative to diatoms and green algae that seem to have a lower salt tolerance.

Several studies using long-term data sets or modeling indicate that the phenology of phytoplankton species, particularly in spring, has been or will be affected by climatic change (e.g. Shatwell et al., 2008). Results generally show earlier onset of thermal stratification and spring diatom blooms, and an expanding temporal mismatch between *Daphnia* populations and spring blooms. The 25-year study by Shatwell et al. (2008) shows that a change in timing of spring plankton events in warm years surprisingly led to lower mean water temperatures during the growth period, favouring cold-adapted diatoms over cyanobacteria. However, under high P:Si ratios, the increased time between phytoplankton and cladoceran peaks opened a loophole for filamentous cyanobacteria in warm years to establish dominance after the diatoms.

Modeling Future Changes

In a review of present and future predictions in shallow lakes of the Netherlands, Mooij et al. (2005) concluded that climate change will likely favour and stabilize cyanobacterial dominance. Mooij et al. (2007) developed both simple and complex models to predict the effect of climate change on shallow lakes (non-stratifying lakes). Cyanobacteria showed the strongest response to temperature. Their models assume that winter precipitation and extreme rainfall events will increase (although in some regions decreased precipitation is predicted) concurrently to water temperature, which will favour phytoplankton (particularly cyanobacteria) against macrophytes through bottom-up (increased nutrient availability and external loading, reduced light availability) and top-down (reduced zooplankton predation) effects. Climate change would therefore mimic the effects of eutrophication or exacerbate ongoing eutrophication (e.g. Moss et al., 2003). The main output parameters of interest for water quality managers are the critical nutrient loading at which the system will switch from clear to turbid and the much lower critical nutrient loading – due to hysteresis – at which the system switches back from turbid to clear. The model predicts that climate change will likely lead to decreased critical nutrient loadings. Combined with an expected increase in the external nutrient loading, this will increase the probability of a shift from a clear to a turbid state.

Several coupled biological-physical models predict that cyanobacterial blooms will be favoured under a warmer climate and that the cyanobacterial dominance is greatest when high water temperature is combined with high nutrient loads (e.g. Jöhnk et al., 2008; Fig. 21). Linking a regional climate model (RCM) with a phytoplankton community model (PROTECH), Elliott et al. (2005) produced simulations of 20 years of data from Bassenthwaite Lake in England. With an increase of 1% in atmospheric CO_2 concentrations until 2100, the model predicted that spring bloom of *Planktothrix* had greater success but declined earlier due to nutrient limitation caused by the increase spring growth.

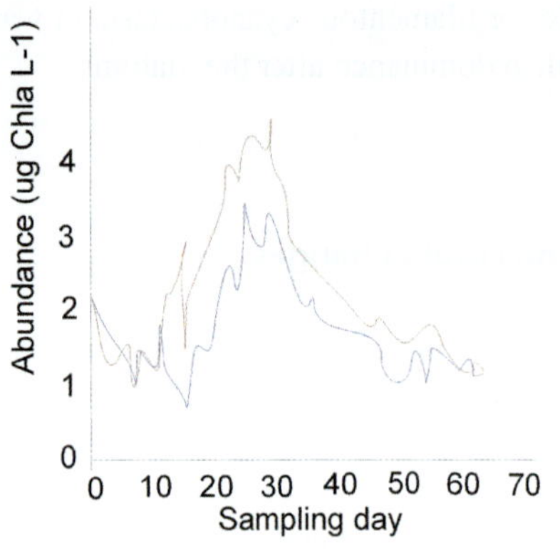

Fig. 21 Abundance of cyanobacteria under colder (*blue line*) or warm (*red line*) scenario. Figure modified from de Senerpont Domis et al. (2007)

Eutrophication

Climate change might affect distribution of cyanobacteria, as previously explained, but it might also affect other primary producers and lead to increased productivity and eutrophication. Changes in climate can affect eutrophication (Fig. 22): (1) by increasing the nutrient input to the lake with increased river runoff and/or organic matter input from soil, (2) by changing the thermal stratification of lakes, (3) by changing the lake level, thus affecting the stratification, (4) by changing the onset of stratification, the ice-free season or the periods of maximum peaks of phytoplankton; or (5) by causing food-web altercations.

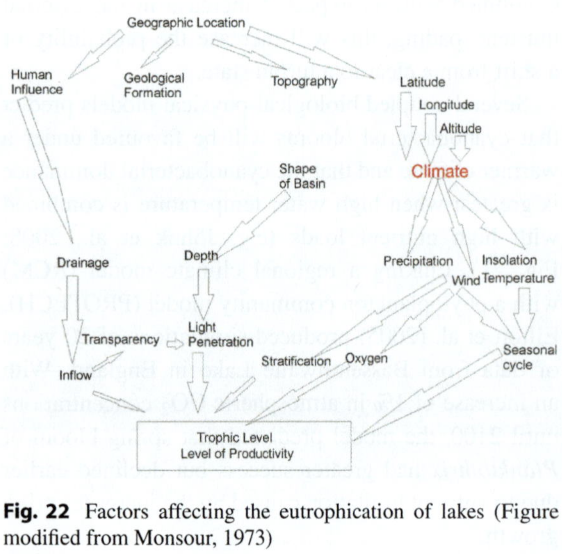

Fig. 22 Factors affecting the eutrophication of lakes (Figure modified from Monsour, 1973)

Nõges et al. (2007) have shown a substantial increase of nutrient loading and dissolved organic matter by river discharge in Lake Võrstjärv, Estonia, with increased precipitation. Moser et al. (2002) found that warmer summer temperatures increased thermal stratification leading to increase diatom-inferred total phosphorus (TP). During the mid-Holocene, lake levels were 8 m lower than today in a boreal lake of northwestern Ontario and lead to an increase of nutrient-rich diatom assemblages, an increase in diatom accumulation and a decrease in the chrysophytes relative to the diatom ratios (Moos et al., 2009). In China, shallow lakes are more eutrophied and prone to algal blooms due to the wind-wave resuspension of material which may result in huge nutrient releases (Qin et al., 2006). Carvahlo et al. (2003) have shown that winter temperature increased from 2.5°C to 3.6°C between 1968–1972 and 1997–2001 leading to a displacement of diatom peaks occurring in February instead of in spring.

Five major changes can occur following an increase in eutrophication: (1) species diversity decreases and the dominant biota changes, (2) plant and animal biomass increase, (3) turbidity increases, (4) the rate of sedimentation increases and (5) anoxic conditions may develop.

Changes Through Time

Eutrophication has been reported in association with the increase of human activities in recent times. Here, we try to report only on eutrophication/lake productivity which occurred in the past and was linked primarily to climate. This is not an easy task since most eutrophication studies are made in lakes where anthropogenic activities have altered the ecosystem. Since many of the algae responsible for eutrophication are not preserved in lake sediment (except as pigments), the reconstructed level of productivity or eutrophication might be underestimated through time.

The Last Glacial Maximum

A decrease in aquatic productivity during cold periods, such as the Younger Dryas, were observed in Ontario, Canada (Yu et al., 2000, 2008) in Spain, (Morellón et al., 2009) and on Easter Island (Sáez et al., 2009).

The Holocene

Warmer periods during the Holocene lead to increases in productivity in France (Loizeau et al., 2000), in Lituania (Stancikaitė et al., 2008), in Finland (Itkonen et al., 1999), in the Canadian arctic (Porinchu et al., 2009) and in Mexico (Bridgwater et al., 1999). Such increases in primary productivity have negative feedbacks on copepods and foraminifera (Cromer et al., 2005).

The Last Millennium

Increases in lake productivity were observed during warmer periods, such as the Medieval Warm Period in the Canadian arctic (Podritske and Gajewski, 2007), in Minnesota (USA) (Bradbury et al., 2002) and in Chile (Moy et al., 2008).

Last Decades

Unprecedented changes in lake productivity were observed on Baffin Island (Thomas et al., 2008) and in Alberta (Canada) (Moser et al., 2002). Increases in productivity associated with warmer climate were observed in British Columbia (Canada) (French and Petticrew, 2007), in Lake Tahoe (USA) (Coats et al., 2006), in northern Russia (Solovieva et al., 2005) and in Uganda (Panizzo et al., 2008). However, O'Reilly et al. (2003) suggested that primary productivity may have decreased by about 20%, implying a roughly 30% decrease in fish yields.

Future Changes

Byron and Goldman (1991) used GCM models and 25 years of limnological data to simulate changes (with doubled atmospheric CO_2) in primary production in a subalpine lake. The primary cause of enhanced productivity was the increased length of the growing season resulting from earlier spring ice-out. On the contrary, Brooks and Zastrow (2002) predicted a decrease in primary production in Lake Michigan by 2090 due to physical/chemical constraints imposed on spring primary production by altered climate conditions. Early stratification would shorten the period of winter-spring mixing, during which time nutrients from the sediment are transported to the productive euphotic zone. The spring bloom was projected to diminish if early stratification capped the nutrient supply, and increased cloud cover reduced light input for photosynthesis. In other Great Lakes, the increased productivity under doubled CO_2 models varied from strong increases in Lake Erie to almost negligable values in Lake Ontario (except near major tributaries) (Lam and Schertzer, 1999). In a mesocosm experiment, Sommer and Lengfellner (2007) showed that experimental temperature elevation had a strong effect on phytoplankton peak biomass (decreasing with temperature), mean cell size (decreasing with temperature) and on the share of microplankton diatoms (decreasing with temperature). All these changes will lead to poorer feeding conditions for copepod zooplankton and, thus, to a less efficient energy transfer to fish production under a warmer climate. In the Arctic, melting of the permafrost will likely increase nutrient input and primary productivity (Wrona et al., 2006). Using a lake model tuned with 30 years of meteorological data Blenckner et al. (2002) predicted that a warmer climate might cause increased nutrient cycling and lake productivity in Lake Erken, Sweden.

In summary, the impact of future climate change on primary production/eutrophication is more complicated than expected and will depend greatly on the physical properties of the lakes. This small overview enhances the need of studying various lakes in different part of the globe.

Oxygen Availability

Oxygen in a lake is produced via photosynthesis in the epilimnion (Fig. 23) while oxygen is used for respiration in the hypolimnion. The oxygen level available for respiration in the hypolimnion depends on two factors: the light penetration and the stratification. A strong thermocline in summer can lead to anoxia when there is no mixing and availability of oxygen in the hypolimnion. If a lake becomes eutrophic, the light penetration in the epilimnion is limited, decreasing the availability of oxygen to the hypolimnion.

With climate warming, algae production is predicted to increase leading to eutrophication (see above) and a decrease of oxygen availability since no light can

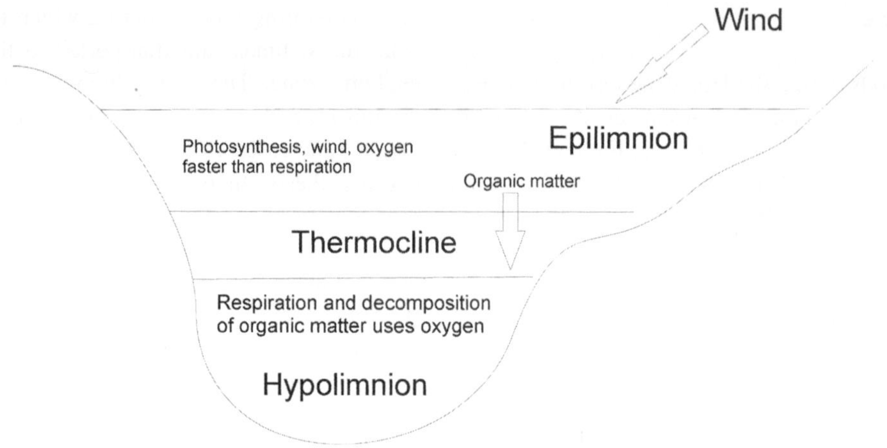

Fig. 23 Production of oxygen in stratified lakes

penetrate the mat of algae and biological production in the deeper part of the lake becomes negligible. Cold climate can also lead to oxygen depletion when the lake ice-cover is thicker and the ice-cover season is longer, creating lower light penetration, low productivity of algae, thus less oxygen in the deeper part of the lake. Variations in oxygen levels through time can be determined using sedimentological, chemical (sulfur) and biological (e.g. chironomids) indicators. In recent times, it is difficult to determine if climate or anthropogenic activities are responsible for the decline in oxygen level.

Changes Through Time

Long-Temporal Scale

Low susceptibility intervals over the last 65,000 cal years BP correlated with cold (glacial-like) conditions and anoxia due to perennial ice-cover in El'gygytgyn Crater Lake, Northeastern Siberia (Asikainen et al., 2007). During the Younger Dryas cold phase increased soil erosion and hypolimnetic oxygen depletion enhanced the nutrient supply to the lake water causing eutrophic conditions in north-eastern Germany (Enters et al., 2010)

The Holocene

Climate warming was associated with the deterioration of hypolimnetic oxygen conditions in various lakes of Europe (Schmidt et al., 2008) and North America (Odegaard et al., 2003).

The Last Millennium

Warm conditions were associated with anoxia and/or decreased hypolimnetic oxygen in lakes of USA (Whitlock et al., 2008) and the Austrian Alps (Schmidt et al., 2008).

Last Decades

Increasing productivity and hypolimnetic anoxia conditions were observed during the twentieth century in Michigan (USA) (Francis, 2001), in UK (Winfield et al., 2008) and in Switzerland (Straile et al., 2003). Low dissolved oxygen conditions during the summer of 2003 was induced by extreme warm temperature in Germany (Wilhelm and Adrian, 2007).

Future Changes

In Lake Zurich, Switzerland, a 50-year model predicted that increased temperatures will lead to a decrease in winter mixing events (Peeters et al., 2002) possibly leading to a decrease in hypolimetic oxygen.

Lake Level Changes

As shown by Street-Perrott and Harrison (1985), instrumental and historical records of lake levels are available for a small number of sites. Unfortunately, these records are of insufficient duration and spatial coverage to identify regional-to-global scale trends. To fill this void in knowledge, lake levels are reconstructed using a combination of biological and geochemical/geophysical methods. On an individual scale, lake level reconstructions have been used to identify local shifts in the water balance caused by glacier melt, ice-damming, changes to vegetative cover in the watershed and redirection of drainage (e.g. Brahney et al., 2008). On a regional-to-global scale, synchronous shifts in lake levels can provide information about high and low frequency shifts in temperature and evaporation (including solar insolation), precipitation (prevailing winds and storm tracks), climate oscillations (e.g. ENSO, AAO), and transitions from glacial to interglacial periods (e.g. Magny, 2004).

Ideal sites for lake level reconstruction include those found in small, topographically closed watersheds (e.g. Shuman et al., 2009). Shuman et al. (2009) and Last et al. (1998) suggest lakes which are recharged by groundwater so that shifts in lake level reflect multiple years of precipitation or evaporation. Therefore, lake basins must have sufficient relief to maintain water levels above the water table even during dry conditions without being over-steepened and risking slope failure (Street-Perrott and Harrison, 1985). Because immediate responses to climate can be dampened by long-term changes in groundwater, Street-Perrott and Harrison (1985) suggest lakes whose runoff equals inputs (amplifier lakes) and those which are sensitive to changes in the atmosphere. Lakes with limited local stream in- and outflows which concentrate the effects of consecutive short-term fluctuations in temperature and precipitation to create larger lake level transgressions or regressions while avoiding problems of incision are preferred (Shuman et al., 2009). Finally, Harrison et al. (1993) and Colman (1998) avoid reconstructing levels from lakes exposed to active tectonics. Unfortunately, even where lakes levels respond to changing climate or environmental conditions, the magnitude of lake level fluctuations (or absolute lake level) is difficult to reconstruct (Harrison et al., 1996). Instead, relative shifts are identified and refer to an increase or decrease relative to a lake's internal range of fluctuations (Street and Grove, 1979). Furthermore, it is not always possible to identify the start and end of regressions and transgressions (Magny, 2004). Once a suitable lake is selected, a variety of biological and geochemical/geophysical methods can be applied to identify past changes in lake level. Descriptions of these methods can be found in various textbooks (e.g. Smol et al., 2001; Francus, 2004).

The Late Glacial

At the Late Glacial to Holocene transition, lake levels in the Americas appear to be strongly influenced by the latitudinal location of the lakes. Higher lake levels in North American sites have been ascribed to an equator-ward shift of the Westerlies due to the presence of the Laurentide ice sheet (and greater winter precipitation), a strengthened south-western monsoon (and enhanced summer precipitation), and reduced evaporation due to cooler temperatures (Killarney-Younger Dryas) (e.g. Lavoie and Richard, 2000). In South America, conflicting lake level results have been published for the Altiplano and Amazonia region during Last Glacial Maximum (Baker et al., 2001). Baker et al. (2001) suggest that a range of climate forcings may have caused high lake levels at Titicaca. A high range of sea-surface temperatures could have strengthened north-east trade winds and enhanced the atmospheric advection of cold water into the Amazon region. Concurrently, high rainy-season solar insolation in the Southern Hemisphere would have hampered precipitation (Baker et al., 2001). For Argentinean lakes Cardiel and Potrok Aike, low lake levels at 11,200 cal year BP have been attributed to Westerlies being displaced north. The transgression of both lakes by 10,230 cal years BP may be related to the return of the Westerlies to their southern position (e.g. Ariztegui et al., 2008).

Low lake levels in northern Europe (18,000–15,000 cal years BP) and higher levels along the Mediterranean (18,000 cal years BP) appear to be caused by a southern displacement of Westerlies, the formation of anticyclones over the ice sheet and cooler high latitude oceans (COHMAP, 1988).

As described by Gasse (2000), many African lake level reconstructions are plagued by weak chronologies (radiocarbon age distortions and low temporal resolution). Furthermore, the use of calibration functions (e.g. biological) and models is rare (Gasse, 2000). Despite these caveats, African lake levels at the Last Glacial Maximum appear to respond to a variety of climate forcings. Gasse (2000) suggests that arid conditions occurred in northern and southern Africa during the Last Glacial Maximum due to low land and sea surface temperatures in the tropical region. For the glacial–interglacial transition, Barker and Gasse (2003) propose different lake level trends above and below the equator in tropical eastern Africa (inter-hemispheric asymmetry). Low levels may be caused by a cool and arid climate initiated by lower sea level and amplified continentality, cooler sea surface temperatures (especially in the North Atlantic), intensified trade winds (and increased evaporation rates), a northward displacement of the Antarctic polar front and a reduced summer monsoon (e.g. Stott et al., 2002). Higher lake levels could be related to a southward displacement of the eastern African rainfall belt (Garcin et al., 2007).

Late Glacial lake level reconstructions for Asia imply widespread aridity, probably related to reduced monsoon precipitation (Sun et al., 2009). On the Tibetan Plateau, low lake levels occurred at Nam Co and Lake Koucha (Mügler et al., 2009; Mischke et al., 2008). In paleolake Zhuyeze, a coarse layer of Aeolian sand was deposited (Zhao et al., 2008). Meanwhile, Tal Chapar and Parihara had an ephemeral existence (Achyuthan et al., 2007).

In Australia, coastal aridity during the Late Glacial appears to be related to a southward displacement of Westerlies and the winter subtropical belt of high pressure and divergence (Harrison, 1993). Alternatively, high lake levels in the interior are possibly due to precipitation originating in the Pacific. As the high pressure belt returned equator-ward, the coastal region became increasingly moist while lakes of the Australian interior regressed (Harrison, 1993).

Holocene Lake Level Trends

Lower lake levels in eastern North America during the early Holocene have been attributed to enhanced evaporation caused by a reduced Laurentide ice sheet and enhanced summertime solar insolation. Coincidentally, periods of higher lake level in parts of southwestern North America are believed to relate to a stronger southerly flow supported by an enhanced monsoon low (e.g. Weng and Jackson, 1999). High lake levels at intermediate latitudes in the Southern Hemisphere may be due to higher temperatures, reduced Andean glacier cover, warm sub-Antarctic sea surface temperatures, north-shifted southern westerly storm-tracks and a greater presence of Easterlies (e.g. Ariztegui et al., 2008). During the Hypsithermal (8,500–5,500 cal year BP), the North American continental interior was dry and warm. Where high lake levels have been reconstructed for eastern North America, they are credited to enhanced precipitation related to a northward shift of the jet-stream (Weng and Jackson, 1999). In temperate South America, episodes of lower lake levels at Lago Cardiel (Argentina) have been ascribed to arid and cooler conditions (Ariztegui et al., 2008). During the second half of the Holocene, summer insolation and monsoon intensity declined in the North American continental interior. These factors, combined with a southward shift of winter Westerlies, increased winter precipitation and probably caused the lake level transgressions in the Midwestern United States at 5,000 cal years BP (Weng and Jackson, 1999; Shuman et al., 2009; Weng and Jackson, 1999). Moister conditions are registered slightly later in eastern North America (3,000 cal years BP), in southern Quebec (4,400 cal years BP), in northern Quebec (3,200 cal years BP; CAN), in the Yukon (2,800–1,300 cal years BP) and in California (2,000 cal years BP to present; USA) (Lavoie and Richard, 2000; Miousse et al., 2003; Brahney et al., 2008; Kirby et al., 2004). In Argentina, modern climate conditions developed after 5,000 cal years BP, characterized by seasonal shifts in westerly storm-tracks (towards the southern pole in austral summer and towards the equator during the austral winter) and an ENSO presence (e.g. Markgraf et al., 2003). Moist conditions at 5,000 cal years BP were followed by increased aridity until 2,500 cal years BP (Stine and Stine, 1990). This is reflected in depressed lake levels at Lago Cardiel and Laguna Potrok Aike (Haberzettl et al., 2008). Finally, the last 2,500 years were characterized by rapid shifts in lake level due to high frequency climate shifts (Stine and Stine, 1990).

The early Holocene was characterized by strong latitudinal gradients in moisture and temperature related to insolation, relic ice cover and a zonal atmospheric circulation in Europe. In Scandinavia, enhanced summer and annual insolation were caused by precession and tilt (maximal at 9,000 cal years BP). At the mid-latitudes, high summer insolation was matched by reduced winter insolation (e.g. Guiot et al., 1993). In Sweden, reconstructions of summer temperatures suggest a warmer climate; however, high lake levels imply that precipitation must have been sufficient to replace summer evaporation (Almquist-Jacobson,

1995). Possible causes for higher precipitation in Scandinavia include higher North Atlantic sea surface temperatures and a northward shift of the sub-tropical anticyclone (Guiot et al., 1993). Modern climate conditions arrived in Sweden by 4,000 cal years BP, including the year-round delivery of moist air by westerly and south-westerly winds and a reduced presence of migratory cyclones (Almquist-Jacobson, 1995). In Holzhauser et al. (2005), lake level fluctuations in west-central Europe are attributed to solar irradiance and the North Atlantic Oscillation.

Low lake levels in Africa at the start of the Holocene have been ascribed to a change in the continental/oceanic temperature gradient caused by the opening of the Northeast Atlantic, the extension of polar waters and a warmer deglacial atmosphere (e.g. Filippi and Talbot, 2005). Transgressions during the early to middle Holocene (e.g. Lake Victoria and Lake Tanganyika) may be caused by high sub-Antarctic sea surface temperatures and a shift in the atmospheric circulation regime of the tropics including a renewed monsoon presence (e.g. Stager et al., 2002).

Shifts between high and low lake levels in the early Holocene imply unstable climatic conditions in Asia, possibly related to the Younger Dryas and the return of the Asian monsoon (e.g. Achyuthan et al., 2007). In Australia, low lake levels between 13,000 and 12,000 cal years BP have been accredited to the Antarctic Cold Reversal as well as the Younger Dryas (Tibby and Haberle, 2007). Conversely, high levels between 11,000 and 7,000 cal years BP have been related to reduced summer insolation in the middle southern latitudes and the northward shift of the subtropical belt of high pressure and divergence (Harrison, 1993). In New Zealand, high lake levels (8,240 and 7,880 cal years BP) may be related to the 8,200 years melt-water event or to the North Atlantic climate anomaly (8,500–7,900 cal year BP) (e.g. Augustinus et al., 2008). Meanwhile, lake levels in China and Mongolia appear to reflect a megathermal episode (7,200–6,000 cal years BP) and a slight warming between 6,500 cal years BP and 5,500 cal years BP, respectively (e.g.Wünnemann et al., 2003). High lake levels between 5,000 cal years BP and 3,000 cal years BP have been attributed to cooler and moister conditions including intense monsoon precipitation (e.g. Mügler et al., 2009). Finally, between 3,000 cal years BP and present, lake levels record the return of arid conditions to certain regions of Austral-Asia (Wünnemann et al., 2003; Khandelwal et al., 2008; Sun et al., 2009; Zhao et al., 2008).

The Last Millennium

In Africa, variations in plants in lake margin vegetation indicate low lake levels in Tanzania, presumably as a result of reduced effective precipitation, contemporary with indications of relatively dry conditions at a prolonged period between ca. 1420 and 1680 AD (Ryner et al., 2008). In eastern Mozambique, high lake levels were reconstructed using diatom assemblages between 900 and 1300 AD, with short low levels at ca. 1100 and 1200 AD. The pollen data reconstructed repeated droughts during the Little Ice Age while higher lake levels are suggested after 1800 AD (Ekblom and Stabell, 2008).

Magnetic susceptibility variations in the sediment of Lake Shkodra (Albania) were attributed to erosion in the watershed related to climate change (Little Ice Age) (van Welden et al., 2008). Based on sedimentary facies, elemental and isotopic geochemistry, and biological proxies (diatoms, chironomids and pollen), shallow lake levels and saline conditions with poor development of littoral environments prevailed during medieval times (1150–1300 AD) in Spain (Morellón et al., 2009). Generally higher water levels and more dilute waters occurred during the LIA (1300–1850 AD). Declining lake levels during the twentieth century, coinciding with a decrease in human impact, are associated with warmer climate conditions. Similarly, using sedimentological, geochemical and palynological analyses Moreno et al. (2008) and all have shown an increase in flood frequency during the LIA and almost no flood during the MWP in Central Iberia.

In a lake in Ontario, the diatom-inferred depth record indicates several periods of persistent low levels during the nineteenth century, from 900 to 1100 AD, and for extended periods prior to 1,500 years ago. Periods of inferred high lake levels occurred from 500–900 AD to 1100–1650 AD (Laird and Cumming, 2009). In the Peace Athabaska Delta, Alberta multiproxy analysis showed that one lake had high levels during the LIA while another lake had low water level (Sinnatamby et al., 2009). These differences were due to the connectivity of the lakes with major tributaries. The lake which was not connected to Lake Athabaska showed low lake levels due to drier climate. Before 730 AD, varved sediment could not be found in Alaska, probably due to the increase of aridity and lowering of the lake level (Bird et al., 2009). Varve thickness was used to infer precipitations (and lake level) in Lago Puyehue (Chile) (Boes and Fagel, 2008). One significant regional peak of winter precipitation (>900 mm) was reconstructed in the mid-twentieth century and a significant period with lower winter precipitation (<400 mm) during the late Medieval Warm Period was inferred.

Last Decades

Longer term changes in some benthic species, the chrysophyte cysts to diatom valve ratio, %C, and C/N ratios suggest declined river inflow and a relative reduction in allochthonous inputs during the last century in Nunavut (Canada) (Stewart et al., 2008). Pollen inferred low lake levels in the late eighteenth and the late nineteenth centuries in Tanzania (Ryner et al., 2008). Declining lake levels in Spain during the twentieth century inferred by sedimentary facies, elemental and isotopic geochemistry, and biological proxies (diatoms, chironomids and pollen) are associated with warmer climate conditions (Morellón et al., 2009).

Future Changes

Kirono et al. (2009) suggested that lake levels in Africa will continue to fall until 2100. In the semi-arid region of China about a 40% annual precipitation change in recent years has caused large lake level decreases and serious drought in the arid basins. In the next few decades, a further decrease of 20–40% in lake levels seems very probable (Yu and Shen, 2010). In Lake Victoria, climate change scenarios were applied to a 75-year lake rainfall inflow series and evaporation data to estimate future water balances of the lake. The scenarios produced a potential fall in lake levels by the 2030s horizon, and a rise by the 2080s horizon (Tate et al., 2005).

UV Penetration in Lakes

Dissolved organic carbon (DOC) acts as a protection against UV penetration: the higher the DOC concentration, the shallower the UV penetration (Fig. 24). Climate and UV penetration in lakes have been shown to be related: climate warming can lead to a decrease in the protective dissolved organic content in lakes and increased UV penetration (Schindler et al., 1996). A decrease in lake level, due to decreased precipitation can have similar effects (Yan et al., 1996). DOC exerts a major control on the penetration of solar ultraviolet radiation (UVR) and photosynthetically available radiation (PAR) in northern lake waters (Laurion et al., 1997). Zhang et al. (2007) found that the UV attenuation increased with chromophoric dissolved organic matter (CDOM), chlorophyll-*a* (Chla) and total suspended matter (TSM) in high altitudinal Tibetan Plateau lakes. In arctic or high elevation lakes, however, the DOC concentration is low (0.2–0.4 mg/L) and UV attenuation is influenced by phytoplankton (Sommaruga, 2001).

In lakes with higher DOC concentration, there is a significant tendency of warmer surface temperature, probably due to greater heat absorption (Gunn et al., 2001; Fig. 25). These results suggest that DOC

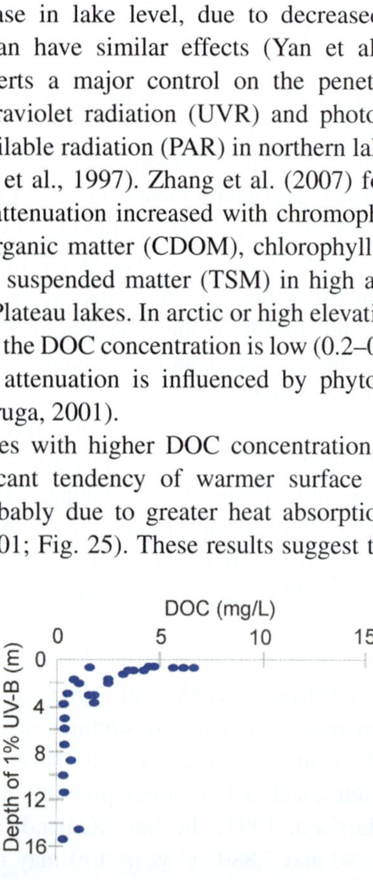

Fig. 24 Depth of UV penetration in relation to DOC concentration. Figure modified from Häder et al. (2007)

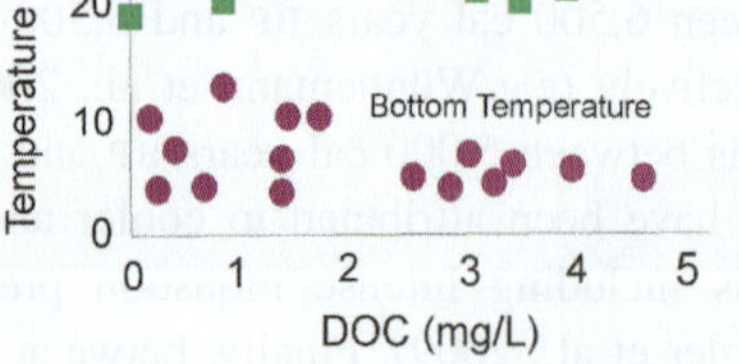

Fig. 25 Relationhsip between water temperature and DOC concentration. Figure modified from Gunn et al. (2001)

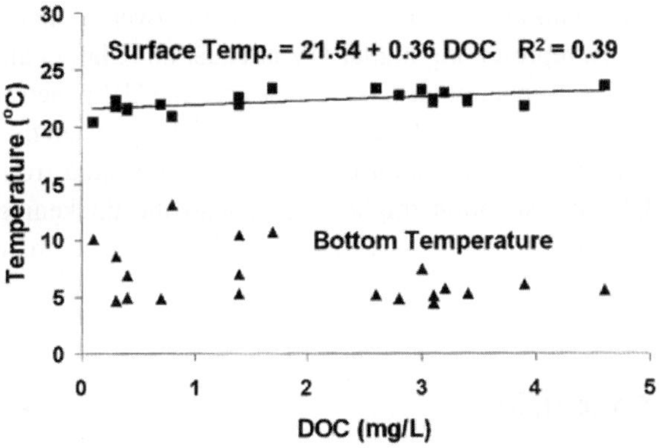

Fig. 26 Changes in thermal stratification with increasing DOC concentration. Figure modified from Gunn et al. (2001)

concentration increases with climate warming but has also a feedback while higher DOC concentration will trap the heat in the surface water. Other lakes parameters affected by DOC concentration are the depth of the 1% light penetration (PAR) and the depth of the 10°C isotherm (Fig. 26).

Exposure to solar UV radiation can reduce productivity, affect reproduction and development, and increase the mutation rate in phytoplankton, macroalgae, eggs and larval stages of fish and other aquatic animals (Häder et al., 2007). Macfadyen et al. (2004) have shown that organisms that depend on DNA repair processes may be less able to survive high UV exposure. Xenopoulos and Frost (2003) have shown that climate warming and increases in phosphorous change the composition of lake phytoplankton, small chrysophyte biomass was reduced by 350% after exposure to UV while others (*Tabellaria flocculosa*) were not affected. Zooplankton were found in shallow depth when Chaoborus was present but needed to stay at deeper depths when UV radiation increased, increasing their predation level (Boeing et al., 2004). Mesocosm experiments have demonstrated that larval chironomids are more sensitive to UVB than sympatric algae and might lead to increases in algae in habitats exposed to UVB due to lower predation from zooplankton (Bothwell et al., 1994). This latest study illustrates the complexity of UV effects on ecosystems. Long temporal scale studies can help in understanding how the lacustrine ecosystem has changed through time with changes in UV radiation. In arctic lakes, Sommaruga (2001) has shown that "understanding the overall impact of UVR on alpine lakes would need to consider synergistic and antagonistic processes."

Various indicators can be used to qualitatively or quantitatively estimate UV penetration. DOC can be quantitatively reconstructed through time based on transfer functions and UV penetration can be estimated (e.g. Dixit et al., 2001). Water transparency for UV radiation can be calculated from the inferred DOC values using the equations in Gibson et al. (2001). However, caution should be applied in such reconstruction since Sommaruga et al. (2006) have shown that in lakes with low DOC (range: 0.12–0.65 mg/L) located above the treeline, the relationship between dissolved organic matter and UV transparency can be insignificant but decrease in UV transparency was significantly correlated with the increase in phytoplankton chlorophyll a. Chlorophyll-*a* preservation was found to be positively related to light penetration (Buchaca and Catalan, 2007). Saulnier-Talbot et al. (2003) used a diatom-based palaeo-optical approach to estimate the depth of UV penetration and Leavitt et al. (2001) used fossil pigments analyses to determine UV penetration in lakes. The pigmentation of cladocerean can also be used to evaluate the UV penetration (e.g. Rautio and Korhola, 2002) or cladoceran assemblages can quantitatively estimate DOC levels (DeSellas et al., 2007). Cyanobacteria have photoprotective compounds which are preserved in sediments (Hodgson et al., 2005).

Changes Through Time

Late Glacial

During the Late glacial, dry cold conditions prevailed with lake-level minima and greater UV radiation was inferred (Hodgson et al., 2006). The mean exposure to UV during the Late Glacial was three times higher than during the Holocene (Hodgson et al., 2005).

The Holocene

At Lake Kachishayoot, in northern Québec (Canada), Saulnier-Talbot et al. (2003) estimated high UV exposure after the initial formation of the lake at ca. 5,400 cal years BP and the exposure decreased to the present. DOC was reconstructed in another lake in northern Québec and stable concentrations were estimated for the last 3,000 years, suggesting constant underwater light conditions (Ponader et al., 2002). Pienitz et al. (1999) inferred higher DOC concentration in mid-Holocene (5,000–3,000 cal years) associated with climate warming in lakes of the Northwestern territories, and this increased DOC was associated with a decrease in UV penetration in Queen's lake (Pienitz and Vincent, 2000).

High pigment concentrations in the early lake history occurred concomitant with low DOC content while this relationship was inversed during the mid-Holocene in Sweden (Reuss et al., 2009). In Austria, DOC concentrations were higher between 4,000 and 5,000 cal years BP, a time frame where the famous iceman was found, and were associated with warm summers (Schmidt et al., 2006).

The Last Millenium

Based on fossil pigments (scytonemins and its derivatives) and siliceous microfossils, Verleyen et al. (2005) reconstructed four well-defined maxima in the UVR proxy centred around 1820–1780, 1580–1490, 790–580 and 680–440 AD.

Last Decades

Optical proxies for UVR attenuation were correlated with chlorophyll-*a* concentration (0–30 m) during typical dry summer months from 1984 to 2002 in Crater Lake, Oregon (USA) (Hargreaves et al., 2007). In the Sudbury area (Ontario, Canada), 19 lakes were studied for their recovery following acidification and although the diatom and chrysophytes have partially recovered, their recovery might no be complete due to factors such as increased exposure to UV-B radiation which might be affecting the assemblages (Dixit et al., 2002).

Future Changes

Experimental research in Nova Scotia has shown that UV attenuation, due to increase DOC, will affect zooplankton. Increase in copepod reproduction, higher species diversity (evenness), and greater variability in the community structure was seen in covered enclosures, representing higher UV attenuation (Clair et al., 2001). Kepner et al. (2000) measured the UV penetration under the ice of Antarctic lakes and suggested that although the microbial organisms are protected from UV, this situation might change with the thickening and disappearance of ice cover predicted with climate warming.

Conclusions

This chapter described the potential of using lacustrine ecosystems to reconstruct climate (e.g. temperature, precipitation) and evaluate the effect of climatic changes on lakes. The findings are the following:

- Statistical methods have been recently developed to allow quantitative reconstruction using biological and sedimentological proxies.
- The differences between inferences and instrumental data are generally lower than the model predictions, allowing their use for reconstructing small climatic variations as observed during the last millennium.
- No consensus on the amplitude of climate change was obtained, however most sites (except one in Alaska) showed a Younger Dryas much colder (-4.5°C) than today's temperatures.
- The timing of the "Holocene Thermal Maximum" differed from sites and proxies used. However the climate was generally warmer after deglaciation until ca. 5,000 cal years BP.
- The 8,200 cal years BP event was present in most studied sites in the northern hemisphere.
- Warm Medieval and cold Little Ice Age periods were reconstructed at most studied sites in the northern hemisphere. The climate of the last century was similar to the MWP.
- In the arctic, Russia and Switzerland, the climate of the past decades is unprecedented during the last millennium. However, many sites showed that present-day temperatures are similar to those of the MWP.
- In the past, maximum biodiversity was associated with warmer climate. However, the future climate might reach the tolerance of many taxa in the arctic leading to a decrease in biodiversity.

- Cyanobacteria will increase with increased temperatures in lakes affected by nutrient increases.
- Lake productivity/eutrophication generally increased with warmer climate in the past; however, the link with climate is not direct. Physical parameters of the lake (especially depth and light penetration) are as important in influencing the bloom of phytoplankton.
- Oxygen depletion increased with warmer climate in the past and increased temperatures in the future might lead to anoxic lakes.
- During the Late Glacial, low lake levels were generally observed in northern Europe, South America, Africa, Asia and coastal Australia while high lake levels were reconstructed in North America, the Mediterranean and the interior of Australia.
- In the early Holocene, most lakes in North America had low lake levels which increased after ca. 5,000 cal years BP with increased precipitations. A similar pattern was observed in Africa. In Austral-Asia, most lakes had high levels in the early Holocene and levels decreased after ca. 3,000 cal years BP.
- In the last millennium and the last decades, warmer climate was generally associated with a decline in lake levels attributed mostly to a decrease in precipitation.
- The climate models applied to lakes in Africa and China suggest that the lake levels will continue to decline with climate warming.
- In arctic sites, UV penetration increased with warmer climate due to a decrease in ice cover. In southern sites, warmer climate lead to increases in DOC and higher UV attenuation.

References

Abrahams C (2008) Climate change and lakeshore conservation: a model and review of management techniques. Hydrobiologia 613:33–43.

Achyuthan H, Kar A, Eastoe C (2007) Late quaternary-Holocene lake-level changes in the eastern margin of the Thar Desert, India. J Paleolimnol 38:493–507.

Allen JRM, Lonh AJ, Ottley CJ, Pearson DG, Huntley B (2007) Holocene climate variability in northernmost Europe. Quat Sci Rev 26:1432–1453.

Almquist-Jacobson H (1995) Lake-level fluctuations at Ljustjärnen, central Sweden and their implications for the Holocene climate of Scandinavia. Palaeogeogr Palaeoclimatol Palaeoecol 118:269–290.

Ampel L, Wohlfarth B, Risberg J, Veres D (2008) Paleolimnological response to millennial and centennial scale climate variability during MIS 3 and 2 as suggested by the diatom record in Les Echets, France. Quat Sci Rev 27:1493–1504.

Anderson NJ, Brodersen KP, Ryves DB, McGowan S, Sander L, Jeppesen E, Leng MJ (2008) Climate versus in-lake processes as controls of community development in an oligotrophic lake in the low-Arctic of West Greenland. Ecosystems 11:307–324.

Ariztegui D, Anselmetti FS, Gilli A, Waldmann N (2008) Late pleistocene environmental change in Eastern Patagonia and Tierra del Fuego – A limnogeological approach. Dev Quat Sci 11:241–253.

Arnaud F, Revel M, Chapron E, Desmet M, Tribovillard N (2005) 7200 years of Rhône river flooding activity in Lake Le Bourget, France: a high-resolution sediment record of NW Alps hydrology. Holocene 15:420–428.

Asikainen CA, Francus P, Brigham-Grette J (2007) Sedimentology, clay mineralogy and grain-size as indicators of 65 ka of climate change from El'gygytgyn Crater Lake, Northeastern Siberia. J Paleolimnol 37:105–122.

Auer I, Böhm R, Jurkovic A, Lipa W, Orlik A, Potzmann R, Schöner W, Ungersbök M, Matulla C, Briffa K, Jones P, Efthymiadis D, Brunetti M, Nanni T, Maugeri M, Mercalli L, Mestre O, Moisselin J-M, Begert M, Müller-Westermeier G, Kveton V, Bochnicek O, Stastny P, Lapin M, Szalai S, Szentimrey T, Cegnar T, Dolinar M, Gajic-Capka M, Zaninovic K, Majstorovic M, Nieplova M (2007) HISTALP – historical instrumental climatological surface time series of the greater Alpine Region. Int J Climatol 27:17–46.

Augustinus P, Bleakley N, Deng Y, Shane P, Cochran U (2008) Rapid change in early holocene environments inferred from Lake Pupuke, Auckland City, New Zealand. J Quat Sci 23:435–447.

Axford Y, Geirsdóttir A, Miller GH, Langdon P (2009) Climate of the little ice age and the past 2000 years in northeast Iceland inferred from chironomids and other lake sediment proxies. J Paleolimnol 41:7–24.

Baker PA, Seltzer GO, Fritz SC, Dunbar RB, Grove MJ, Tapia PM, Cross SL, Rowe HD, Broda JP (2001) The history of South American tropical precipitation for the past 25,000 years. Science 291:640–643.

Barker P, Gasse F (2003) New evidence for a reduced water balance in East Africa during the last glacial maximum: implications for model-data comparison. Quat Sci Rev 22:823–837.

Barley EM, Walker IR, Kurek J, Cwynar LC, Mathewes RW, Gajewski K, Finney B (2006) A northwest North America training set: distribution of freshwater midges in relation to air temperature and lake depth. J Paleolimnol 36:295–314.

Battarbee RW (2000) Palaeolimnological approaches to climate change, with special regard to the biological record. Quat Sci Rev 19:107–124.

Besonen MR, Patridge W, Bradley RS, Francus P, Stoner JS, Abbott MB (2008) A record of climate over the last millennium based on varved lake sediments from the Canadian high arctic. Holocene 18:169–180.

Bigler C, Hall RI (2003) Diatoms as quantitative indicators of July temperature: a validation attempt at century-scale with meteorological data from northern Sweden palaeogeography, palaeoclimatology, palaeoecology. 189:147–160.

Bird BW, Abbott MB, Finney BP, Kutchko B (2009) A 2000 year varve-based climate record from the central Brooks Range, Alaska. J Paleolimnol 41:25–41.
Blass A, Bigler C, Grosjean M, Sturm M (2007b) Decadal-scale autumn temperature reconstruction back to AD 1580 inferred from the varved sediments of Lake Silvaplana (southeastern Swiss Alps). Quat Res 68:184–195.
Blass A, Grosjean M, Troxler A, Strum M (2007) How stable are twentieth-century calibration models? A high-resolution summer temperature reconstruction for the eastern Swiss Alps back to AD 1580 derived from proglacial varved sediments. Holocene 17:51–63.
Blenckner T, Omstedt A, Rummukainen M (2002) A Swedish case study of comtemporary and possible future consequences of climate change on lake function. Aquat Sci 64:171–184.
Boeing WJ, Leech DM, Williamson CE, Cooke S, Torres L (2004) Damaging UV radiation and invertebrate predation: conflicting selective pressures for zooplankton vertical distribution in the water column of low DOC lakes. Oecologia 138:603–612.
Boes X, Fagel N (2008) Timing of the Late Glacial and Youmger Dryas cold reversal in Southern Chile varved sediments. J Paleolimnol 39:267–281.
Bordon A, Peyron O, Lézone A-M, Brewer S, Fouache E (2009) Pollen-inferred Late-Glacial and Holocene climate in southern Balkans (Lake Maliq). Quat Int. 200:19–30.
Bothwell ML, Sherbot DMJ, Pollock CM (1994) Ecosystem response to Solar Ultraviolet-B Radiation: Influence of trophic-level interactions. Science 265:97–100.
Bradbury P, Cumming B, Laird K (2002) A 1500-year record of climatic and environmental change in Elk Lake, Minnesota III: measures of past primary productivity. J Paleolimnol 27:321–340.
Bradley R (Ed) (1999) Paleoclimatology (2nd ed). San Diego, CA: Academic Press.
Brahney J, Clague JJ, Menounos B, Edwards TWD (2008) Geochemical reconstruction of late Holocene drainage and mixing in Kluane Lake, Yukon Territory. J Paleolimnol 40:489–505.
Brandriss ME, OŃeil JR, Edlund MB, Stoermer EF (1998) Oxygen isotope fractionation between diatomaceous silica and water. Geochim Cosmochim Acta 62:1119–1125.
Briand JF, Leboulanger C, Humbert JF, Bernard C, Dufour P (2004) Cylindrospermopsis raciborskii (Cyanobacteria) invasion at mid-latitudes: selection, wide physiological tolerance, or global warming? J Phycol 40:231–238.
Bridgwater ND, Heaton THE, O'Hara SL (1999) A late Holocene palaeolimnological record from central Mexico, based on faunal and stable-isotope analysis of ostracod shells. J Paleolimnol 22:383–397.
Broecker WS (2001) Was the medieval warm period global? Science 291:1497–1499.
Broecker WS (2006) Was the Younger Dryas triggered by a flood? Science 312:1146–1148.
Brooks SJ, Langdon PG, Heiri O (2007) The Identification and Use of Palaearctic Chironomidae Larvae in Palaeoecology (276 pp). QRA technical guide No. 10. Quaternary Research Association, London. ISBN 0907 780 717.
Brooks AS, Zastrow JC (2002) The Potential influence of climate change on offshore primary production in Lake Michigan. J Great Lakes Res 28:597–607.
Broxton WB, Abbott MB, Finney BP, Kutchko B (2009) A 2000 year varved-based climate record from the central Brooks Range, Alaska. J Paleolimnol 41:25–41.
Buchaca T, Catalan J (2007) Factors influencing the variability of pigments in the surface sediments of mountain lakes. Freshw Biol 52:1365–1379.
Bunbury J, Gajewski K (2009) Postglacial climate inferred from a lake at treeline, southwest Yukon Territory. Quat Sci Rev 28:354–369.
Büntgen U, Esper J, Frank DC, Nicolussi K, Schmidhalter M (2005) A 1052-year tree-ring proxy for Alpine summer temperatures. Clim Dyn 25:141–153.
Büntgen U, Frank DC, Nievergelt D, Esper J (2006) Summer temperature variations in the European Alps, AD 755–2004. J Clim 19:5606–5623.
Byron ER, Goldman CR (1991) The potential effects of global warming on the primary prodictivity of a subalpine lake. J Am Water Resour Assoc 26:983–989.
Carvahlo L, Kirika A (2003) Changes in shallow lake functioning: Response to climate change and nutrient reduction. Hydrobiologia 506–509:789–796.
Caseldine C, Langdon P, Holmes N (2006) Early Holocene climate variability and the timing and extent of the Holocene thermal maximum (HTM) in northern Iceland. Quat Sci Rev 25:2314–2331.
Castenholz RW, Garcia-Pichel F (2000) Cyanobacterial responses to UV radiation. In BA Whitton, M Potts (Eds), The Ecology of Cyanobacteria (pp. 591–611). Dordrecht, The Netherlands: Kluwer Academic Publishers.
Casty C, Wanner H, Luterbacher J, Esper J, Böhm R (2005) Temperature and precipitation variability in the European Alps since 1500. Int J Climatol 25:1855–1880.
Chase M, Bleskie C, Walker IR, Gavin DG, Hu FS (2008) Midge-inferred Holocene summer temperatures in Southeastern British Columbia, Canada. Palaeogeogr Palaeoclimatol Palaeoecol 257:244–259.
Cherapanova MV, Snyder JA, Brigham-Grette JB (2007) Diatom stratigraphy of the last 250 ka at Lake El'gygytgyn, northeast Siberia. J Paleolimnol 37:155–162.
Chmura GL, Stone PA, Ross MS (2006) Non-pollen microfossils in everglades sediments. Rev Palaeobot Palynol 141:103–119.
Clair TA, Ehrman J, Kaczmarska I, Locke A, Tarasick DW, Day KE, Maillet G (2001) Will reduced summer UV-B levels affect zooplankton populations of temperate humic and clearwater lakes? Hydrobiologia 462:75–89.
Coats R, Perez-Losada J, Schladow G, Richards R, Goldman C (2006) The warming of Lake Tahoe. Clim Change 76:121–148.
Cohen AS (Ed) (2003) Paleolimnology: The History and Evolution of Lake Systems. New York, NY: Oxford University Press.
COHMAP Members (1988) Climate changes of the last 18,000 years: Observations and model simulations. Science 241:1043–1052.
Coleman SM (2007) Conventional wisdom and climate history. PNAS 104:6500–6501.
Colman SM (1998) Water-level changes in Lake Baikal, Siberia: tectonism versus climate. Geology 26:531–534.
Cromer L, Gibson JAE, Swadling KM, Ritz DA (2005) Faunal microfossils: Indicators of Holocene ecological change in a

saline Antarctic lake. Palaeogeography, Palaeoclimatology, Palaeoecology. 221:83–97.

Cwynar LC, Levesque AJ (1995) Chironomid evidence for Late-Glacial climatic reversals in maine. Quat Res 43:405–413.

Daryin AV, Kalugin IA, Maksimova NV (2005) The scanning XRF-microanalysis of annual layers in microsection of bottom sediments of Lake Telteskoe Nucl Instrum Methods Phys Res A 543:255–258.

Davis TW, Berry DL, Boyer GL, Gobler CJ (2009) The effects of temperature and nutrients on the growth and dynamics of toxic and non-toxic strains of *microcystis* during cyanobacteria blooms. Harmful Algae 8:715–725.

De Senerpont Domis LN, Mooij WM, Huisman J (2007) Climate-induced shifts in an experimental phytoplankton community: A mechanistic approach. Hydrobiologia 584:403–413.

DeSellas AM, Paterson AM, Sweetman JN, ` JP (2007) Cladocera assemblages from the surface sediments of southcentral Ontario (Canada) lakes and their relationships to measured environmental variables. Hydrobiologia 600:105–119.

Dieffenbacher-Krall AC, Birkel S, Borns HW, Cwynar L, Doner LA, Fastook J, Jaconson JL Jr, Chase Langley G, Nurse A. 2009. A multi-proxy reconstruction of Younger Dryas environmental conditions in Northern Maine and the validity of chironomid-inferred temperatures. Northeastern Section – 44th annual meeting (22–24 March 2009).

Dixit SS, Dixit AS, Smol JP (2002) Diatom and chrysophyte transfer functions and inferences of post-industrial acidification and recent recovery trends in Killarney lakes (Ontario, Canada). J Paleolimnol 27:79–96.

Dixit SS, Keller W, Dixit AS, Smol JP (2001) Diatom-inferred dissolved organic carbon reconstructions provide assessments of past UV-B penetration in Canadian Shield lakes. Can J Fish Aquat Sci 58:543–550.

Downing JA, Prairie YT, Cole JJ, Duarte CM, Tranvik LJ, Striegl RG, McDowell WH, Kortelainen P, Caraco NF, Melack JM, Middelburg JJ (2006) The global abundance and size distribution of lakes, ponds, and impoundments. Limnol Oceanogr 51:2388–2397.

Dupuis AP, Hann BJ (2009) Warm spring and summer water temperatures in small eutrophic lakes of the Canadian Prairies: potential implications for phytoplankton and zooplankton. J Plankton Res 31:489–502.

Eggermont H, Heiri O, Russell J, Vuille M, Audenaert L, Verschuren D (2010) Paleotemperature reconstruction in tropical Africa using fossil chironomidae (Insecta: Diptera). J Paleolimnol 43:413–435.

Ekblom A, Stabell B (2008) Paleohydrology of Lake Nhaucati (Southern Mozambique), 400 AD to present. J Paleolimnol 40:1127–1141.

Elias SA (Ed) (2006a) Encyclopedia of Quaternary Science, Elsevier, London, 3576p.

Elias SA (Ed) (2006b) Quaternary beetle research: The state of the art. Quat Sci Rev 25(15–16):1731–2024.

Elliott JA, Thackery SJ, Huntingford C, Jones RG (2005) Combining a regional climate model with a phytoplankton community model to predict future changes in phytoplanton in lakes. Freshw Biol 50:1404–1411.

Ellison CRW, Chapman MR, Hall IR (2006) Surface and deep ocean interactions during the cold climate event 8200 years ago. Science 312:1929–1932.

Enters D, Kirilova E, Lotter AF, Lücke A, Parplies J, Jahns S, Kuhn G, Zolitschka B (2010) Climate change and human impact at sacrower See (NE Germany) during the past 13,000 years: a geochemical record. J Paleolimnol 43:719–737.

Esper J, Niederer R, Bebi P, Frank D (2008) Climate signal age effects – Evidence from young and old trees in the Swiss Engadin. For Ecol Manage 255:3783–3789.

EPA (1999) Climate change and public lands United States Environmental Protection Agency. EPA-236-F-99-004.

Fallu M-A, Pienitz R, Walker IR, Lavoie M (2005) Paleolimnology of a shrub-tundra lake and response of aquatic and terrestrial indicators to climatic change in arctic Quе́bec, Canada. Palaeogeogr Palaeoclimatol Palaeoecol 215:183–203.

Feurdean A, Klotz S, Brewer S, Mosbrugger V, Tãmas T, Wohlfarth B (2008) Late glacial climate development in NW Romania – Comparative results from three quantitative pollen-based methods. Palaeogeogr Palaeoclimatol Palaeoecol 265:121–133.

Filippi ML, Talbot MR (2005) The palaeolimnology of northern Lake Malawi over the last 25 ka based upon the elemental and stable isotopic composition of sedimentary organic matter. Quat Sci Rev 24:1303–1328.

Finkelstein SA, Gajewski K (2008) Responses of Fragilariod-dominated diatom assemblages in a small Arctic lake to Holocene climatic changes. Russell Island, Nunavut, Canada. J Paleolimnol 40:1079–1095.

Fletcher WJ, Boski T, Moura D (2007) Palynological evidence for environmental and climatic change in the lower Guadiana valley, Portugal, during the last 13 000 years. Holocene 17:481–494.

Folke C, Carpenter S, Walker B, Scheffer M, Elmqvist T, Gunderson L, Holling CS (2004) Regime shifts, resilience, and biodiversity in ecosystem management. Annu Rev Ecol Evol Syst 35:557–581.

Fortin M-C, Gajewski K (2010) Holocene climate change and its effect on lake ecosystem production on Northern Victoria Island, Canadian Arctic. J Paleolimnol 43:219–234.

Francis D (2001) A record of hypolimnetic oxygen conditions in a temperate multi-depression lake from chemical evidence and chronomid remains. J Paleolimnol 25: 351–365.

Francus P (Ed) (2004) Image Analysis, Sediments and Paleoenvironments (300 pp). Dordrecht, The Netherlands: Kluwer Academic Publishers.

French TD, Petticrew EL (2007) Chlorophyll a seasonality in four shallow eutrophic lakes (northern British Columbia, Canada) and the critical roles of internal phosphorus loading and temperature. Hydrobiologia 575:285–299.

Gajewski K, Bouchard G, Wilson SE, Kurek J, Cwynar LC (2005) Distribution of chironomidae (Insecta: Diptera) head capsules in recent sediments of Canadian Arctic lakes. Hydribiologia 549:131–143.

Garcin Y, Vincens A, Williamson D, Buchet G, Guiot J (2007) Abrupt resumption of the African Monsoon at the Younger Dryas – Holocene climatic transition. Quat Sci Rev 26: 690–704.

Gasse F (2000) Hydrological changes in the African tropics since the Last Glacial maximum. Quat Sci Rev 19:189–211.

Gibson JAE, Vincent WF, Pienitz R (2001) Hydrologic control and diurnal photobleaching of CDOM in a subarctic. Arch Hydrobiol 152:143–159.

Goosse H, Arzel O, Luterbacher J, Mann ME, Renssen H, Riedwyl N, Timmermann A, Xoplaki E, Wanner H (2006) The origin of the European "medievalwarm period." Clim Past 2:99–113.

Guilizonni P, Lami A, Manca M, Musazzi S, Marchetto A (2006) Palaeoenvironmental changes inferred from biological remains in short lake sediment cors from the Central Alps and Dolomites. Hydrobiologia 562:167–191.

Guiot J, Harrison SP, Prentice IC (1993) Reconstruction of Holocene precipitation patterns in Europe using Pollen and Lake-Level Data. Quat Res 40:139–149.

Gunn JM, Snucins ED, Yan ND, Arts MT (2001) Use of water clarity to monitor the effects of climate change and other stressors on Oligotrophic Lakes. Environ Monit Assess 67:69–88.

Haberzettl T, Kück B, Wulf S, Anselmetti F, Ariztegui D, Corbella H, Fey M, Janssen S, Lücke A, Mayr C, Ohlendorf C, Schäbitz F, Schleser GH, Wille M, Zolitschka B (2008) Hydrological variability in southeastern Patagonia and explosive volcanic activity in the southern Andean Cordillera during Oxygen Isotope Stage 3 and the Holocene inferred from lake sediments of Laguna Potrok Aike, Argentina. Palaeogeogr Palaeoclimatol Palaeoecol 259:213– 229.

Haltia-Hovi E, Saarinen T, Kukkonen M (2007) A 2000-year record of solar forcing on varved lake sediment ine astern Finland. Quat Sci Rev 26:678–689.

Hamilton PB, Ley LM, Dean S, Pick FR (2005) The occurrence of the cynobacterium Cylindrospermopsis raciborskii in Constance Lake: an exotic cynoprokaryote new to Canada. Phycologia 44:17–25.

Hargreaves BR, Girdner SF, Buktenica MW, Collier RW, Urbach E, Larson GL (2007) Ultraviolet radiation and bio-optics in Crater Lake, Oregon developments in hydrobiology long-term limnological research and monitoring at Crater Lake, Oregon. In GL Larson, R Collier, MW Buktenica (Eds), A Benchmark Study of a Deep and Exceptionally Clear Montane Caldera Lake (pp. 107–140). Dordrecht, The Netherlands: Springer.

Harrison SP, Prentice IC, Guiot J (1993) Climatic controls on Holocene lake-level changes in Europe. Clim Dyn 8:189–200.

Häder D-P, Kumar HD, Smith RC, Worrest RC (2007) Effects of solar UV radiation on aquatic ecosystems and interactions with climate change. Photochem Photobiol Sci 6:267–285.

Harrison SP, Yu G, Tarasov PE (1996) Late quaternary lake-level record from Northern Eurasia. Quat Res 45:138–159.

Heiri O, Millet L (2005) Reconstruction of Late Glacial summer temperatures from chironomid assemblages in Lac Lautrey (Jura, France). J Quat Sci 20:33–44.

Heiri O, Filippi M-L, Lotter AF (2008) Lateglacial summer temperature in the Trentino area (Northern Italy) as reconstructed by fossil chironomid assemblages in Lago di Lavarone (1100 m a. s.l.). Studi Tent Sci Nat Acta Geol 82:299–308.

Hense I (2007) Regulative feedback mechanisms in cyanobacteria-driven systems: a model study. Mar Ecol Prog Ser 339:41–47.

Herczeg AL, Smith AK, Dighton JC (2001) A 120 year record of changes in nitrogen and carbon cycling in Lake Alexandrina, South Australia: C:N, $\delta^{15}N$ and $\delta^{13}C$ in sediments. Appl Geochem 16:73–84.

Hodell DA, Schelske CL (1998) Production, sedimentation, and isotopic composition of organic matter in Lake Ontario. Limnol Oceanogr 43:200–214.

Hodgson DA, Vyverman W, Verleyen E, Leavitt PR, Sabbe K, Squier AH, Keely BJ (2005) Late Pleistocene record of elevated UV radiation in an Antarctic lake. Earth Planet Sci Lett 236:765–772.

Hodgson DA, Verleyen E, Sabbe K, Squier AH, Keely BJ, Leng MJ, Saunders KM, Vyverman W (2006) Late Quaternary climate-driven environmental change in the Larsemann Hills, East Antarctica, multi-proxy evidence from a lake sediment core. Quat Res 64:83–99.

Holmes JA, Cook ER, Yang B (2009) Climate change over the past 2000 years in Western China. Quat Int 194:91–107.

Holzhauser H, Magny M, Zumbühl HJ (2005) Glacier and lake-level variations in west-central Europe over the last 3500 years. Holocene 15:789–801.

Hostetler SW (1995) Hydrological and thermal response of lakes to climate: Description and modelling. In A Lerman, DM Imboden, JR Gat (Eds), Physics and Chemistry of Lakes (pp. 63–82). New York, NY: Springer.

Huisman J, Sharples J, Stroom J, Visser PM, Kardinaal WEA, Verspagen JMH, Sommeijer B (2004) Changes in turbulent mixing shift competition for light between phytoplankton species. Ecology 85:2960–2970.

IPCC (2007) Climate Change 2007. UK: Synthesis Report. An Assessment of the Intergovernmental Panel on Climate Change. Cambridge: Cambridge University Press.

Ilyashuk B, Gobet E, Heiri O, Lotter AF, van Leeuwen JFN, van der Knaap WO, Ilyashuk E, Oberli F, Ammann B (2009) Lateglacial environmental and climatic changes at the Maloja Pass, central Swiss Apls, as recorded by chironomids and pollen. Quat Sci Rev 28:1340–1353.

Itkonen A, Marttila V, Meriläinen JJ, Salonen V-P (1999) 8000-year history of palaeoproductivity in a large boreal lake. J Paleolimnol 21:271–294.

Jacquet S, Briand J-F, Leboulanger C, Avois-Jacquet C, Oberhaus L, Tassin B, Vincon-Leite B, Paolini G, Druart J-C, Ammeville O, Humbert J-F (2005) The proliferation of the toxic cyanobacterium planktothrix rubescens following restoration of the largest natural French lake (Lac du Bourget). Harmful Algae 4:651–672.

Jones VJ, Leng MJ, Solovieva N, Sloane HJ, Tarasov P (2004) Holocene climate of the Kola Peninsula; evidence from the oxygen isotope record of diatom silica. Quat Sci Rev 23: 833–839.

Jöhnk KD, Huisman J, Sharples J, Sommeijer B, Visser PM, Stroom JM (2008) Summer heatwaves promote blooms of harmful cyanobacteria. Glob Chang Biol 14:495–512.

Juillet Leclerc A, Labeyrie L (1987) Temperature dependence of the oxygen isotopic fractionation between diatom silica and water. Earth Planet Sci Lett 84:69–74.

Kamenik C, Schmidt R (2005 Chrysophyte resting stages: A tool for reconstructing winter/spring climate from Alpine lake sediments. Boreas 34:477–489.

Kardinaal WEA, Tonk L, Janse I, Hol S, Slot P, Huisman J, Visser PM (2007) Competition for light between toxic and nontoxic strains of the harmful cyanobacterium *microcystis*. Appl Environ Microbiol 73:2939–2946.

Kattel GR, Battarbee RW, Mackay AW, Birks HJB (2008) Recent ecological change in a remote Scottish mountain loch: An evaluation of a Cladocera-based temperature transfer-function Palaeogeogr Palaeoclimatol Palaeoecol 259 51–76.

Kennett DJ, Kennett JP, West A, Mercer C, Que Hee SS, Bement L, Bunch TE, Sellers M, Wolbach WS (2009) Nanodiamonds in the Younger Dryas boundary sediment layer. Science 323:94.

Kepner Jr RL, Wharton Jr. RA, Collier RD, Cockell CS, Jeffrey WH. (2000) UV radiation and potential biological effects beneath the perennial ice cover of an antarctic lake. Hydrobiologia 427:155–165.

Khandelwal A, Mohanti M, Garcia-Rodriguez F, Scharf BW (2008) Vegetation history and sea level variations during the last 13,500 years inferred from a pollen record at Chilika Lake, Orissa, India. Veg Hist Archaeobot 17:335–344.

Kienel U, Schwab MJ, Schettler G (2005) Distinguishing climatic from direct anthropogenic influences during the past 400 years in varved sediments from Lake Holzmaar (Eifel, Germany). J Paleolimnol 33:327–347.

Kirby ME, Poulsen CJ, Lund SP, Patternson WP, Reidy L, Hammond DE (2004) Late Holocene lake level dynamics inferred from magnetic susceptibility and stable oxygen isotope data: Lake Elsinore, Southern California (USA). J Paleolimnol 31:275–293.

Kirono DGC, Jones RN, Kent D, Leahy P 2009. Modelling lake levels under climate change conditions: Three closed lakes in Western Victoria. 18th World IMACS/MODSIM Congress, Cairns, Australia, 13–17 July 2009.

Korhola A, Sorvari S, Rautio M, Appleby PG, Dearing JA, Hu Y, Rose N, Lami A, Cameron NG (2002) A multi-proxy analysis of climate impacts on the recent development of subarctic Lake Saanajärvi in Finnish Lapland. J Paleolimnol 28:59–77.

Kurek J, Cwynar LC, Ager TA, Abbott MB, Edwards ME (2009) Late Quaternary paleoclimate of western Alaska inferred from fossil chironomids and its relation to vegetation histories. Quat Sci Rev 28:799–811.

Labeyrie L (1974) New approach to surface seawater palaeotemperatures using $^{18}O/^{16}O$ ratios in silica of diatom frustules. Nature 248:40–42.

Lacoul P, Freedman B (2006) Relationships between aquatic plants and environmental factors along a steep Himalayan altitudinal gradient. Aquat Bot 84:3–16.

Laird K, Cumming B (2009) Diatom-inferred lake level from near-shore cores in a drainage lake from the Experimental Lakes Area, northwestern Ontario, Canada. J Paleolimnol 42:65–80.

Lam DCL, Schertzer WM (1999) Potential Climate Change Effects on Great Lakes Hydrodynamics and Water Quality (232 pp). ASCE Publications, Boston, USA. ISBN 0784404135, 9780784404133.

Larocque I, Hall RI, Grahn E (2001) Chironomids as indicators of climate change: A 100-lake training set from a subarctic region of Northern Sweden (Lapland). J Paleolimnol 36:307–322.

Larocque I, Hall RI (2003) Chironomids as quantitative indicators of mean July air temperature: Validation by comparison with century-long meteorological records from Northern Sweden. J Paleolimnol 29:475–493.

Larocque I, Rolland N2006 Le guide visuel des chironomids sub-fossiles, du Québec à l'île d'Ellesmere, INRS rapport de Recherche R-900, ISBN 2-89146-430-3.

Larocque I (2008) Nouvelle fonction de transfert pour reconstruire la température à l'aide des chironomides préservés dans les sédiments lacustres. INRS Rapport de Recherche R1032, ISBN 978-2-89146-587-588.

Larocque I, Pienitz R, Rolland N (2006) Factors influencing the distribution of chironomids in lakes distributed along a latitudinal gradient in northwestern Quebec, Canada. Can J Fish Aquat Sci 63:1286–1297.

Larocque I, Finsinger W (2008) Late-glacial chironomid-based temperature reconstructions for Lago Piccolo di Avigliana in the Southwestern Alps (Italy). Palaeogeogr Palaeoclimatol Palaeoecol 257:207–223.

Larocque I, Grosjean M, Heiri O, Bigler C, Blass A (2009a) Comparison between chironomid-inferred July temperatures and meteorological data AD 1850–2001 from varved Lake Silvaplana, Switzerland. J Paleolimnol 41:329–342.

Larocque I, Grosjean M, Heiri O, Trachsel M (2009b) High-resolution mean July air temperature history since AD 1580 as reconstructed by chironomids preserved in varved Lake Silvaplana, Engadine, Switzerland. Holocene 18.

Larocque I, Heiri O, Wehrli M (2010) Late Glacial and Holocene temperature changes at Egelsee, Switzerland, reconstructed by chironomids (non-biting midges). J Paleolimnol 44:243–251.

Larocque-Tobler I, Grosjean M, Heiri O, Trachsel M, Kamenik C (2010) Thousand years of climate change reconstructed from chironomid subfossils preserved in varved-lake Silvaplana, Engadine, Switzerland. Quat Sci Rev.

Last WM, Vance RE, Wilson S, Smol JP (1998) A multi-proxy limnologic record of rapid early-Holocene hydrologic change on the northern Great Plains, southwestern Saskatchewan, Canada. Holocene 8:503–520.

Laurion I, Vincent WF, Lean DR (1997) Underwater ultraviolet radiation: development of spectral models for northern high latitude lakes. Photochem Photobiol 65:107–114.

Lavoie M, Richard PJH (2000) Postglacial water-level changes of a small lake in southern Québec, Canada. Holocene 10:621–634.

Leavitt PR, Hodgson DA (2001) Sedimentary pigments. In JP Smol, HJB Birks, WM Last (Eds), Tracking Environmental Changes Using Lake Sediments (vol. 3, pp. 295–325) Kluwer Academic Publishers, Boston, USA.

Leng MJ, Barker PA (2006) A review of the oxygen isotope composition of lacustrine diatom silica for palaeoclimate reconstruction. Earth Sci Rev 75:5–27.

Lennox B, Spooner I, Jull T, Patterson WP (2010) Post-glacial climate change and its effect on a shallow dimictic lake in Nova Scotia, Canada. J Paleolimnol 43:15–27.

Lim DSS, Smol JP, Douglas MSV (2008) Recent environmental changes on Banks Island (N.W.T., Canadian Arctic) quantified using fossil diatom assemblages. J Paleolimnol 40:385–398.

Loizeau J-L, Span D, Coppee V, Dominik J (2000) Evolution of the trophic state of Lake Annecy (eastern France) since the last glaciation as indicated by iron, manganese and phosphorus speciation. J Paleolimnol 25:205–214.

Loso MG, Anderson RS, Anderson SP, Reimer PJ (2006) A 1500-year record of temperature and glacial response inferred from varved Iceberg Lake, southcentral Alaska. Quat Res 66:12–14.

Loso MG (2009) Summer temperatures during the medieval warm period and little ice age inferred from varved proglacial

lake sediments in southern Alaska. J Paleolimnol 41: 117–128.
Luoto TP, Nevalainen L, Sarmaja-Korjonen K (2008) Multiproxy evidence for the 'Little Ice Age' from Lake Hamptra¨sk, Southern Finland. J Paleolimnol 40: 1097–1113.
Lücke A, Schleser GH, Zolitschka B, Negendank JFW (2003) A Lateglacial and Holocene organic carbon isotope record of lacustrine palaeoproductivity and climatic change derived from varved lake sediments of Lake Holzmaar, Germany. Quat Sci Rev 22:569–580.
MacDonald GM, Moser KA, Bloom AM, Porinchu DF, Potito AP, Wolfe BB, Edwards TWD, Petel A, Orme AR, Orme AJ (2008) Evidence of temperature depression and hydrological variations in the eastern Sierra Nevada during the Younger Dryas stade. Quat Res 70:131–140.
Macfadyen EJ, Williamson CE, Grad G, Lowery M, Jeffrey WH, Mitchell DL (2004) Molecular response to climate change: temperature dependence of UV-induced DNA damage and repair in the freshwater crustacean Daphnia pulicaria. Glob Chang Biol 10:408–416.
Magny M (2004) Holocene climate variability as reflected by mid-European lake-level fluctuations and its probable impact on prehistoric human settlements. Quat Int 113:65–79.
Markgraf V, Bradbury JP, Schwalb A, Burns SJ, Stern C, Ariztegui D, Gilli A, Anselmetti FS, Stine S, Maidana N (2003) Holocene palaeoclimates of southern Patagonia: Limnological and environmental history of Lago Cardiel, Argentina. Holocene 13:581–591.
Massol F, David P, Gerdeaux D, Jarne P (2007) The influence of trophic status and large-scale climatic change on the structure of fish communities in Perialpine lakes. J Anim Ecol 76: 538–551.
Mayr C, Lücke A, Maidana NI, Wille M, Habersettl T, Corbella H, Ohlendorf C, Schäbitz F, Fey M, Janssen S, Zolitschka B (2009) Isotopic fingerprints on lacustrine organic matter from Laguna Potrok Aike (southern Patagaonia, argentina) reflect environmental changes during the last 16,000 years. J Paleolimnol 42:81–102.
McGowan S, Juhler RK, Anderson NJ (2008) Autotrophic response to lake age, conductivity and temperature in two West Greenland lakes. J Paleolimnol 39:301–317.
McKay NP, Kaufman DS, Michelutti N (2008) Biogenic silica concentration as a high-resolution, quantitative temperature proxy at Hallet Lake, south-central Alaska. Geophys Res Lett 35. doi:10.1029/2007GL032876.
Metcalfe SE, Davies SJ (2007) Deciphering recent climate change in central Mexican lake records. Clim Change 83:169–189.
Millet L, Arnaud F, Heiri O, Magny M, Verneaux V, Desmet M (2009) Late-Holocene summer temperature reconstruction from chironomid assemblages of Lake Anterne, northern French Alps. Holocene 19:317–328.
Miousse L, Bhiry N, Lavoie M (2003) Isolation and water-level fluctuations of Lake Kachishayoot, Northern Québec, Canada. Quat Res 60:149–161.
Mischke S, Kramer M, Zhang C, Shang H, Herzschuh U, Erzinger J (2008) Reduced early Holocene moisture availability in the Bayan Har Mountains, northeastern Tibetan Plateau, inferred from a multi-proxy lake record. Palaeogeogr Palaeoclimatol Palaeoecol 267:59–76.
Monsour V (1973) Eutrophication. A Natural Process (21p). Fort Lauderdale, FL: Broward County School Board.
Mooij WM, Hülsmann S, De Senerpont Domis LN, Nolet BA, Bodelier PLE, Boers PCM, Pires LMD, Gons HJ, Ibelings BW, Noordhuis R, Portielje R, Wolfstein K, Lammens EHRR (2005) The impact of climate change on lakes in the Netherlands: A review. Aquat Ecol 39:381–400.
Mooij WM, Janse JH, De Senerpont Domis LN, Hülsmann S, Ibelings. BW (2007) Predicting the effects of climate change on temperate shallow lakes with the ecosystem model PCLake. Hydrobiologia 584:443–454.
Moos MT, Laird KR, Cumming BF (2009) Climate-related eutrophication of a small boreal lake in northwestern Ontario: a palaeolimnological perspective. Holocene 19:359–367.
Morellón M, Valero-Garcés B, González-Sampériz P, Vegas-Vilarrubia T, Rubio E, Rieradevall M, Delgado-Huertas A, Mata P, Romero O, Engstrom DR, Lopez-Vicente M, Navas A, Soto J (2009) Climate changes and human activities recorded in the sediments of Lake Estanya (NE Spain) during the medieval warm period and little ice age. J Paleolimnol doi: 10.1007/s10933-009-9346-3.
Moreno A, Valero-Garceés BL, González-Sampériz P, Rico M (2008) Flood response to rainfall variability during the last 2000 years inferred from the Taravilla Lake record (Central Iberian Range, Spain). J Paleolimnol 40:943–961.
Moschen R, Lücke A, Schleser GH (2005) Sensitivity of biogenic silica oxygen isotopes to changes in surface water temperature and palaeoclimatology. Geophys Res Lett 32. doi: 10.1029/2004GL022167.L07708.
Moschen R, Lücke A, Parplies J, Radtke U, Schleser GH (2006) Transfer and early diagenesis of biogenic silica oxygen isotope signals during settling and sedimentation of diatoms in a temperate freshwater lake (Lake Holzmaar, Germany). Geochim Cosmochim Acta 70:4367–4379.
Moschen R, Lücke A, Parplies J, Radtke U, Schleser GH (2009) Controls on the seasonal and interannual dynamics of organic matter stable carbon isotopes in mesotrophic Lake Holzmaar, Germany. Limnol Oceanogr 54:194–209.
Moser KA, Smol JP, MacDonald GM, Larsen CPS (2002) 19th Century eutrophication of a remote boreal lake a consequence of climate warming? J Paleolimnol 28:269–281.
Moss B, McKee D, Atkinson D, Collings SE, Eaton JW, Gill AB, Harvey I, Hatton K, Heyes T, Wilson D (2003) How important is climate? Effects of warming, nutrient addition and fish on phytoplankton in shallow lake microcosms. J Appl Ecol 40:782–792.
Moy CM, Dunbar RB, Moreno PI, Francois J-P, Villa-Martínez R, Mucciarone DM, Guilderson TP, Garreaud RD (2008) Isotopic evidence for hydrologic change related to the westerlies in SW Patagonia, Chile, during the last millennium. Quat Sci Rev 27:1335–1349.
Mügler I, Gleixner G, Günther F, Mäusbacher R, Daut G, Schütt B, Berking J, Schwalb A, Schwark L, Xu B, Yao T, Zhu L, Yi C (2009) A multi-proxy approach to reconstruct hydrological changes and Holocene climate development of Nam Co, Central Tibet. J Paleolimnol 1–24.
Nara F, Tani Y, Soma Y, Soma M, Naraoka H, Watanabe T, Horiuchi K, Kawai T, Osa T, Nakamura T (2005) Response of phytoplankton productivity to climate change recorded by sedimentary photosynthetic pigments in Lake

Hovsgol (Mongolia) for the last 23,000 years. Quat Int 136: 71–81.
Nesje A, Bjune AE, Bakke J, Dahl SO, Lie Ø, Birks HJB (2006) Holocene palaeoclimate reconstructions at Vanndalsvatnet, western Norway, with particular reference to the 8200 cal yr BP event. Holocene 16:717–729.
Nõges P, Kägu M, Nõges T (2007) Role of climate and agricultural practice in determining matter discharge into large, shallow Lake Võrstjäv, Estonia. Hydrobiologia 581:125–134.
Odegaard C, Rea DK, Moore TC Jr (2003) Stratigraphy of the mid-Holocene black bands in Lakes Michigan and Huron: Evidence for possible basin-wide anoxia. J Paleolimnol 29:221–234.
Ojala AEK, Heinsalu A, Kauppira T, Alenius T, Saarnisto M (2008a) Characterizing changes in the sedimentary environment of a varved lake sediment records in southern central Finland around 8000 cal yr BP. J Quat Sci 23:765–775.
Ojala AEK, Alenius T, Seppä H, Giesecke T (2008b) Integrated varve and pollen-based temperature reconstruction from Finland: evidence for Holocene seasonal temperature patterns at high latitudes. Holocene 18:529–538.
Ortu E, Peyron O, Bordon A, de Beaulieu JL, Siniscalco C, Caramiello R (2008) Lateglacial and Holocene climate oscillations in the South-western Alps: an attempt at quantitative reconsruction. Quat Int 190:71–88.
Overpeck JT, Webb T, Prentice IC (1985) Quantitative interpretation of fossil pollen spectra: Dissimilarity coefficients and the method of modern analogs. Quat Res 23:87–108.
O'Reilly CM, Alin SR, Plisnier P-D, Cohen AS, McKee BA (2003) Climate change decreases aquatic ecosystem productivity of Lake Tanganyika, Africa. Nature 424:766–768.
Paerl H, Huisman J (2009a) Blooms like it hot. Science 320:57.
Paerl HW, Huisman J (2009b) Climate change: A catalyst for global expansion of harmful cyanobacterial blooms. Environ Microbiol Rep 1:27–37.
Panizzo VN, Mackay AW, Ssemmanda I, Taylor R, Rose N, Leng MJ (2008) A 140-year record of recent changes in aquatic productivity in a remote, tropical alpine lake in the Rwenzori Mountain National Park, Uganda. J Paleolimnol 40:325–338.
Paus A, Velle G, Larsen J, Nesje A, Lie Ø (2006) Lateglacial nunataks in central Scandinavia: Biostratigraphical evidence for ice thickness from Lake Flåfattjønn. Tynset, Norway. Quat Sci Rev 25:1228–1246.
Peeters F, Livingstone DM, Goudsmith G-H, Kipfer R, Forster R (2002) Modeling 50 years of historical temperature profiles in a large central European lake. Limnol Oceanogr 47: 186–197.
Peros MC, Gajewski K (2009) Pollen-based reconstructions of late Holocene climate from the central and western Canadian arctic. J Paleolimnol 41:161–175.
Perren BB, Douglas MSV, Anderson NJ (2009) Diatoms reveal complex spatial and temporal patterns or recent limnological change in West Greenland. J Paelolimnol 42:233–247.
Pienitz R, Smol JP, MacDonald GM (1999) Paleolimnological reconstruction of Holocene climatic trends from two boreal lakes, Northwest Territories, Canada. J Paleolimnol 13:21–49.
Pienitz R, Vincent WF (2000) Effect of climate change relative to ozone depletion on UV exposure in subarctic lakes. Nature 404:484–487.
Pla S, Catalan J (2005) Chrysophyte cysts from lake sediments reveal the sub-millennial winter/spring climate variability in the northwestern Mediterranean region throughout the Holocene. Clim Dyn 24:263–278.
Poberezhnaya AE, Fedotov AP, Sitnikova TYa, Semenov My, Ziborova GA, Otiniva EL, Khabuev AV (2006) Palaeoecological and paleoenvironmental record of the Late Pleistocene record of Lake Khubsugul (Mongolia) based on ostracod remains. J Paleolimnol 36:133–149.
Podritske B, Gajewski K (2007) Diatom community response to multiple scales of Holocene climate variability in a small lake on Victoria Island, NWT, Canada. Quat Sci Rev 26:3179–3196.
Ponader K, Pienitz R, Vincent W, Gajewski K (2002) Limnological conditions in a subarctic lake (northern Québec, Canada) during the late Holocene: analyses based on fossil diatoms. J Paleolimnol 27:353–366.
Porinchu DF, MacDonald G, Rolland N (2009) A 2000 year midge-based paleotemperature reconstruction from the Canadian Arctic archipelago. J Paleolimnol 41:177–188.
Prasad S, Brauer A, Rein B, Negendank JFW (2006) Rapid climate change during the early Holocene in western Europe and Greenland. Holocene 16:153–158.
Pražáková M, Veselý J, Fott J, Majer V, Kopáček J (2006) The long-term succession of cladoceran fauna and paleoclimate forcing: A 14,600-year record from Plesne Lake, the Bohemian Forest. Biologia Bratislava 61:S387–S399.
Qin B, Yang L, Chen F, Zhu G, Zhang L, Chen Y (2006) Mechanism and control of lake eutrophication. Chin Sci Bull 51:2401–2412.
Quinlan R, Douglas MSV, Smol JP (2005) Food web changes in arctic ecosystems related to climate warming. Glob Chang Biol 11:1381–1386.
Racca JM, Racca R, Pienitz R, Prairie YT (2007) PaleoNet: New software for building, evaluating and applying neural network based transfer functions in paleoecology. J Paleolimnol 38:467–472.
Rautio M, Korhola A (2002) Effects of ultraviolet radiation and dissolved organic carbon on the survival of subarctic zooplankton. Polar Biol 25:469–473.
Reuss N, Leavitt PR, Hall RI, Bigler C, Hammarlund D (2009) Development and application of sedimentary pigments for assessing effects of climatic and environmental changes on subarctic lakes in northern Sweden. J Paleolimnol doi: 10.1007/s10933-009-9323-x.
Robson BJ, Hamilton DP (2003) Summer flow event induces a cyanobacterial bloom in a seasonal Western Australian estuary. Mar Freshw Res 54:139–151.
Rolland N, Larocque I, Francus P, Pienitz R, Laperrière L (2008) Holocene climate inferred from biological (Diptera:Chironomidae) analyses in a Southampton Island (nunavut, Canada) lake. Holocene 18:229–241.
Rolland N, Larocque I, Francus P, Pienitz R, Laperrière L (2009) Evidence for a warmer period during the 12th and 13th centuries AD from chironomid assemblages in Southampton Island, Nunavut, Canada. Quat Res 72:27–37.
Romero-Viana L, Keely BJ, Camacho A, Vicente E, Miracle MR (2009) Photoautotrophic community changes in Lagunilla

del Tojo (Spain) in response to lake level fluctuation: Two centuries of sedimentary pigment records. Org Geochem 40:376–386.

Rosquist GC, Rietti-Shati M, Shemesh A (1999) Late Glacial to middle Holocene climatic record of lacustrine biogenic silica oxygen isotopes from a Southern Ocena island. Geology 27:967–970.

Rosèn P, Dåbbak E, Renberg I, Hall RI (2000) Near-infrared spectrometry (NIRS): A new tool for inferring past climatic changes from lake sediments. Holocene 10:161–166.

Rull V, López-Sáez JA, Vegas-Vilarrubia T (2008) Contribution of non-pollen palynomorphs to the paleolimnological study of a high-altitude Andean lake (Laguna Verde Alta, Venezuela). J Paleolimnol 40:399–411.

Ryner M, Holmgren K, Taylor D (2008) A record of vegetation dynamics and lake level changes from Lake Emakat, northern Tanzania, during the last c. 1200 years. J Paleolimnol 40:583–601.

Sáez A, Valero-Garcés BL, Giralt S, Moreno A, Bao R, Pueyo JJ, Hernández A, Casas D (2009) Glacial to Holocene climate changes in the SE Pacific. The Raraku Lake sedimentary record (Easter Island, 27°S). Quat Sci Rev 28:2743–2759.

Sarmaja-Korjonen K, Seppä H (2007) Abrupt and consistent responses of aquatic and terrestrial ecosystems to the 8200 cal yr cold event: a lacustrine record from Lake Arapisto, Finland. Holocene 17:457–467.

Saulnier-Talbot E, Pienitz R, Vincent W (2003) Holocene lake succession and palaeo-optics of a subarctic lake, northern Québec (Canada). Holocene 13:517–526.

Saulnier-Talbot E, Larocque I, Pienitz R (submitted) Holocene summer temperatures in northernmost Ungava (Canada): A regional assessment of inferences from two chironomid-based transfer functions. Quat Res Sci.

Schindler DW, Jefferson Curtis P, Parker BR, Stainton MP (1996) Consequences of climate warming and lake acidification for UV-B penetration in North American boreal lakes. Nature 379:705–70.

Schleser GH, Helle G, Lücke A, Vos H (1999) Isotope signals as climate proxies: The role of transfer functions in the study of terrestrial archives. Quat Sci Rev 18:927–943.

Schmidt R, Kamenik C, Tessadri R, Koinig KA (2006) Climatic changes from 12,000 to 4,000 years ago in the Austrian Central Alps tracked by sedimentological and biological proxies of a lake sediment core. J Paleolimnol 35:491–505.

Schmidt R, Kamenik C, Roth M (2007) Siliceous algae-based seasonal temperature inference and indicator pollen tracking ca. 4,000 years of climate/land use dependency in the southern Austrian Alps. J Paleolimnol 38:541–554.

Schmidt R, Roth M, Tessardi R, Weckström K (2008) Disentangling late-Holocene climate and land use impacts on an Austrian alpine lake using seasonal temperature anomalies, ice-cover, sedimentology, and pollen tracers. J Paleolimnol 40:453–469.

Seppä H, MacDonald GM, Birks HJB, Gervais BR, Snyder JA (2008) Late-Quaternary summer temperature changes in the northern-European tree-line region. Quat Res 69:404–412.

Shatwell T, Köhler J, Nicklisch A (2008) Warming promotes cold-adapted phytoplankton in temperate lakes and opens a loophole for Oscillatoriales in spring. Glob Chang Biol 14:2194–2200. doi: 10.1111/j. 1365–2486.2008.01630.x.

Shemesh A, Burckle LH, Hays JD (1995) Late Pleistocene oxygen isotope records of biogenic silica from the Atlantic sector of the Southern Ocean. Paleoceanography 10: 179–196.

Shuman B, Henderson AK, Colman SM, Stone JR, Fritz SC, Stevens LR, Power MJ, Whitlock C (2009) Holocene lake-level trends in the Rocky Mountains, USA. Quat Sci Rev 28:1861–1879.

Sinnatamby RN, Yi Y, Sokal MA, Clogg-Wright KP, Asada T, Vardy SR, Karst-Riddoch TL, Last WM, Johnston JW, Hall RI, Wolfe BB, Edwards TWD (2009) Historical and paleolimnological evidence for expansion of Lake Athabasca (Canada) during the Little Ice Age. J Paleolimnol doi: 10.1007/s10933-009-9361-4.

Smol J, Birks HJB, Last WM (Eds) (2001) Tracking Environmental Change Using Lake Sediments. Dordrecht, The Netherlands: Kluwer Academic Publishers.

Smol JP (2002) Pollution of Lakes and Rivers. A Paleoenvironmental Perspective (280 pp). New York, NY: Oxford University Press.

Solovieva N, Jones VJ, Nazarova L, Brooks SJ, Birks HJB, Grytnes J-A, Appleby PG, Kauppila T, Kondratenok B, Renberg I, Ponomarev V (2005) Palaeolimnological evidence for recent climatic change in lakes from the northern Urals, arctic Russia. J Paleolimnol 33:463–482.

Soma Y, Tani Y, Soma M, Mitake H, Kurihara R, Hashomoto S, Watanabe T, Nakamuara T (2007) sedimentary steryl chlroin esters (SCEs) and other photosynthetic pigments as indicators of paleolimnological change over the last 28,000 years from the Buguldeika Saddle of Lake Baikal. J Paleolimnol 37:163–175.

Sommaruga R (2001) The role of solar UV radiation in the ecology of alpine lakes. J Photochem Photobiol B Biol 62:35–42.

Sommaruga R, Augustin G (2006) Seasonality in UV transparency of an alpine lake is associated to changes in phytoplankton biomass. Aquat Sci 68(2006):129–141.

Sommer U, Lengfellner K (2007) Climate change and the timing, magnitude, and composition of the phytoplankton spring bloom. Glob Chang Biol 14:1199–1208.

St-Jacques J-M, Cumming BF, Smol JP (2008) A 900-year pollen-inferred temperature and effective moisture record from varved Lake Mina, west-central Minnesota, USA. Quat Sci Rev 27:781–796.

Stager JC, Mayewski PA, Meeker LD (2002) Cooling cycles, Heinrich event 1, and the desiccation of Lake Victoria. Palaeogeogr Palaeoclimatol Palaeoecol 183:169–178.

Stančikaitė M, Šinkūnas P , Šeirienė V, Kisielienė D (2008) Patterns and chronology of the Lateglacial environmental development at Pamerkiai and Kašučiai, Lithuania. Quat Sci Rev 27:127–147.

Stewart KA, Lamoureux SF, Finney BP (2008) Multiple ecological and hydrological changes recorded in varved sediments from Sanagak Lake, Nunavut, Canada. J Paleolimnol 40:217–233.

Stine S, Stine M (1990) A record from Lake Cardiel of climate change in southern South America. Nature 345: 705–707.

Stott L, Poulsen C, Lund S, Thunell R (2002) Super ENSO and global climate oscillations at millennial time scales. Science 297:222–226.

Stow C 2008 Analysis of water level changes in lakes Michigan and Huron. 51st annual conference of the international association for Great Lakes Research, Trent University, Peterborough, Ontario, Canada, May 19–23.

Straile D, Jöhnk K, Rossknecht H (2003) Complex effects of winter warming on the physicochemical characteristics of a deep lake. Limnol Oceanoghr 48:1432–1438.

Street FA, Grove AT (1979) Global maps of lake-level fluctuations since 30000 BP. Quat Res 12:83–118.

Street-Perrott FA, Harrison SP (1985) Lake levels and climate reconstruction. In AD Hecht (Ed), Paleoclimate Analysis and Modeling (pp. 291–340). New York, NY: Wiley.

Sun Q, Wang S, Zhou J, Shen J, Cheng P, Xie X, Wu F (2009) Lake surface fluctuations since the late glaciation at Lake Daihai, North central China: A direct indicator of hydrological process response to East Asian monsoon climate. Quat Int 194:45–54.

Talbot MR, Johannessen T (1992) A high resolution palaeoclimatic record for the last 27,500 years in tropical West Africa from the carbon and nitrogen isotopic composition of lacustrine organic matter. Earth Planet Sci Lett 110:23–37.

Tarasov PE, Bezrukova EV, Krivonogov SK (2009) Late Glacial and Holocene changes in vegetation cover and climate in southern Siberia derived from a 15 Kyr long pollen record from lake Kotokel. Clim Past Discuss 5:127–151.

Tate E, Sutcliffe J, Conway D, Farquharson F (2005) Water balance of Lake Victoria: Update to 2000 and climate change modeling to 2100. Hydrol Sci 49:563–574.

Teranes JL, Bernasconi SM (2005) Factors controlling δ^{13}C values of sedimentary carbon in hypertrophic Baldeggersee, Switzerland, and implications for interpreting isotope excursions in lake sedimentary records. Limnol Oceanogr 50: 914–922.

Thomas EK, Axford Y, Briner. JP (2008) Rapid 20th century environmental change on northeastern Baffin Island, Arctic Canada inferred from a multi-proxy lacustrine record. J Paleolimnol 40:507–517.

Tibby J, Haberle SG (2007) A Late Glacial to present diatom record from Lake Euramoo, wet tropics of Queensland, Australia. Palaeogeogr Palaeoclimatol Palaeoecol 251:46–56.

Toivonen HTT, Mannila H, Korhola A, Olander H (2001) Applying Bayesian statistics to organusm-based environmental reconstruction. Ecol Appl 11:618–630.

Tonk L, Bosch K, Visser PM, Huisman J (2007) Salt tolerance of the harmful cyanobacterium *microcystis aeruginosa*. Aquat Microb Ecol 46:117–123.

Trachsel M, Eggenberger U, Grosjean M, Blass A, Sturm M (2008) Mineralogy-based quantitative precipitation and temperature reconstructions from annually laminated lake sediments (Swiss Alps) since AD 1580. Geophys Res Lett 35 L13707, 6pp.

Uwins HK, Teasdale P, Stratton H (2007) A case study investigating the occurrence of geosmin and 2-methylisoborneol (MIB) in the surface waters of the Hinze Dam, Gold Coast, Australia. Water Sci Technol 55:231–238.

Verleyen E, Hodgson DA, Sabbe K, Vyverman W (2005) Late Holocene changes in ultraviolet radiation penetration recorded in an East Antarctic lake. J Paleolimnol 34: 191–202.

Viau AE, Gajewski K, Sawada MC, Fines P (2006) Millennial-scale temperature variations in North America during the Holocene. J Geophys Res 111:D09102. doi: 10.1029/2005JD006031.

Väliranta MM (2006) Long-term changes in aquatic plant species composition in North-eastern European Russia and Finnish Lapland, as evidenced by plant macrofossil analysis. Aquat Bot 85:224–232.

van Welden A, Beck C, Reyss J-L, Bushati S, Koci R, Jouanne F, Mugnier J-L (<yr>2008</yr>) The last 500 year of sedimentation in Shkodra Lake (Albania/Montenegro): paleoenvironmental evolution and potential for paleoseismicity studies. J Paleolimnol 40:619–633.

Wagner B, Bennike O, Bos J, Cremer H, Lotter A, Melles M (2008) A multidisciplinary study of Holocene sediment records from Hjort Sø on Store Koldewey, Northeast Greenland. J Paleolimnol 39:381–398.

Walker IC, Cwynar LC (2006) Midges and palaeo- temperature reconstruction – The North American experience. Quat Sci Rev 25:1911–1925.

Waters MN, Piehler MF, Rodriguez AB, Smoak JM, Bianchi TS (2009) Shallow lake trophic status linked to late Holocene climate and human impacts. J Paleolimnol 42:51–64.

Weckström J, Korhola A, Erästo P, Holmström L (2006) Temperature patterns over the past eight centuries in Northern Fennoscandia inferred from sedimentary diatoms. Quat Res 66:78–86.

Weng C, Jackson ST (1999) Late Glacial and Holocene vegetation history and paleoclimate of the Kaibab Plateau, Arizona. Palaeogeogr Palaeoclimatol Palaeoecol 135:179–201.

Wetzel RG (1983) Limnology (767p). Philadelphia, PA: Saunders College Publishing House.

Whitehead PG, Wilby RL, Battarbee RW, Kernan M, Wade AJ (2009) A review of the potential impacts of climate change on surface water quality. Hydrol Sci 54:101–123.

Whitlock C, Dean W, Rosenbaum J, Stevens L, Fritz S, Bracht B, Power M (2008) A 2650-year-long record of environmental change from northern Yellowstone National Park based on a comparison of multiple proxy data. Quat Int 188:126–138.

Whitney BS, Vincent JH, Cwynar LC (2005) A midge-based Late-Glacial temperature reconstruction from southwestern Nova Scotia. Canadian. J Earth Sci 42:2051–2057.

Wiedner C, Nixdorf B, Heinze R, Wirsing B, Neumann U, Weckesser J (2007) Regulation of cyanobacteria and microcystin dynamics in polymictic shallow lakes. Arch Hydrobiol 155:383–400.

Wilhelm S, Adrian R (2007) Long-term response of Dreissena polymorpha larvae to physical and biological forcing in a shallow lake. Oecologia 151:104–114.

Winfield IJ, James JB, Fletcher JM (2008) Northern pike (*Esox lucius*) in a warming lake: Changes in population size and individual condition in relation to prey abundance. Hydrobiologia 601:29–40.

Wrona FJ, Prowse TD, Reist JD, Hobbie JE, Levesque LMJ, Vincent WF (2006) Climate change effects on aquatic biota, ecosystem structure and function. Ambio 35:359–369.

Wünnemann B, Chen F, Riedel F, Zhang C, Mischke S, Chen G, Demske D, Ming J (2003) Holocene lake deposits of Bosten Lake, southern Xinjiang, China. Chin Sci Bull 48: 1429–1432.

Xenopoulos MA, Frost PC (2003) UV radiation, phosphorus, and their combined effects on the taxonomic composition of phytoplankton in a boreal lake. J Phycol 39:291–302.

Yan ND, Keller W, Scully NM, Lean DRS, Dillon PJ (1996) Increased UV-B penetration in a lake owing to drought-induced acidification. Nature 81:141–143.

Yu Z (2000) Ecosystem response to Late Glacial and early Holocene climate oscillations in the Great Lakes region of North America. Quat Sci Rev 19:1723–1747.

Yu G, Shen H (2010) Lake water changes in response to climate change in northern China: Simulations and uncertainty analysis. Quat Int 212:44–56.

Yu Z, Walker KN, Evenson EB, Hajdas I (2008) Late Glacial and early Holocene climate oscillations in the Matanuska Valley, south-central Alaska. Quat Sci Rev 27: 148–161.

Zhang YL, Zhang EL, Liu ML, Wang X, Qin BQ (2007) Variation of chromophoric dissolved organic matter and possible attenuation depth of ultraviolet radiation in Yunnan Plateau lakes. Limnology 8:311–319.

Zhao Q, Wang N, Li X (2008) Lacustrine strata sedimentology and lake-level history in ancient Zhuyeze Lake since the Last Deglaciation. Front Earth Sci China 2: 199–208.

Rivers

Emmanuel Gandouin and Philippe Ponel

Abstract Quaternary insect subfossils preserved in river deposits allow for the reconstruction of past environmental and climate changes. During the last glacial period, warm climate periods such as Dansgaard/Oeschger events or the Late-Glacial Interstadial have been recorded by the reappearance of temperate faunas, typical of very shallow lakes or low-energy anastomosing rivers. Inference models indicate temperatures reaching values similar to those of the present day. During cold periods such as the Younger Dryas, insect faunas were characterized by the presence of numerous cold-adapted species. Summer air temperatures ranged between 9 and 11°C. Winter temperatures were close to −11 to −12°C. During the early Holocene, temperatures rapidly increased to reach a thermal optimum (about 18–20°C) at around 8,000 year ^{14}C BP. From the Middle Holocene, climate reconstructions are biased by increasing local impact of human activities on river water bodies. In France, cultivation and pastoralism led to the clearance of alluvial forest around 5,000–4,000 BP. This induced the disappearance of riffle beetles due to the rise in load in fines in the stream resulting from soil erosion. The fossil fauna from 2,000 BP to the present is mainly made of ruderal grassland associated taxa, dung-beetles and synanthropic taxa associated with food stored products.

Keywords Chironomids · Coleoptera · River · Climate change · Flood impact · Human impact

E. Gandouin (✉)
Institut Méditerranéen d'Ecologie et de Paléoécologie, IMEP-CNRS, UMR 6116, Bâtiment Villemin, Domaine du Petit Arbois, Université Paul Cézanne, 13545 Aix-en-Provence Cedex 04, France
e-mail: e.gandouin@univ-cezanne.fr

The potential of biological remains in palaeoecological investigations of alluvial sedimentary deposits: with special emphasis to Insect data from France.

Introduction

Why study past climatic and environmental changes in river systems?

1. River floodplains provide good palaeoecological archives in regions where continuous lacustrine and peat bog records are rare, as in the case for seasonally dry regions, such as around the Mediterranean.
2. To improve our knowledge on both natural and non-natural (anthropogenic) processes that shaped the present fluvial hydrosystems.

The achievement of these aims needs analyses performed at different timescales:

(i) Long-term geological timescale: global climatic changes such as glacial cooling or interglacial warming lead to geomorphological changes that affect the hydrosystems sensu stricto (main channel and secondary arms) as well as flood plain landscapes. These changes have important consequences on dynamics of biological species that inhabit these ecosystems, and thus on the biodiversity and functions related to changing environments.
(ii) Historical timescale (500–100 year timescale): increased human impacts through organic or

J. Dodson (ed.), *Changing Climates, Earth Systems and Society*, International Year of Planet Earth,
DOI 10.1007/978-90-481-8716-4_7,

chemical contamination, river regulation and exploitation of water resources have profoundly modulated present hydrosystems (Petts, 1989). It is necessary to better document the mechanisms that lead to this anthropogenic environment in a perspective of ecological conservation or restoration aims. In addition, the sedimentary deposits of fluvial terraces and alluvial plains are important archives for archaeological studies since all over the world, rivers have attracted populations and are often the starting point of civilizations. Hence, the knowledge of the history of the catastrophic flood events or of the alternation between phases of aridity or wetness that affect economic activities (fisheries, navigation, water-supply for agriculture) may be very informative concerning the historical expansion of human populations and their trade linkages. In spite of their technological advances, modern societies are still exposed to these natural catastrophes or to the desertification related to both global warming and increased water consumption.

(iii) At 50–10 years timescale: the rapid growth of human demography and urbanization leads to degradation of natural river processes and dynamics. This very short term observation supports the conclusion of the previous historical timescale and emphasizes the necessity to improve the management of Large Alluvial Rivers Network, and thus at continental spatial scales. This requires comprehensive international cooperation in scientific programs as well as in politic or economic development strategies, since many catchments cross national boundaries.

Why Fossil Insects Are a Cost Efficient Tool to Analyse Past River Changes

Invertebrate faunas are directly impacted by changing ecological factors in rivers. Water temperature, chemistry, flow and discharge rates are particularly important. Moreover, macroinvertebrates play a key role in the trophic structure of alluvial ecosystems. In addition, responses of invertebrates to environmental changes are often very rapid. These taxonomic groups can be considered as functional biological descriptors for both climate and environmental reconstruction purposes. This allows detection of abrupt changes and thus permits work at high resolution timescales. This is particularly interesting in the present context of rapid global warming.

Recent palaeocological studies performed during the last few decades have shown that coleopteran (e.g. Buckland and Coope, 1991; Elias, 1994, 2009) and chironomids (e.g. Engels et al., 2008a, b, c; Gandouin et al., 2005, 2006) assemblage changes recorded from rivers and floodplain sequences are very efficient tools to reconstruct past temperatures, human impact, vegetational and hydrological changes, for both prehistorical and historical periods. In the present section, our aim is to provide a review of recent studies (with an emphasis on France) dealing with palolimnological research in river and floodplain hydrosystems.

Chironomid and Coleopteran Remains as Efficient Palaeoecological Tools

Chironomids

Among the invertebrate faunal groups colonizing aquatic environments, the Chironomidae (Diptera) inhabit all kinds of water bodies from river main channels, cut-off features such as oxbow lakes and to the most isolated, temporary aquatic environments (Wilson, 1980). Chironomids are abundant in freshwater sediments, where they are often the predominant macroinvertebrate taxon. Klink (1989) has reported, for example, that this group accounts for 90% of the entire macroinvertebrate population in the Lower Rhine region. As the chitinous remains of chironomid larvae are relatively well preserved in Quaternary sediments (Hofmann, 1986), chironomids are widely used as palaeoecological indicators in lake environments, in order to reconstruct high resolution past environmental dynamics. In addition the use of chironomid larvae for palaecological purposes is facilitated by relatively simple laboratory methods (Hofmann, 1986). The recent key for determination for European chironomid remains (Brooks et al., 2007) gives a review of the main type of head capsules encountered in Quaternary freshwater sediments for the region.

In lakes, chironomid species distribution follows the depth gradient (Barley et al., 2006) so they can be used as indicators of past lake level fluctuations. Some taxa such as *Limnophyes* and *Smittia* include terrestrial or semi-terrestrial species. The littoral part of the lake is colonized by taxa such as *Glyptotendipes*, *Dicrotendipes* and *Polypedilum* which contain species associated with macrophytes. The profundal zone of lakes is dominated by taxa, such as *Chironomus,* tolerant to low oxygen concentrations because of the presence of haemoglobin in their hemolymph.

Chironomids are widely used as indicators of past changes in lake water quality since a classification of lake trophic types has been established by Brundin (1958). A short list of useful European chironomid species is provided by Brooks et al. (2007). Recent studies have shown that chironomids can be used to reconstruct total phosphorus (Brooks et al., 2001) and chlorophyll a concentrations (Brodersen and Lindegaard, 1999). Chironomids are also sensitive to many other environmental variables such as pH (Mousavi, 2002), dissolved oxygen concentration (e.g. Quinlan and Smol, 2002), salinity (Heinrichs et al., 2001) and the nature of the substratum (e.g. Larocque et al., 2001). Chironomids have also been used to reconstruct air and/or water temperatures for various regions (e.g. Bedford et al., 2004; Caseldine et al., 2006; Stefanova et al., 2003). And, based on studies of the modern distribution of lacustrine chironomids along an altitudinal gradient, it is possible to develop modern chironomid data calibration sets (temperature-chironomid transfer functions), that allows inferences of T° July with prediction errors of the models in the order of 1°C (Brooks, 2006). However, the differences between instrumental data and inferences can be much smaller (see Larocque and Hall, 2003; Larocque et al., 2009).

Coleoptera

The majority of Quaternary fossil coleopteran species are morphologically identical with their modern counterparts, and their ecology has thus not changed significantly on this time scale (Coope, 1986; Buckland and Coope, 1991), at least in central and northern Europe. Since many fossil beetle fragments can be identified to species level, a number of ecological parameters may be drawn from any fossil assemblage, using the extensive entomological literature devoted to modern insects that has been published recently.

Coleopteran remains are widely used in investigations concerning archaeological sedimentary deposits because they present a very wide range of ecological requirements. They occur everywhere in terrestrial and aquatic ecosystems and thus are able to provide a diversity of palaeoecological information concerning agricultural and pastoral human activities, human habitats, climate and environmental changes.

In the context of floodplain sedimentary sequences, the analysis of fossil beetle assemblages may provide additional information concerning a wider area than the aquatic ecosystem itself, since an important part of the many species contained in the fossil records are terrestrial insects carried into the place of deposit either by active (flight) or passive (drift) transport. For example, huge assemblages of beetles mixed with plant detritus often accumulate in dead arms or slow meandering rivers during flood periods. Such assemblages recovered from sedimentary sequences are often extremely rich in taxa, most of them very well preserved due to the resistant chitinous exoskeleton. This material is extremely valuable since it may yield a wide range of data that are not necessarily provided by other proxies. For example, the vegetation that constitutes the riverine forest may be analysed in detail, since many phytophagous coleoptera are monophagous or oligophagous insects associated with one or very few tree or herbaceous plant species.

Saproxylophagous beetles may provide data concerning the age and the general condition of health of the riverine forest, since many species are associated with dead wood, declining trees, bracket fungi, tree mould and more. The nature of the river banks may also be interpreted from the fossil beetle record, since many ground or rove beetles are associated with a particular type of substratum: shingle, gravels, sand or clay. The landscape beyond the river bed may be reconstructed using beetle species associated with particular types of environments, such as beetles associated with steppic or forest vegetation, dry open substratum or marshy environments. Coprophagous coleoptera are often present in number in Quaternary deposits. Although this information is not unequivocal since this category of insects may be associated either with cattle or with wild mammals, it is often possible to infer from large quantities of fossil dung beetles that pastoralism was practised in the vicinity of the depositional site, in association with other categories of insects. Another important group of beetles in the context of Human impact and anthropogenic change is the occurrence of synanthropic insects, which are associated with stored food products (such as cereals), cattle shelters or human dwellings. Most of these interesting species cannot live in the wild and provide a wealth of data concerning human living conditions and activities.

Concerning the river itself, water beetles may provide objective data concerning water temperature, current speed and the nature of the river bottom. There is a large body of data obtained from the analysis of fossil beetle assemblages to reveal palaeoclimates. From a qualitative point of view, it is possible to infer past climatic conditions from modern analogue distribution of fossil insects: for example, the occurrence of arcti-alpine taxa in a fossil record suggests that low temperatures prevailed then, conversely the occurrence of comparatively southern taxa indicates warmer temperatures. This principle has been refined by Atkinson et al. (1986) in order to provide quantitative temperature estimates and was first applied in Britain by Atkinson et al. (1987). In brief, the MCR (Mutual Climatic Range) method is based on the ranges of climates corresponding to the total geographical area occupied today by the modern counterpart of the species present in a fossil assemblage. The temperature estimates for T_{max} and T_{min} for the whole assemblage are obtained by the overlap of the specific climatic ranges of the taxa identified in this assemblage. This method is now extensively used in Europe. French data were obtained for the Late glacial and for the Holocene, from floodplain sedimentary sequences, by Ponel et al. (2005) and Ponel et al. (2007).

Climate Reconstructions

Pleistocene Sedimentary Sequences: Last Glacial and Late Glacial Periods

It is well known that rivers have responded to main Pleistocene climate changes, particularly to the interglacial-glacial successions (Starkel, 1995; Huisink, 2000). Channel changes have also responded to factors such as global sea level, and thus to the river base level, topography (slope, size of the watershed, number of outlets, etc.) and tectonic and glacio-hydro-eustatic processes. Despite the importance of these factors, climate is one of the more important factors driving the geomorphological processes of European rivers (Huisink, 2000). Antoine et al. (2000) have shown that upstream sections are more influenced by climate than those downstream, with the latter mainly controlled by sea level fluctuations. Hence, in streams and rivers from lowland temperate areas, both interglacial and interstadial warm periods are usually characterized by low energetic hydrosystems such as meandering or anastomosed channels, whereas cold glacial periods are characterized by higher energetic braided channels. (Huisink, 2000; Berendsen and Stouthamer, 2000). In temperate periods, climate improvement leads to the development of the vegetation and to the stabilization of soils over the watershed. These induce a decrease of peak discharges of the rivers, a decrease of bedload and an increase in load of fines (Berensen and Stouthamer, 2000). During cold periods, the sediment load from the watershed due to the substrate gelifraction and the higher eolian activity may lead to a rapid infilling of river channels, and thus braided channels may form.

Up to now, very few palaeoecological studies based on river fossil insect evidence from the Pleistocene period were available, especially those based on chironomid assemblages. In The Netherlands, at Dinkel Valley (Ran, 1990), chironomids were mainly used as indicators of running water. Engels et al. (2008b) have inferred past climate conditions from two floodplain lacustrine sediments (SR-X1 and LM-8 records) intercalated in Weichselian Early Glacial and Early Pleniglacial fluvial and aeolian sediments from eastern Germany. Two Dansgaard/Oeschger like climate events dated to approximately 80 ka BP (OSL dating) and 45 ka BP (radiocarbon dating) have been detected. Chironomid fauna and lithology were typical of very shallow lakes for the SR-X1 record and of a low-energetic anastomosing-river deposit for LM-8, an abundance of lentic chironomids often associated with macrophytes, such as *Dicrotendipes* and *Glyptotendipes* were present. From these and other records, inference models indicated mean July air temperatures of ca. 15°C, stable for SR-X1 and abruptly declining from 15–16°C to ca. 13°C for LM-8. In northern Finland, at Sloki, Engels et al. (2008c) have studied a floodplain lacustrine record covering the earlier part of OIS3 around 45–50 ka BP (OSL and radiocarbon dating). Because of the absence of a regional transfer function, the Norwegian calibration data set (153 Norwegian lakes) from Brooks and Birks (2001), spanning a mean July air temperature range of 3.5–15.6°C, was used. During this period, the chironomid inferred July air temperatures of between 10.5 and 14°C are surprisingly high for a glacial period, reaching values similar to the current regional temperature. Botanical inferred temperatures from the same site were slightly lower but, like the chironomid data, provided evidence for a period of high (minimum) July air temperatures at the onset of the sequence. The authors discussed the factors that should influence the faunal assemblages such as lake depth or decoupling water and air temperatures. Engels et al. (2008c) concluded that although there are some differences between the chironomid assemblages of floodplain lakes and of isolated lakes, these differences did not have a major effect on chironomid-based temperature reconstructions.

To our knowledge, the late glacial climate reconstructions based on river chironomid communities are under-represented. Gandouin et al. (2007) have described Younger Dryas assemblages at St Omer basin in northern France. The base of the St Omer sequence is composed of lotic assemblages typical of shallow and oligothrophic streams with slow water flow, under very cold climate conditions. Based on the modern distributions of chironomid species along a thermal gradient from northern Italy (Rossaro, 1991), Gandouin et al. (2007) estimated water temperatures during Younger Dryas spanned a range of 9–11°C. According to the strong correlation between surface water temperature and $T°_{jul}$ (Olander et al., 1999) these estimations are consistent with temperature data inferred from insect data (coleopteran and chironomids) in Britain (e.g. Bedford et al., 2004; Coope et al., 1998) and in The Netherlands (Bohncke et al., 1987). They also corresponded with inferred temperatures from botanical data in a central and north-western Europe data set (Isarin and Bohncke, 1999; Bogaart, 2003). At St Omer, a coleopteran analysis (Ponel et al., 2007) performed in parallel with chironomid analysis has provided T_{max} (the mean July temperature) of 10°C and T_{min} (the mean temperature of January/February) close to −11/−12°C, using the Mutual Climatic Range method.

In addition to these valuable results concerning temperatures, a chironomid analysis at St Omer allowed for climate reconstructions based on the positive correlation between river hydrological regimes and precipitation (Kiely, 1999; Rîmbu et al., 2002). Hence, the Younger Dryas has been sub-divided into an earlier colder and wetter phase (characterized by the dominance of cold and rheophilous fauna) followed by a relatively warmer and drier phase (rise in abundances of warmer and lenitophilous fauna). This sub-division is in agreement with several palaeoecological studies from central and northwestern Europe (e.g. Isarin et al., 1998; Magny, 2001; Bogaart, 2003).

For the Late glacial period, a synthesis of beetle data from northern France (Paris and St-Omer Basins) is given by Ponel et al. (2007) (see Fig. 1). The northern France climatic trend is similar to those reconstructed from late glacial coleopteran assemblages from the British Isles (Coope and Brophy, 1972; Atkinson et al., 1987; Walker et al., 1993; 2003). The T_{max} values showed that the warming at the beginning of the Bølling was intense, reaching figures similar to those of the present day. During the Allerød phase, the climate in Britain was rather cooler but still warm enough to permit expansion of forest cover. A moderate rise in temperature seems to have taken place towards the end of the Allerød in northern France. Nevertheless, French data are at present rather incomplete. In northern France and Britain, the intense Younger Dryas cooling was characterized by the presence of numerous cold-adapted beetle species, some of which have exclusively arctic distributions today. This indicates that T_{max} figures were as low or lower than 10°C and T_{min} values were even more depressed.

The Holocene

Based on river chironomid communities and coleopteran assemblages, the Holocene period in northern France is better documented than the Late glacial. Gandouin et al., (2005, 2006, 2007) and Ponel et al. (2007) reconstructed climate changes at the St-Omer Basin over a period spanning to ca. 9,500–3,000 ^{14}C uncal. BP. Chironomid river communities are in equilibrium with water temperatures because of the major role of this parameter on midge adult emergence (Rossaro, 1991). Based on this hypothesis, Gandouin et al. (2007) have shown that it is possible to estimate ranges for water summer temperatures

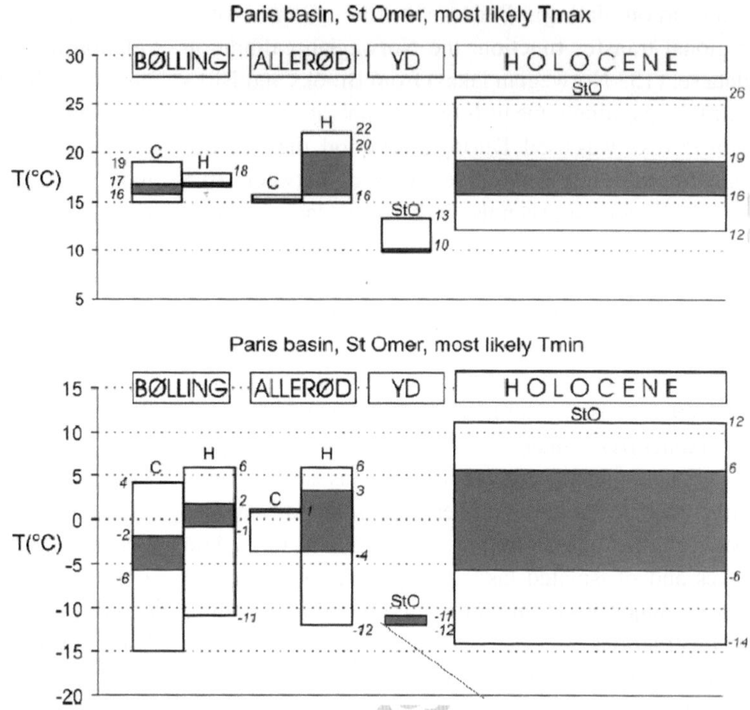

Fig. 1 Synthetic MCR beetle reconstruction from several sites in Paris Basin and St-Momelin data set, after Ponel et al. (2007). Shaded areas indicate most likely values for T_{max} and T_{min}. C = Conty, H = Houdancourt, StO = St-Omer

for the early Holocene and the mid-Holocene periods (Fig. 2), completed by the MCR method from beetle analysis (Fig. 3) given by Ponel et al. (2007). After the Younger Dryas, during the early Holocene, temperatures rapidly increased until values were close to those of the present day. The correspondence analysis derived from St-Omer chironomid assemblages (Gandouin et al., 2007) suggests that a thermal optimum was reached around 8,000 years ^{14}C uncal. BP and that mid-Holocene water temperatures were lower after 6,000 years ^{14}C uncal. BP, may be related to a slight decrease in atmospheric thermal conditions (cf. Neoglacial period). Several other hypotheses were also discussed in Gandouin et al. (2007), and may be related to (i) possible cold running water, (ii) water-table flooding and probable thermal buffer effect and (iii) shading effects of the riverine forest. Pollen analysis carried out on the same alluvial deposits has shown a late Atlantic cooling which persisted over the Subboreal period, as suggested by the regular occurrences of *Abies*, and by the decrease in *Tilia* percentages from regional pollen assemblages (Gandouin et al., 2009).

If climate reconstruction is more difficult in floodplain environments than for lakes because of the lack of temperature transfer function and the complexity of the river processes, one of the main interests of river chironomid palaeoecological analyses consist in the reconstruction of the level of connectivity that existed during floods between the main channel and the various waterbodies of the alluvial plain. Based on a first

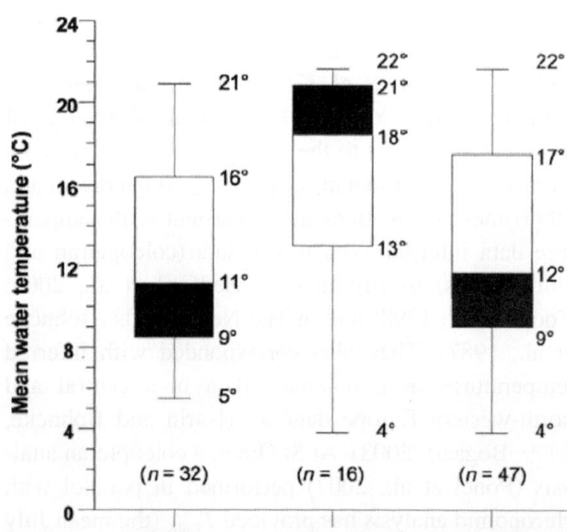

Fig. 2 Possible mean water temperatures (*black area*) at St-Momelin during the Late glacial and The Holocene (after Gandouin et al., 2007)

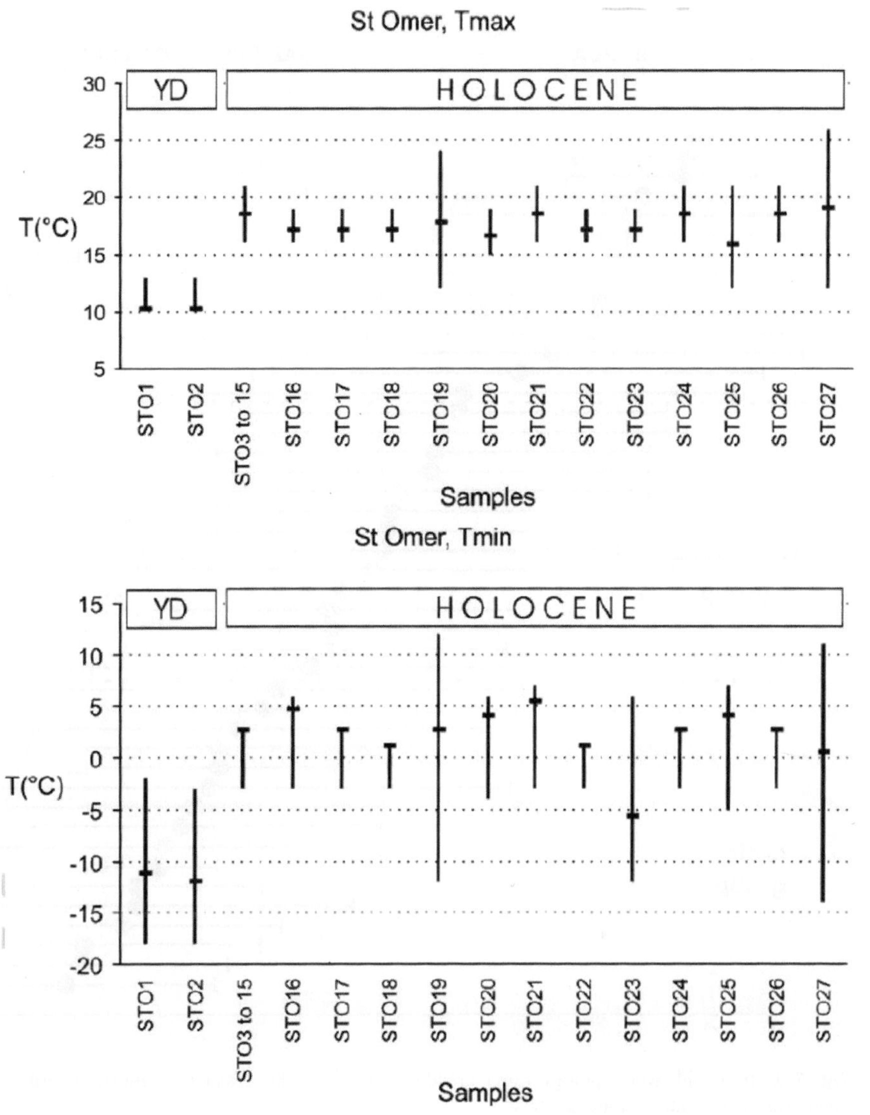

Fig. 3 Temperature reconstruction from Younger Dryas to Sub-Boreal at St-Momelin using the MCR method. *Horizontal lines* on the *bars* indicate the most likely position for the actual mean temperature (after Ponel et al., 2007)

river chironomid typology provided by Gandouin et al. (2006) from a gradient of lotic to lentic habitats over a river floodplain of France (Fig. 4), it is now possible to obtain descriptive palaeoecological information about past hydrological changes in river systems directly comparable to lake level changes (Fig. 5) over large spatial scales (such as continental scale) and thus to complete the knowledge of past atmospheric humidity schemes and better understand the mechanisms and processes leading, for example, to understanding the functioning of the northern Atlantic Wersterlies. In view of the positive correlations observed between precipitation and the flow rates of waterways (e.g. Kiely, 1999; Rîmbu et al., 2002), it should be very interesting now to develop elaborate mathematical models (cf., transfer functions) that allow inferences of river discharge and precipitation rates based on the modern distribution of chironomid species throughout river floodplains, and therefore to improve the quantification of past environmental change.

Historical Period (Anthopocene): Local Impacts of Human Activities on River Waterbodies

Although Ruiz et al. (2006) have underlined the potential importance of chironomids to help trace temporal changes in water quality related to human activities,

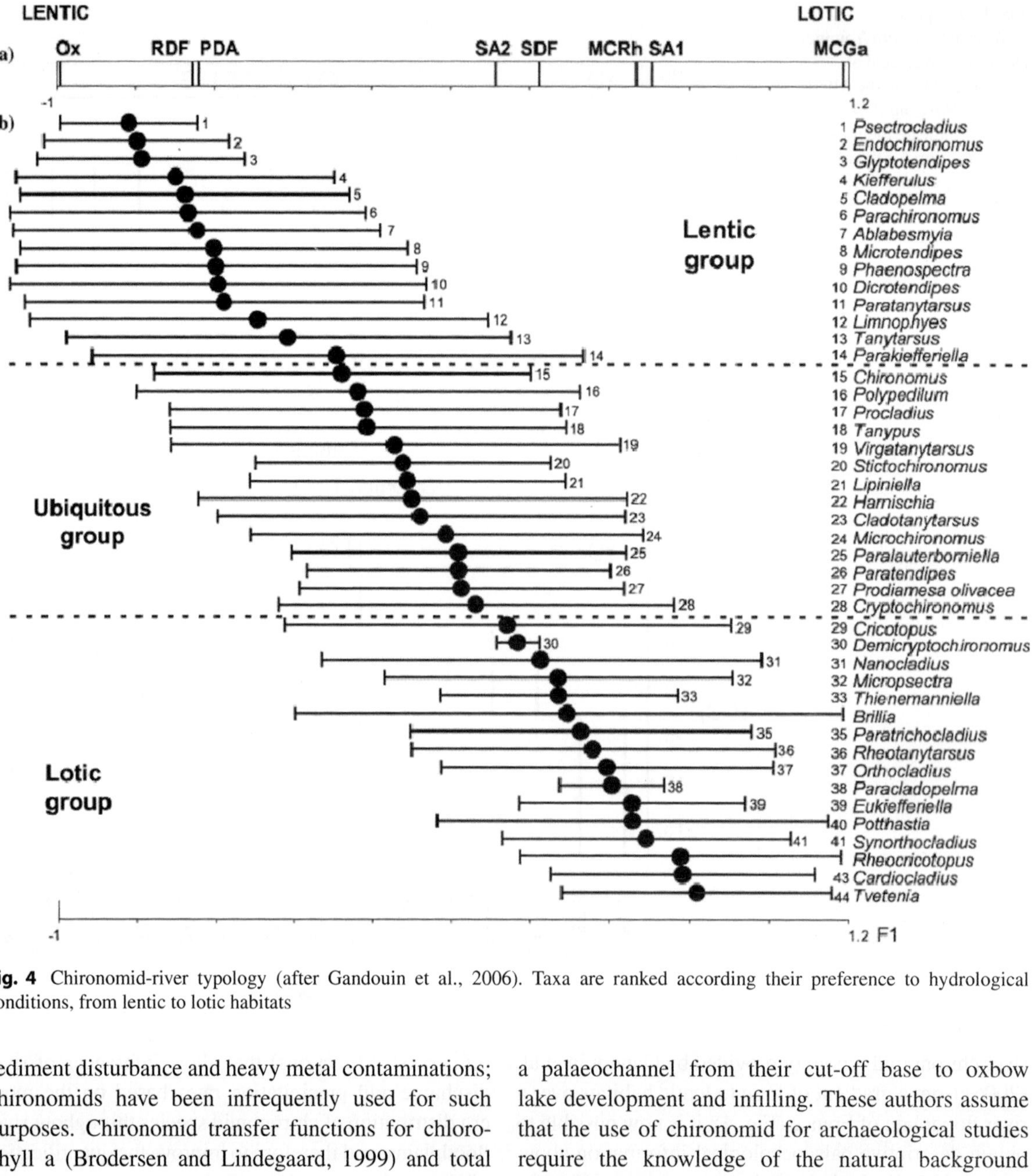

Fig. 4 Chironomid-river typology (after Gandouin et al., 2006). Taxa are ranked according their preference to hydrological conditions, from lentic to lotic habitats

sediment disturbance and heavy metal contaminations; chironomids have been infrequently used for such purposes. Chironomid transfer functions for chlorophyll a (Brodersen and Lindegaard, 1999) and total phosphorus (e.g. Brooks et al., 2001) are available and allow the reconstruction of past water quality. Nevertheless, as these functions have been elaborated in lacustrine conditions, their extrapolation to river environments has difficulties, according to the first observations by Ruiz et al. (2006) made for a palaeochannel at West Cotton (British West Midlands). At West Cotton, interpretations about water quality were difficult because of the poor knowledge concerning natural succession of chironomid taxa in a palaeochannel from their cut-off base to oxbow lake development and infilling. These authors assume that the use of chironomid for archaeological studies require the knowledge of the natural background conditions of the site, and thus to work with other palaeoecological data (such as pollen and coleoptera) or historical tools.

Concerning the coleopteran record in floodplain environments, the data obtained from France are currently scarce, especially when compared with the numerous British studies available (see, for example, the works by Smith et al., 2005; Smith and Howard, 2004; Elias et al., 2009). One of the most striking examples of faunal changes induced by human action

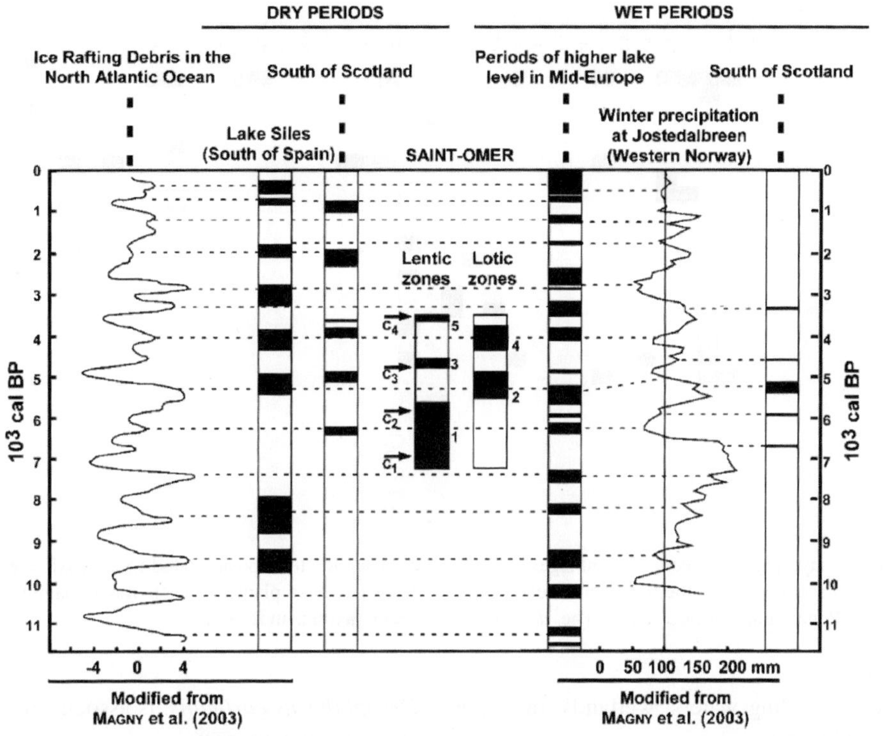

Fig. 5 Comparison between the North-Atlantic Ice Rafting Discharge (IRD) values (Bond et al., 2001) and various hydrological records detected in Europe from Gandouin et al. (2005). Dry periods are in *black* on the *left*: South of Spain (Carrión, 2002 in Magny et al., 2003), South of Scotland (Langdon et al., 2003) and Saint-Omer (lentic zones: SMch-1, 3 and 5). Wet periods are shown in *black* on the *right*: Saint-Omer (lotic zones: SMch-2 and 4), Mid-Europe (Magny et al., 2003), West of Norway (Nesje et al., 2000) and south of Scotland (Langdon et al., 2003). C1–C4: peaks in the dry periods detected at Saint-Omer [cf. maxima of limnophilous taxa (Fig. 4)]. These peaks and zones have been dated with the mean rate of sedimentation (Rs) calculated from the calibrated ages obtained on the part of core analysed. Estimated dates in cal years BP are [SAINT-OMER (SMch) 1] 7,320–5,640, (SAINT-OMER 2) 5,540–4,900, (SAINT-OMER 3) 4,800–4,590, (SAINT-OMER 4) 4,370–3,700, (SAINT-OMER 5) 3,600–3,500, (C1) 6,920, (C2) 5,850, (C3) 4,800 and (C4) 3,530

in river ecosystems in the Holocene is certainly the disappearance of representatives of the Elmidae family ("riffle beetles"), a particular category of water beetles associated today with clear and well oxygenated streams with coarse sediment bottoms, where they live hanging on the lower part of the stones and gravels by means of their strong claws. Evidence for such changes was reported by Osborne (1988) from the River Avon in Britain. A Late Bronze Age assemblage containing several species and many individuals of Elmidae, including the very rare *Stenelmis canaliculata*, is described. A comparison with the modern living fauna inhabiting the rivers in the same area has shown that this Elmid fauna is today almost entirely eradicated. Such major changes may be explained by the deposition of mud and fine silt at the bottom of the river, clogging the interstices between stones and making the biotope unsuitable for riffle beetles (Smith, 2000). It is likely that this process is a consequence of the destruction of the surrounding forest cover, leading to increased soil erosion and thence load in fines in the stream.

Evidence for Human Impact at Neolithic on the Alluvial Forests

Downstream of the Seine, in a meander near Rouen (France), a Holocene sedimentary sequence was extracted from the bank of the river. The rich beetle assemblages in the sediment profile contain a wide range of categories of coleoptera, suggesting the presence of a diversity of environments in the vicinity

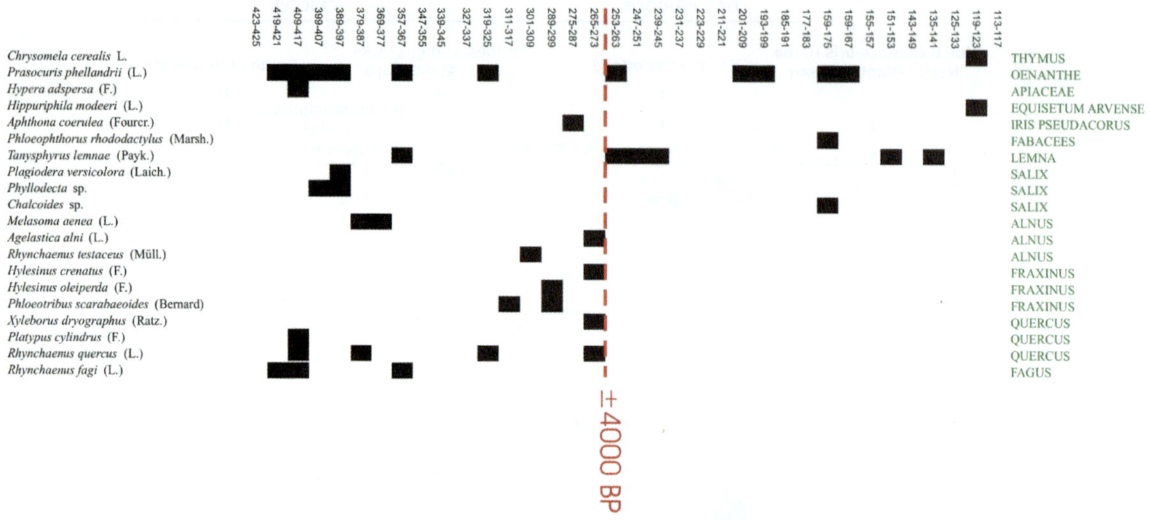

Fig. 6 Changes in phytophagous beetle assemblages from an alluvial sedimentary sequence from the river Seine near Rouen, Northern France (Ponel, unpublished data). The *dotted line* corresponds to deposits dated to approximately 4,000 BP. The preferred host plant for each monophagous or oligophagous beetle taxon is indicated

of the site: e.g. standing waters, wetlands, dry open steppic areas and forests.

The analysis of the tree-dependent coleoptera group showed that there was a diverse range of ecological niches present. Some of these insects are monophagous or oligophagous taxa associated with one plant species or genus. This is the case for phyllophagous insects such as leaf-beetles (chrysomelids) and weevils (curculionids) and for wood borers such as scolytids: *Plagiodera versicolora*, *Phyllodecta*, *Chalcoides*, live on willows, *Melasoma aenea*, *Agelastica alni*, *Rhynchaenus testaceus*, live on alders, *Hylesinus crenatus*, *Hylesinus oleiperda*, *Phloeotribus scarabaeoides*, live on ash, *Xyleborus dryographus*, *Platypus cylindrus*, *Rhynchaenus quercus*, live on oak and *Rhynchaenus fagi* lives on beech.

Other coleoptera are saproxylophagous insects living in rotten wood, decaying stumps or on old trees. This is the case for the members of the Eucnemidae family such as *Melasis buprestoides* and *Dromaeolus barnabita*. The latter is today a very rare species associated with rotten stumps in dark old growth forests, and these have a very scattered distribution. Many other species are also dependent on old dead wood in various stades of decay: *Cerylon* live under loose bark, *Hedobia imperialis*, *Grynobius planus*, *Hypophloeus unicolor* and *Stenoscelis submuricatus* are associated with old dry wood. Another saproxylophagous weevil, *Dryophthorus corticalis*, is associated with old forests and is today very rare.

The study of the distribution within the sedimentary sequence of these tree-dependent and primeval forest species shows that most of them, if not all, are restricted to the lower half of the sequence, approximately from sample 4 to sample 20. Conversely, the species associated with open environments and grasslands persist through to the present. Such distribution suggests than a major ecological change took place approximately at the transition sample 20–sample 21, corresponding to about 4,000 BP (see Fig. 6). It is likely that such change was not induced by climate, but rather by the large scale forest clearance at time of spread of cultivation and pastoralism during the Neolithic period.

At Saint-Omer (northern France), only pollen data (Gandouin et al., 2009) showed this post 4,000 BP deforestation (because of the lack of sediment spanning this period for coleopteran analysis). Before 4,000 BP, the coleopteran record (Ponel et al., 2007) did not contain any species associated with cultivated grounds; on the contrary several species typical of pristine, undisturbed primeval forests were identified, for example, *Mycetina cruciata*, *D. barnabita* and *Dirrhagus lepidus*. The occurrence of *M. cruciata* is especially significant since this saproxylophagous beetle is associated with old forests with very low

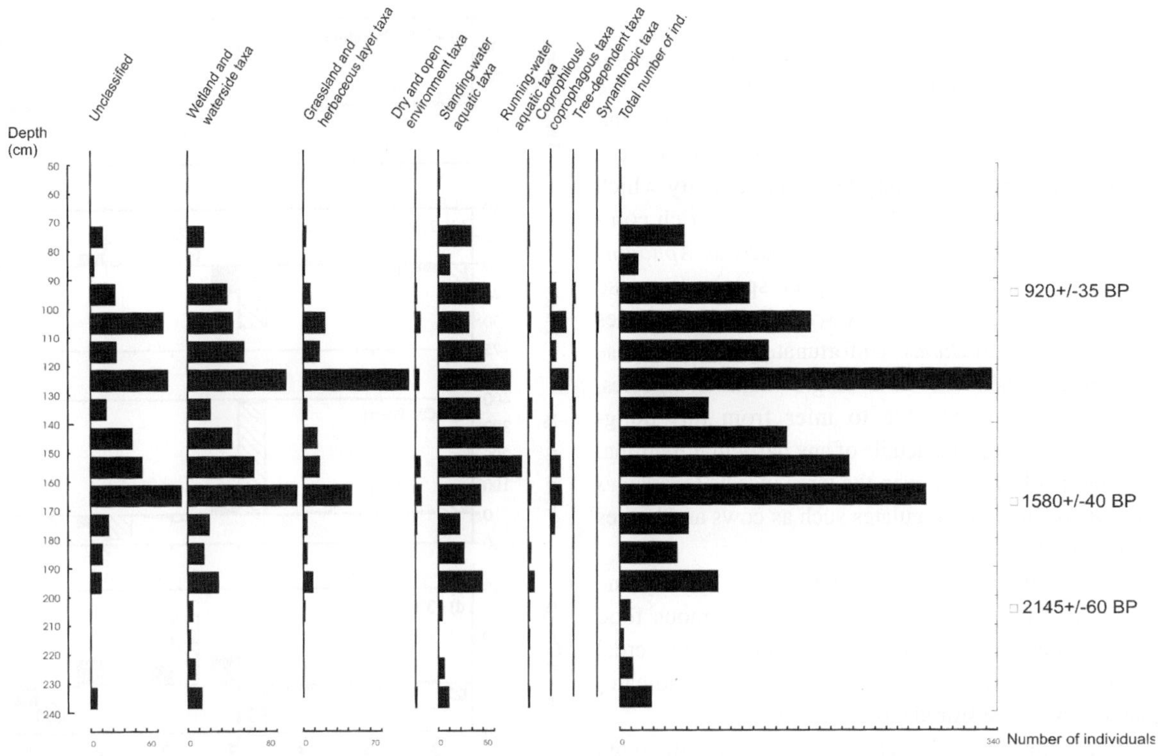

Fig. 7 Main ecological categories of Coleoptera at La Calade

human impact ("urwald" forests), where it is found in wet rotten wood attacked by fungus (Ponel et al., 2007).

The proximity of the Rhône river that feeds the groundwater and prevents the periodic drying up of the site in summer enabled Andrieu-Ponel et al. (2000) to obtain, from the site of La Calade near Arles, a very rich pollen and beetle assemblage succession dated to the last two millenia. This record is especially interesting because in the Mediterranean region, palaeoecological sequences are rare due to the dry seasonal climatic conditions which are prejudicial to the preservation of fossil biological material. The exceptional site of La Calade offered an opportunity to reconstruct, for the first time, the landscape of the floodplain of the Lower Rhone valley from 2,000 BP to the present.

As expected in such depositional environment, the wetland and waterside taxa are by far the dominant group of coleoptera. However, some other categories of beetle are very informative. This is the case for the tree dependent coleoptera which are extremely underepresented: four taxa and four individuals only (Fig. 7). The leaf-beetle *A. alni* and the weevil *Lepyrus palustris* suggest respectively that *Alnus glutinosa* or *A. incana* and *Salix* were present. The occurrence of these riparian trees is not surprising, but their low representation is unexpected. Two other weevil species (*Curculio nucum, Neocoenorhinus aeneovirens*) indicated that *Corylus* and/or *Quercus* grew on the site or in its close vicinity. This insignificant representation of tree-dependent taxa (0.2% of the total number of individuals) is a strong indication for an almost treeless landscape, not only on the site itself but certainly also within a wider area. By contrast with this scarcity of arboreal trees, two other categories were much better represented. grassland taxa, some of which are today very common in cultivated areas (*Agriotes* spp.) and the Alticinae taxa (*Sitona* spp., *Ceuthorhynchus* sensu lato), which may become pests for domestic plants. Several ground-beetles are also frequently found in cultivated fields: the predatory *Metallina properans*, *Phyla obtusum*, *Poecilus cupreus* but also the phytophagous *Zabrus ignavus* and *Z. tenebrionides*. The latter species is regularly reported to cause damage to cereals in south-east Europe. These grassland and herbaceous layer taxa identified throughout the sequence indicated that, at all times, cultivated and ruderal plants grew on the site. Another important ecological category is the dung-beetle community which is well represented with 19 taxa. There is a rich community of small size dung-beetles such as *Aphodius* and *Onthophagus* but some bigger species are also recorded, especially *Sisyphus schaefferi,* and species of *Geotrupes* and *Bubas*. Unfortunately none of these species is dependent on a single mammal species, thus it seems impossible to infer from this dung-beetle assemblage the details of any particular mammal species involved. However, the large insects *Geotrupes* and *Bubas* hint that ungulates such as cows and horses were raised on the site.

The synanthropic taxa are represented by only one specimen of *Sitophilus granarius*. This famous food product pest is associated with unprocessed cereal grains and is unable to survive outdoors. It indicates that cereals were stored close to the site.

In conclusion, insect data (in agreement with pollen data) suggested there was a heavy focus of pastoral activities at La Calade during the last two millenia. Remarkably, the importance of pastoralism in the nearby Plaine de la Crau is attested as soon as in the first century in ancient texts by Strabo and Pliny and was confirmed by the discovery of huge sheep-folds. It is possible that the flocks wintering in the "Plaine de la Crau" moved in spring to the Lower Rhône Valley to feed in humid depressions?

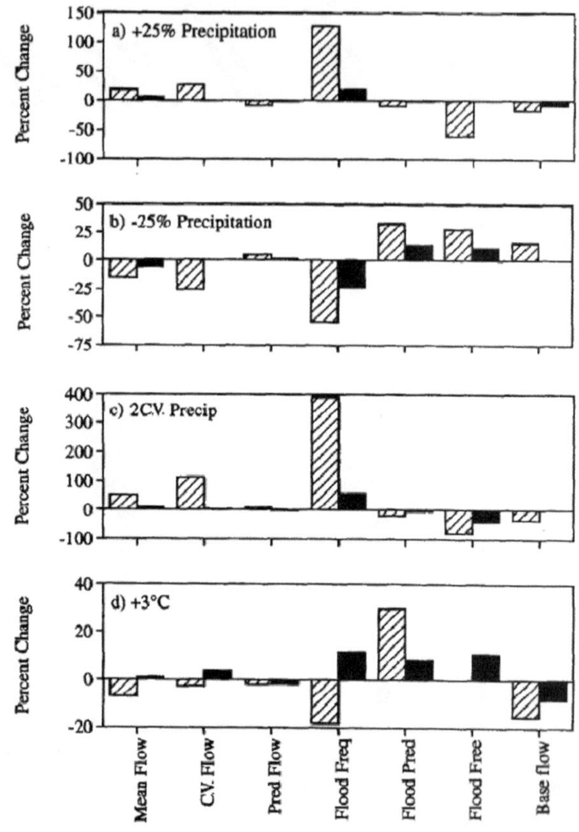

Fig. 8 Responses of seven hydrological variables in the Little Patuxent (hatched bars) and Independence Rivers (*solid bars*) to four scenarios of climate change: 25% increase in precipitation; 25% decrease in precipitation; doubling of the Coefficient of Variation (C.V.) in precipitation; temperature increase of 3°C. Variables are mean annual flow, C.V. of daily flow, predictability of flow, flood frequency, flood predictability, flood-free period, and baseflow stability. Note differences in scale for *y*-axis. According to Poff et al. (1996)

Implications for the Future

Insects have great potential for paleaeocological investigations of alluvial sedimentary deposits. In the context of global climate warming, their use can be very helpful for the understanding of mechanisms relating climate and river processes. Poff et al. (1996) have shown that, in general, the rainfall-dominated streams exhibit rapid response to climate change scenarios: for example, in rainfall-dominated streams, a temperature increase of 3°C would induce a significant decrease in flood frequency (Fig. 8). This will have major consequences on human societies, especially those from subhumid mediterranean regions, where water resource management is the challenge of the coming decades. Moreover, floodplain ecosystems will be affected too, with a decrease in the frequence of connectivities between main channel and others waterbodies, leading to a fall in landscape heterogeneity, followed by a biodiversity decrease in these areas. Intermittent rivers and mires will be the first victim of this direction in climate change. The intermittent flow tributaries play a major role in maintaining the taxonomic richness in catchments, highly impacted by anthropogenic activities (Maasri et al., 2008). The preservation of these hydrosystems should be an important step in catchment management.

References

Andrieu-Ponel V, Ponel P, Bruneton H, Leveau P, Beaulieu J-L (2000) Palaeoenvironments and cultural landscapes of the last 2000 years reconstructed from pollen and Coleopteran records in the Lower Rhône Valley, southern France. Holocene 10:341–355.

Antoine P, Lautridou JP, Laurent M (2000) Long-term fluvial archives in NW France: Response of the Seine and Somme rivers to tectonic movements, climatic variations and sea-level changes. Geomorphology 33:183–207.

Atkinson TC, Briffa KR, Coope GR, Joachim JM, Perry DW (1986) Climatic calibration of Coleopteran data. In BE Berglund (Ed), Handbook of Holocene Palaeoecology and Palaeohydrology (pp. 851–858). Chichester: Wiley.

Atkinson TC, Briffa KR, Coope GR (1987) Seasonal temperatures in Britain during the past 22,000 years, reconstructed using beetle remains. Nature 325:587–592.

Barley EM, Walker IR, Kurek J, Cwynar LC, Mathewes RW, Gajewski K, Finney BP (2006) A northwest North American training set: Distribution of freshwater midges in relation to air temperature and lake depth. J Paleolimnol 36:295–314.

Bedford A, Jones RT, Lang B, Brooks S (2004) A Late-glacial chironomid record from Hawes Water, N.W. England. J Quat Sci 19:281–291.

Berendsen HJA, Stouthamer E (2000) Late Weichselian and Holocene palaeogeography of the Rhine-Meuse delta, The Netherlands. Palaeogeogr Palaeoclimatol Palaeoecol 161:311–335.

Bogaart PW (2003). Process-based modelling of the fluvial response to rapid climate change – With reference to the river Maas during the Last glacial–interglacial transition. Ph.D. thesis, Vrije Universiteit Amsterdam.

Bohncke S, Vanderberghe J, Coope GR, Reiling R (1987) Geomorphology and palaeoecology of the Mark valley (southern Netherlands): Palaeoecology, palaeohydrology and climate during the Weichselian Late Glacial. Boreas 16:69–85.

Bond G, Kromer B, Beer J, Muscheler R, Evans MN, Showers W, Hoffmann S, Lotti-Bond R, Hadjas I, Bonani G (2001) Persistent solar influence on North Atlantic climate during the Holocene. Science 294:2130–2136.

Brodersen KP, Lindegaard C (1999) Classification, assessment and trophic reconstruction of Danish lakes using chironoimids. Freshw Biol 42:143–157.

Brooks SJ (2006) Fossil midges (Diptera: Chironomidae) as palaeoclimatic indicators of the Eurasian region. Quat Sci Rev 25:1894–1910.

Brooks SJ, Birks HJB (2001) Chironomid-inferred air temperatures from Lateglacial and Holocene sites in north-west Europe: Progress and problems. Quat Sci Rev 20:1723–1741.

Brooks SJ, Bennion H, Birks HJB (2001) Tracing lake trophic history with a chironomid-total phosphorus inference model. Freshw Biol 46:1–21.

Brooks SJ, Langdon PG, Heiri O (2007) The Identification and Use of Paleoarctic Chironomidae Larvae in Palaeoecology. Technical Guide No. 10. London: Quaternary Research Association.

Brundin L (1958) The bottom faunistical lake type system and its application to the southern hemisphere. Moreover a theory of glacial erosion as a factor of productivity in lakes and oceans. Verhandlungen der Interntionalen Vereinigung für Theoretische und Angewandte Limnologie 13:288–297.

Buckland PC, Coope GR (1991) A Bibliography and Literature Review of Quaternary Entomology. University of Sheffield: J.R. Collis Publications, Department of Archaeology & Prehistory.

Carrión JS (2002) Patterns and processes of Late Quaternary environmental change in a montane region of southwestern Europe. Quat Sci Rev 21:2047–2066.

Caseldine CJ, Langdon PG, Holmes N (2006) Early Holocene climate variability and the timing and extent of the Holocene Thermal Maximum (HTM) in Northern Iceland. Quat Sci Rev 25:2314–2331.

Coope GR (1986) Coleoptera analysis. In BE Berglund (Ed), Handbook of Holocene Palaeoecology and Palaeohydrology (pp. 703–713). Chichester: Wiley.

Coope GR, Brophy JA (1972) Late glacial environmental changes indicated by a coleopteran succession from NorthWales. Boreas 1:97–142.

Coope GR, Lemdahl G, Lowe JJ, Walkling A (1998) Temperature gradients in northern Europe during the last glacial-Holocene transition (14–9 ^{14}C kyr BP) interpreted from coleopteran assemblages. J Quat Sci 13:419–433.

Elias SA (1994) Quaternary Insects and Their Environments. Washington, DC: Smithsonian Institution Press.

Elias SA, Webster L, Amer M (2009) A beetle's eye view of London from the Mesolithic to Late Bronze Age. Geol J DOI 10.1002/gj.1158.

Engels S, Bohncke SJP, Heiri O, Nyman M (2008a) Intraregional variability in chironomid-inferred temperature estimates and the influence of river inundations on lacustrine chironomid assemblages. J Paleolimnol 40:129–142.

Engels S, Bohncke SJP, Bos JAA, Heiri O, Vandenberghe J, Wallinga J (2008b) Environmental inferences and chironomid-based temperature reconstructions from fragmentary records of the Weichselian Early Glacial and Pleniglacial periods in the Niederlausitz area (eastern Germany). Palaeogeogr Palaeoclimatol Palaeoecol 260: 405–416.

Engels S, Bohncke SJP, Bos JAA, Brooks SJ, Heiri O, Helmens KF (2008c) Chironomid-based palaeotemperature estimates for northeast Finland during Oxygen Isotope Stage 3. J Paleolimnol 40:49–61.

Gandouin E, Franquet E, Van Vliet-Lanoë B (2005) Chironomids (Diptera) in river floodplains: Their status and potential use for palaeoenvironmental reconstruction purposes. Archiv für Hydrobiologie 162:511–534.

Gandouin E, Maasri A, Van Vliet-Lanoë B, Franquet E (2006) Chironomid (Insecta: Diptera) assemblages from a gradient of lotic and lentic waterbodies in river floodplains of France: A methodological tool for palaeoecological applications. J Paleolimnol 35:149–166.

Gandouin E, Ponel P, Franquet E, Van Vliet-Lanoë B, Andrieu-Ponel V, Keen DH, Brulhet J, Brocandel M (2007) Chironomid responses (Insect: Diptera) to younger Dryas and Holocene environmental changes in a river floodplain from northern France (St-Momelin, St-Omer basin). Holocene 17:333–349.

Gandouin E, Ponel P, Andrieu-Ponel V, Guiter F, Beaulieu J-L, Djamali M, Franquet E, Van Vliet-Lanoë B, Alvitre M, Meurisse M, Brocandel M, Brulhet J (2009) 10,000 years of

vegetation history of the Aa palaeoestuary, St-Omer Basin, northern France. Rev Palaeobot Palynol 156:307–318.
Heinrichs ML, Walker IR, Mathewes RW (2001) Chironomid-based paleosalinity records in southern British Columbia, Canada: A comparison of transfer functions. J Paleolimnol 26:147–159.
Hofmann W (1986) Chironomid analysis. In BE Berglung (Ed), Handbook of Holocene Palaeoecology and Palaeohydrology (pp. 715–727). Chichester: Wiley.
Huisink M (2000) Changing river styles in response to Weichselian climate changes in the Vecht valley, eastern Netherlands. Sediment Geol 133:115–134.
Isarin RFB, Bohncke SJP (1999) Mean July Temperatures during the Younger Dryas in Northwestern and Central Europe as Inferred from Climate Indicator Plant Species. Quat Res 51:158–173.
Isarin RFB, Renssen H, Vandenberghe J (1998) The impact of the North Atlantic Ocean on the Younger Dryas climate in northwestern and central Europe. J Quat Sci 13:447–453.
Kiely G (1999) Climate change in Ireland from precipitation and streamflow observations. Adv Water Resour 23: 141–151.
Klink A (1989) The lower Rhine: Palaeoecological analysis. In GE Petts, H Moller, AL Roux (Eds), Historical Change of Large Alluvial Rivers: Western Europe (pp. 183–201). London: Wiley.
Langdon PG, Barber KE, Hughes PDM (2003) A 7500-year peat-based palaeoclimatic reconstuction and evidence for an 1100-year cyclicity in bog surface wetness from Temple Hill Moss, Pentland Hills, southeast Scotland. Quat Sci Rev 22:259–274.
Larocque I, Grosjean M, Heiri O, Bigler C, Blass A (2009) Comparison between chironomid-inferred July temperatures and meteorological data AD 1850–2001 from varved Lake Silvaplana, Switzerland. J Paleolimnol 41:329–342.
Larocque I, Hall RI (2003) Chironomids as quantitative indicators of mean July air temperature: Validation by comparison with century-long meteorological records from northern Sweden. J Palaeolimnol 29:475–493.
Larocque I, Hall RI, Grahn E (2001) Chironomids as indicators of climate change: A 100-lake training set from a subarctic region of northern Sweden (Lapland). J Palaeolimnol 26:307–322.
Maasri A, Dumont B, Claret C, Archambaud-Suard G, Gandouin E, Franquet E (2008) Tributaries under Mediterranean climate: Their role in macrobenthos diversity maintenance. Comptes Rendus Biologie 331:547–558.
Magny M (2001) Palaeohydrological changes as reflected by lake level fluctuations in the Swiss Plateau, the Jura Mountains and the northern French Pre-Alps during the Last Glacial–Holocene transition: A regional synthesis. Global Planet Change 30:85–101.
Magny M, Bégeot C, Guiot J, Peyron O (2003) Contrasting patterns of hydrological changes in Europe in response to Holocene climate cooling phases. Quat Sci Rev 22:1589–1596.
Mousavi SK (2002) Boreal chironomid communities and their relations to environmental factors - the impact of lake depth, size and acidity. Boreal Environ Res 7:63–75.
Nesje A, Lie O, Dahl SO (2000) Is the North Atlantic Oscillation reflected in Scandinavian glacier mass balance records? J Quat Sci 15:587–601.
Olander H, Birks HJB, Korhola A, Blom T (1999) An expanded calibration model for inferring lakewater and air temperatures from fossil chironomid assemblages in northern Fennoscandia. Holocene 9:279–294.
Osborne P (1988) A Late Bronze Age insect fauna from the river Avon, Warwickshire, England: Its implications for the terrestrial and fluvial environment and for climate. J Archaeol Sci 15:715–727.
Petts GE (1989) Historical analysis of fluvial hydrosystems. In GE Petts, H Moller, AL Roux (Eds), Historical Change of Large Alluvial Rivers: Western Europe (pp. 1–18). London: Wiley.
Poff NL, Tokar S, Johnson P (1996) Stream hydrological and ecological response to climate change assessed with an artificail neural network. Limnol Oceanogr 41:857–863.
Ponel P, Coope GR, Antoine P, Limondin-Lozouet L, Leroyer C, Munaut A-V, Pastre J-F, Guiter F (2005) Late-glacial Palaeoenvironments and Palaeoclimates from Conty and Houdancourt, Northern France, reconstructed from Beetle remains. Quat Sci Rev 24:2449–2465.
Ponel P, Gandouin E, Coope RG, Andrieu-Ponel V, Van Vliet-Lanöe B, Franquet E, Brocandel M, Brulhet J (2007) Insect evidence for environmental and climate changes from Younger Dryas to Sub-Boreal in a river floodplain at St-Momelin (St-Omer basin, northern France), Coleoptera and Trichoptera. Palaeogeogr, Palaeoclimatol, Palaeoecol 245:483–504.
Quinlan R, Smol JP (2002) Regional assessment of long-term hypolimnetic oxygen changes in Ontario (Canada) shield lakes using subfossil chironomids. J Paleolimnol 27:249–260.
Ran ETH (1990) Dynamics of vegetation and environment during the middle pleniglacial in the Dinkel valley (the Netherlands). Mededelingen Rijks Geologische Dienst 44:141–205.
Rîmbu N, Boroneant C, Buta C, Dima M (2002) Decadal variability of the Danube river flow in the Lower basin and its relation with the North Atlantic Oscillation. Int J Climatol 22:1169–1179.
Rossaro B (1991) Chironomid and water temperature. Aquat Insects 13:87–98.
Ruiz Z, Brown AG, Langdon PG (2006) The potential of chironomid (Insecta: Diptera) larvae in archaeological investigations of floodplain and lake settlements. J Archaeol Sci 33:14–33.
Smith DN (2000) Disappearance of elmid 'riffle beetles' from lowland river systems- the impact of alluviation. In AR Nicholson, TP O'Connor (Eds), People as an Agent of Environmental Change (pp. 75–80). Oxford: Oxbow Books.
Smith DN, Howard AJ (2004) Identifying changing fluvial conditions in low gradient alluvial archaeological landscapes: Can coleoptera provide insights into changing discharge rates and floodplain evolution? J Archaeol Sci 31:109–120.
Smith DN, Roseff R, Bevan L, Brown AG, Butler S, Hughes G, Monckton A (2005) Archaeological and environmental investigations of a Lateglacial and Holocene river sedimentary sequence on the River Soar at Croft, Leicestershire, UK. Holocene 15:353–377.
Starkel L (1995) Palaeohydrology of the Temperate Zone. In KJ Gregory, L Starkel, VR Backer (Eds), Global Continental Palaeohydrology (pp. 233–257). New York, NY: Wiley.

Stefanova I, Ognjanova-Rumenova N, Hofmann W, Ammann B (2003) Late glacial and Holocene environmental history of the Pirin Mountains (SW Bulgaria): A paleolimnological study of Lake Dalgoto (2310 m). J Paleolimnol 30: 95–111.

Walker MJC, Coope GR, Lowe JJ (1993) The Devensian (Weichselian) Lateglacial palaeoenvironmental record from Gransmoor, East Yorkshire, England. Quat Sci Rev 12: 659–680.

Walker MJC, Coope GR, Sheldrick C, Turney CSM, Lowe JJ, Blockley SEP, Harkness DD (2003) Devensian Lateglacial environmental changes in Britain: A multi-proxi environmental record from Llanilid, South Wales, UK. Quat Sci Rev 22:475–520.

Wilson RS (1980) Classifying rivers using Chironomid pupal exuviae. In DA Murray (Ed), Chironomidae Ecology, Systematics, Cytology and Physiology (pp. 209–216). Oxford: Pergamon Press.

Climate Change and Desertification with Special Reference to the Cases in China

Xiaoping Yang

Abstract This chapter briefly reviews the terminological origin, development and the latest status of desertification and demonstrates the significant impact of the climatic variations in the processes of desertification and its rehabilitations. Two typical arid regions at Asia's middle latitudes are selected as cases for studies, i.e., the oases in the hyper-arid Taklamakan Desert in the west and the semi-arid Hunshandake Sandy Land in the east (Fig. 1). It shows that humans set up cultivation bases in the center of the Taklamakan Desert over 2,000 years ago and the later abandonment of these bases was probably caused by decrease in runoffs associated with droughts. In shaping the landscape changes between soil formation and aeolian sedimentation in the Hunshandake, the role played by the Holocene changes of East Asia summer monsoons surpasses the impact of human activities since the Neolithic time, meaning that the climatic background and its potential changes should be given greater attention in the aims and schemes of combating desertification. The rapid vegetation regeneration in recent years in the Hunshandake reconfirms the great importance of climatic background in ecological rehabilitations in desertified lands. In the course of global warming, the risks of desertification are likely to increase in both areas due to different reasons.

Keywords Desertification · Climatic change · Oasis · Arid zone · Semiarid area · China

X. Yang (✉)
Institute of Geology and Geophysics, Chinese Academy of Sciences, Beijing 100029, China
e-mail: xyang@mail.igcas.ac.cn; xpyang@263.net.cn

Introduction

Deserts have long been recognized as distinct in their landscapes and have been inhabited in some parts of the world by humans for millennia. The term desertification, however, is new and its definition and hence use has been not always consistent. The word was first used by Aubréville (1949) to describe the clearing and burning of forests in parts of Africa in order to cultivate land and the replacement of tropical forest by secondary savanna and scrub. Following the severe drought and associated famine in the Sahel region of northern Africa between 1968 and 1974, desertification was conceptualized as an issue in need of global attention. In response to the resulting disaster, the United Nations Conference on Desertification (UNCOD) was organized by the United Nations Environment Program (UNEP) in 1977, attended by ca. 300 geographers, botanists, and social scientists. UNEP (1977) defined desertification as the reduction or destruction of the biological potential of land which can ultimately lead to desert-like conditions, attributing the causes solely to human activities. As El-Baz (2008) commented, this conference delivered a sense of emergency and a resounding indictment of the practices of local people. Desertification is now defined in the United Nations Convention to Combat Desertification (UNCCD) as "land degradation in the arid, semi arid and dry sub-humid areas resulting from various factors, including climatic variations and human activities" (UNCCD, 1999). The revision to include climate factors as cause of desertification is indeed a key step toward a better understanding of desertification because water shortage due to droughts is often the main cause of land degradation in drylands. Meteorological data of the

J. Dodson (ed.), *Changing Climates, Earth Systems and Society*, International Year of Planet Earth,
DOI 10.1007/978-90-481-8716-4_8,

Sahel region show that precipitation during the past four decades was substantially less than in the period from 1900 to 1965 and there is a 7-year cycle of alternating wet and dry intervals (El-Baz, 2008). It may be mentioned that some scientists think desertification is a buzzword and should be replaced by land degradation or landscape degradation (e.g., Seuffert, 2001).

Under the auspices of the United Nations, the year 2006 was celebrated as the International Year of Deserts and Desertification and raised public awareness about desertification and the need to protect drylands biodiversity and the knowledge and traditions of the people whose everyday lives are affected by desertification (Conacher and Gisladottir, 2006). However, the assessments about the severity of desertification on both regional and global scales differ considerably. UNCCD (1999) reported that 70% of global drylands are affected by desertification that threatens the livelihood of some one billion people. Such an assessment is, however, questionable as it is mainly derived from data with different levels of authenticity and consistency (Thomas and Middleton, 1994). Based on regional monitoring and assessments of annually integrated Normalized Difference Vegetation Index (NDVI) and annual rainfall, Helldén and Tottrup (2008) argue that systematic desertification trends do not occur at the regional–global level over the past 20 years in any of the regions of the Mediterranean basin, the Sahel and the drylands of Southern Africa, China, Mongolia, and South America. On the contrary, a "greening-up" tendency was reported in several of these regions (Helldén and Tottrup, 2008).

The time scale is another key issue causing disputes in the understanding of desertification. Although it was emphasized that the encroachments of the deserts in Sahara of northern Africa during the Last Glacial Maximum should not be seen as desertification (Mensching, 1990), deserts formed by past Earth System processes are sometimes included in the data of desertification. For example, the large dune fields formed by atmospheric circulation changes in relation to the Cenozoic uplifting of Tibetan Plateau in western China must have been included while one quarter of China was reported to be affected by desertification, as described in Liu and Diamond (2005). Globally, the desert areas have undergone great changes due to climatic fluctuations during the Quaternary. The aeolian sand that is typical in sand seas today originated often from erosion and transportation by wind and surface water. Groundwater in deserts was often mainly recharged during wetter epochs. Geomorphological and paleoclimatic studies have shown that it was particularly arid in the Sahara during the last glacial maximum, accompanied by shrinkage of lakes, disintegration of drainage systems, and southward migration of dunes (Goudie, 2002). Owing to the absence of any human impacts, the landscape degradation associated with such long-time scale climate changes should not be treated as desertification.

On the other hand, humans have become an increasingly significant driver in changing the Earth's environment. Consequently, the time since the beginning of the nineteenth century was suggested to be defined as the Anthropocene (Crutzen et al., 2002). Looking back at the long history of human impact in China, Liu (2004) suggested that the Anthropocene should cover the entire Holocene. It has been also argued that rice farming in China over 5,000 years ago probably had already caused an increase in the greenhouse enhancing gas (methane) concentrations, in turn triggering an increase in temperature (Ruddiman et al., 2008). In this context, the processes of desertification are here traced back to the middle Holocene.

As one of the oldest civilizations of the world, China has a long history of human activities that have left strong fingerprints on the landscape, particularly in the drylands. Arid and semi-arid regions cover one-third of China, and they include deserts, desert steppe, steppe, and scrub-woodland environments distributed in a wide range of geomorphological and tectonic settings, making China very prone to desertification (Yang et al., 2004). Based on case studies from China, it is intended here to examine the relationships between climate change and desertification. The meridian at ca. 106°E, geomorphologically marked by the N–S striking range named Helan Mountains (Fig. 1) and climatologically identical with the 200 mm isoline of mean annual precipitation, marks a boundary between arid and semi-arid regions of China. Although the hyper-arid regions like the Tarim Basin in northwestern China should be excluded for the topic of desertification (Zhu, 1998; Conacher and Gisladottir, 2006), the degradation of oases there can be assessed as desertification as they are strongly affected by both human activities and climate changes (Yang, 1998). Oases, dependant upon rivers and groundwater, are unique

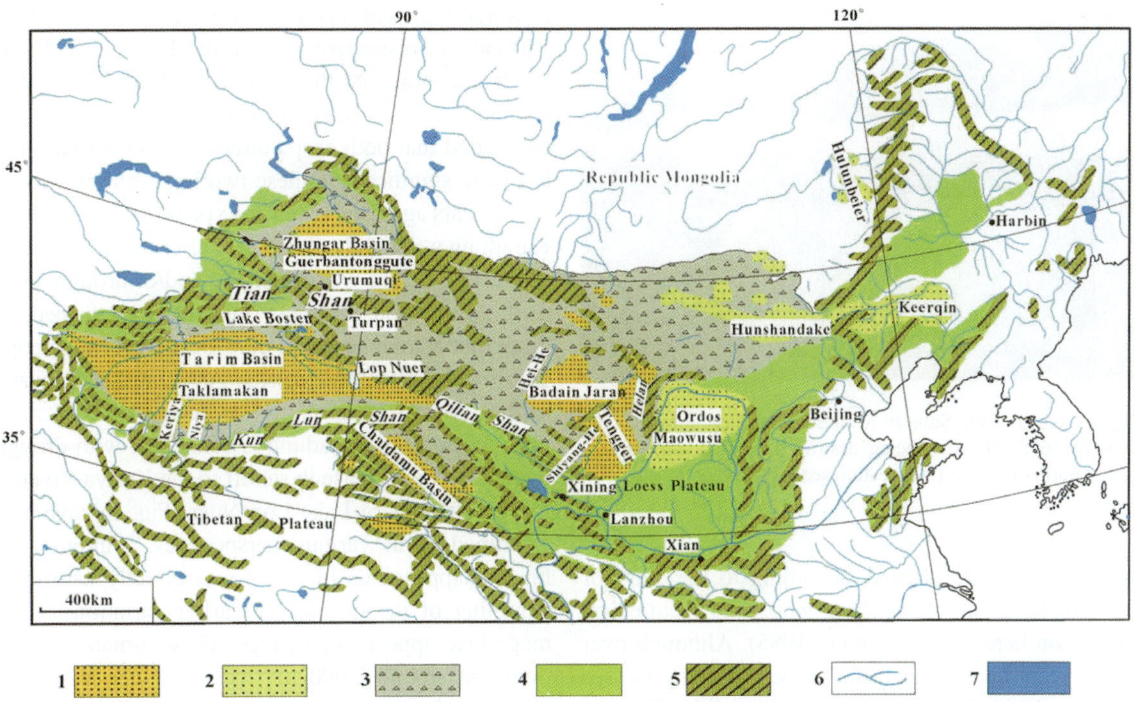

Fig. 1 Distribution map of desert landscapes in northern China. 1, sandy deserts; 2, sandy lands; 3, gravel desert (Gobi); 4, loess; 5, mountain ranges; 6, rivers; 7, lakes

landscapes in arid regions of the world. Owing to water originating from surrounding mountains, oases with varying extensions are established in the fluvial delta and alluvial plains at the edges of alluvial fans and along the river courses in the Tarim Basin. Thus, the Taklamakan Desert, the largest sand sea in China, is encircled by a series of oases and divided by several green corridors developed along the rivers flowing into it. The degradation of oases and shrinkage of wetlands in hyper-arid regions impose serious environmental challenges on the economic development in western China and in other countries of Central Asia as well. Therefore, the case of Tarim Basin is selected here to demonstrate the serious harm caused by desertification in arid regions. As a comparison, the degradation of sandy lands in the areas east of Helan Mountains is presented here to show the level of desertification and the achievable success of rehabilitation efforts in regions with more favorable climate conditions. On the basis of paleoenvironmental knowledge and modern trends, the future tendencies of desertification and the interactions between climate change and desertification are briefly discussed.

Desertification in the Past

One of the most serious difficulties in studying dryland degradation and global climate change in general is to separate the effect of human activities from the impacts of climatic fluctuations (Dodson et al., 2004). The earlier history of human settlements in the Tarim Basin is ideal for investigating the society's response to hydrological changes that are caused by climatic fluctuations. The settlements and agricultural activities in the Taklamakan Desert have been strongly dependent on the runoff of rivers with their headwaters in the surrounding mountains.

Large ruined settlements are robust indicators of desertification of oases in the Tarim Basin where the mean annual precipitation is less than 50 mm nowadays. For example, Subashi was a famous city on the northern margin of the Taklamakan Desert during the Tang Dynasty (618–907 AD). Nowadays, the only remnant of the once large settlement is the 20,000 m^2 Buddhism monastery (Fig. 2). There were 5,000 monks in this monastery and rice, apples, apricots,

Fig. 2 Relics of the Subashi Monastery constructions of Tang Dynasty (618–907 AD), being destroyed by flooding, located 20 km north of Kuche (for location see Fig. 4b)

and peaches were planted, according to the diary of the famous monk Xuan Zang who gave lectures on Buddhism here at that time (Ji, 1985). Although over 1,000 years have passed, the shapes of many houses associated with the monastery are still visible. The precise cause of the abandonment of the Subashi is not known, but might have been related to catastrophic flooding as the monastery was incised by a flood channel (Fig. 2).

In the Tarim Basin, paleo-river courses are much more extensive than the active channels at present time, indicating a more recent trend of aridification in this hyper-arid environment. In the center of the Taklamakan Desert, two large ruined settlements confirm the degradation of the ecosystem in the largest sand sea of China. One is Karadung, and the other Yuansha, located north of Karadung (Fig. 3). The region of these two former cities is characterized by relatively flat terrain, different from the common dune landscape of this desert. A Sino-French Expedition Team (1977) excavated the ruined city Yuansha and found that it was constructed in the form of an old Chinese city, surrounded by a 3–4-m-high and ca. 1-km-long city wall. Yuansha was established before the Western Han Dynasty (206 BC to 25 AD) and was abandoned in the Eastern Han Dynasty (25–220 AD) according to cultural remains and a radiocarbon date from wood. Karadung was a large dwelling area with a diameter of ca. 2 km and with clear evidence of an administration structure, and the bronze coins collected from this site were dated to the first, fourth, and fifth centuries AD (Huang, 1958). Timbers found in Karadung were dated to 2,150 ± 120 and 2,141 ± 130 cal BP, respectively (Zhu and Lu, 1991). To the south and north of the settlement, former irrigation channels with a width of 1 m are still preserved. It was concluded that both crop cultivation and animal grazing were significant in these two large settlements ca. 2,000 years ago (Sino-French Expedition Team, 1977). Karadung was totally abandoned after the fifth century, suggesting severe water shortage and/or drought since then. Although it is not yet objectively confirmed, we think that Karadung was a new location for the people who lived earlier in Yuansha. As Yuansha is located further down stream along the Keriya River, a move from Yuansha to Karadung is consistent with drought that caused a decrease in runoff to the Keriya River.

The changes of Lake Lop Nuer (Fig. 1) have been examined from various perspectives. Although the final disappearance of water was directly caused by damming of the inflow, the earlier variations of the mega lake appear to be triggered by climatic fluctuations. At around 2,000 BP, the Lop Nuer was called "Puchang Sea" with a diameter of 150–200 km according to historical records (Yang et al., 2006). The recharge of the lake was described to be from two origins, i.e., Pamir Plateau and the Kunlun Mountains. This earlier observation was probably correct as it is consistent with current hydrological patterns. In historical records of the Qing Dynasty (1644–1911 AD), Lop Nuer was described as "Tarim Lake" occurring only in the south part of a much larger lake in earlier times, the northern part of the former lake was occupied by extensive marshes. From the digitized maps with information of the historical descriptions, one can see that the extension of Lake Lop Nuer has decreased dramatically in the past 2,000 years (Fig. 4). Before this time it was a single large lake; then it changed to a group of small lakes by the Qing Dynasty, leaving large areas of bare earth around the shorelines. A climatic shift toward drier conditions was probably the main cause of this lake level decrease (Berg, 1907; Yang et al., 2006), although human impact must have played a role in this process also (Xia et al., 2007). Distinct variations of carbon isotopes of the organic matter (Luo et al., 2008) and vegetation shifts (Yan et al., 1998) found in the lacustrine sediments of Lop Nuer show a change in moisture toward an arid climate during the late Pleistocene. Even in this present-day hyper-arid zone, the vegetation was desert steppe between 20 and 15 ka BP rather than desert as seen

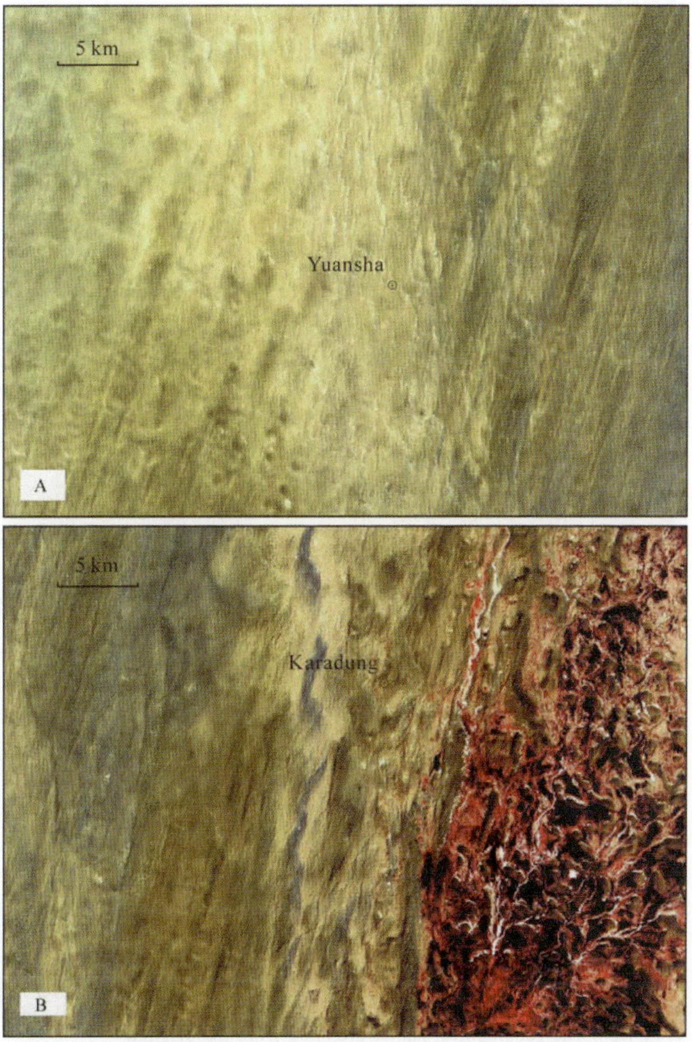

Fig. 3 Formerly large settlements in the lower reaches of the Keriya River (Fig. 4a) in the center of the Taklamakan, covered by flat sand sheets and paleo-channels. (**a**) Yuansha; (**b**) Karadung. The *red color* in (**b**) shows the present delta formed by the same river that formerly provided water to Yuansha and later to Karadung

today according to the palynological evidence (Yan et al., 1998).

Great attention has been given to desertification in China partly because the reworking of formerly stable dunes in the sandy lands east of Helan Mountains (Fig. 1) has been intensive, causing an increase in dust storms, economic losses in agriculture, animal husbandry, and even the transportation industry. Thus, the formation of aeolian landforms has been applied as one of the main indictors for evaluating the degree of desertification in China (Research Group "Study on Combating Desertification/Land Degradation in China", 1998). The initial formation of the dunes in the sandy lands might be linked to the early aridification in northern China while the dry and cold East Asian winter monsoon formed, but geological evidence for this assumption is not yet proved. As sandy lands are mostly located on the marginal zones of the summer monsoon, they are very sensitive to the changes of the intensity of the summer monsoon moisture.

During the drier periods of the last millennia, sand seas occurred in the areas of the present-day sandy lands (Fig. 1). These dunes may become stabilized by denser vegetation cover when it becomes more humid. Owing to the occurrence of massive aeolian sands, the dunes can be easily reactivated while the vegetation cover is deteriorated either due to climate change toward drier conditions or due to unsustainable human activities such as overgrazing or intensification of agriculture. The crucial role of climate in the processes

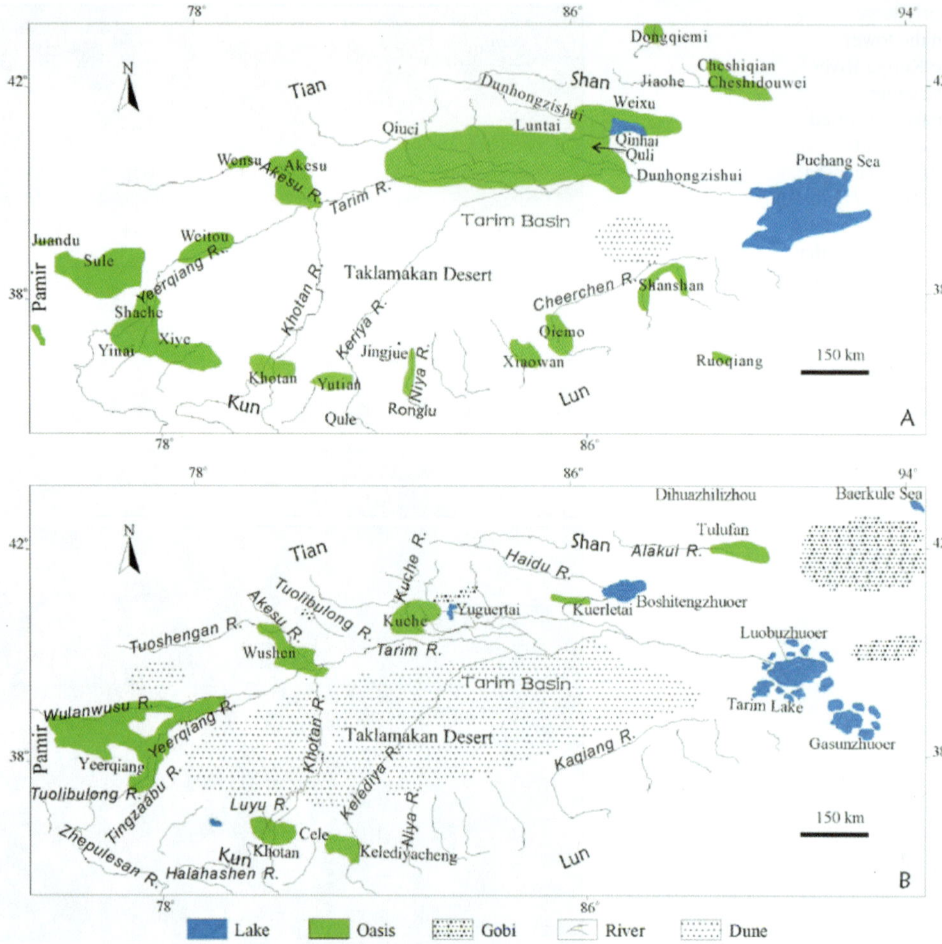

Fig. 4 Area of lakes and oases in the Tarim Basin during the Han Dynasties (206 BC to 220 AD) and the Qing Dynasty (1644–1911 AD). The geographical names are cited from the original text and they are often different from the present ones (modified from Yang et al., 2006)

of desertification is well illustrated by the Holocene changes in the Hunshandake Sandy Land (Fig. 1).

Hunshandake Sandy Land is ca. 21,400 km^2 in area and located between the latitudes 42–44°N and the longitudes 113–118°E in the eastern part of Asia's mid-latitude desert belt. The climate is typically controlled by the East-Asia monsoon system, with cold winters and hot summers. The mean annual precipitation declines from ca. 350 mm in the east to ca. 250 mm in the west and the mean annual temperature is ca. 2°C (Yang et al., 2007). Due to the relatively high amount of precipitation, most of the dunes here are stable or semi-stable, reaching a height of 15–20 m, and occasionally around 30 m. The predominant plants in this sandy land are grasses with some shrubs like *Artemisia* species and *Salix microstachya*, and trees such as *Ulmus* and *Pinus* species. Close to the top of the sediment sequences of the dunes a ca. 80-cm kastanozem paleosol is present, indicating much denser vegetation cover in the past. This layer of soil is thought to be an indicator of the climate optimum during the middle Holocene. The climatic conditions at that time would most likely have been warmer and more humid than at present. According to radiocarbon dates of the paleosols and the optically stimulated luminescence (OSL) dates of aeolian sand, it was concluded that a warmer and wetter period began ca. 5 ka BP and lasted until ca. 3 ka BP (Yang et al., 2008). The aeolian sand overlying this paleosol suggests that the vegetation was degraded after ca.

3 ka BP. The initial beginning of desertification at ca. 3 ka BP must have been triggered by the natural climate change because it occurred simultaneously as the temperature decreased in China according to historical notes based on plant phenology (Zhu, 1972).

Interestingly, all excavation sites of Hongshan Culture of Neolithic Age are close to the eastern margin of the Hunshandake Sandy Land. This Culture was suggested to be prosperous from 6,700 to 5,000 BP (Suo and Li, 2007), broadly simultaneous with Banpo Culture discovered in Xi'an of China. The Hongshan Culture was characterized by permanent settlements with agricultural cultivation and limited animal husbandry. The reasons of the disappearance of this culture in this region are not yet known. Besides, the known ages of the Hongshan culture are based on a very limited samples of physical dating, and the precise dates need to be obtained from future research. Due to the short distance between this sandy land and several excavation sites of the Hongshan Culture, we assume that humans had a certain level of impact in this sandy land from the middle Holocene. But the available ages of the paleosols suggests that the shift from soil formation to sand accumulation occurred later after the culture period, pointing to climatic variation rather than human influences as the main trigger for initiation of the aeolian sedimentation on the formerly stable dunes.

It should be mentioned that the Holocene climate changes in the Sandy Land are still not well understood and require further detailed investigations. Inconsistencies and disagreements do exist in the reconstruction of Holocene climate histories in this sandy land probably due to lack of precise chronology and the use of unsuitable paleoclimatic proxies. The pollen assemblages of a sediment core from a lake located ca. 300 km southwest of Hunshandake suggested that the Holocene Climatic Optimum occurred between ca. 7,900 and 4,450 cal years BP (Xiao et al., 2004). Based on the evidence of pollen and algal data from another lake located immediately on the southern margin of this sandy land, it was suggested that the wettest climate occurred between ca. 10,500 and 6,500 cal years BP (Jiang et al., 2006). The results from the studies of the dune stratigraphy by OSL dating are not consistent either. For example, the Holocene Optimum was suggested to be from 10 to 3.6 ka BP by Li et al. (2002), but between 7 and 2.4 ka BP by Lu et al. (2005), both based on the OSL chronology of selected dune profiles. The records of algal fossils of *Pediastrum* and pollen in an inter-dune lake of this sandy land, however, show that the wettest period of the last 5,000 years occurred between 3,500 and 1,500 cal years BP (Xu et al., 2004). For a correct assessment on the relationship between climate change and desertification in this region, a much more reliable reconstruction of paleoclimate is urgently needed.

At present, desertification is not a real issue in the well-protected areas of the Hunshandake as plants can regenerate rapidly due to relatively high mean annual rainfall (Fig. 5). Reactivation of the former dormant dunes (Fig. 6) occurs mainly in the non-protected central and southern region, ca. 11,000 km^2 in area. Using digital processing of the LANDSAT data, it was found out that the percentage of active dunes (referring to dunes with vegetation coverage under 10% in this study) has varied between 4 and 8%, and the wetlands between 0.8 and 2.9% from 1975 to 2001 (Yang et al., 2007). The expansion of active dunes and the shrinkage of wetlands has taken place in the years of reduced precipitation and increased temperature (Yang et al., 2007), suggesting that both human impact and climate are key factors triggering desertification in this region. Compared with desertification assessment system applied in China (Research Group "Study on combating desertification/land degradation in China", 1998), the active dune field is consistent with the area defined as strong desertification. The areas with light and moderate desertification are supposed to have a vegetation coverage of 50–30% and 30–10%,

Fig. 5 The dunes in the southern Hunshandake Sandy Land with little desertification due to relatively high annual rainfall and the elimination of human activities via fencing

Fig. 6 Strongly desertified land in the southern Hunshandake Sandy Land, with reworked dunes and some remnant trees

respectively (Research Group "Study on combating desertification/land degradation in China", 1998). This research group reported that the desertification occurred in an area of 664,000 km^2 in arid, semi-arid, and dry sub-humid areas of China, accounting for ca. 6.9% of the total land surface in China.

Assessments carried out by others reach different conclusions. For example, Liu and Wang (2007) studied the desertification processes in the southern region of Hunshandake, using the entire administrative county in the south as a monitoring unit. Their results confirm that the degree of desertification varied between serious progression and rehabilitation from the 1970s to 2005. But they claimed unsympathetic human activities, not climate caused desertification in the Hunshandake. In fact, the earlier surveys prior to 1980 concluded that active dunes accounted for 2% of all dunes whereas fixed and semi-fixed ones accounted for 98% in the Hunshandake (Zhu et al., 1980). Here the unsolved scientific question is how high the vegetation cover would be in the sandy land if there was not any human activity. Indeed, the occurrence of a small percentage of active dunes might be inevitable in the drier parts of Hunshandake under present climate conditions. In the historical books of the Beiwei (386–534 AD) and Yuan Dynasty (1279–1368 AD) aeolian landforms were recorded in the Hunshandake (Wang and Ren, 2007). Of course, the rapid population increase has no doubt had a negative impact on the vegetation in this ecologically sensitive region.

Future Trend of Desertification

The future trends of desertification are of great concern in China due to occurrence of extensive drylands and high population densities. The glaciers in the mountains of western China have retreated rapidly in recent decades, causing a particular threat to the sustainability of the oases in the drylands. It was estimated that the area of glaciers in dryland regions of northwestern China has decreased by 1,400 km^2 since 1960s, while the snowline was elevated by 30–60 m (Qin, 2002). The melting of glaciers would accelerate if global warming continued. As the oases in drylands of western China are mainly dependant on the water from surrounding mountains, the future extent of the glaciers would be crucial to the fortune of the communities depending on the oases. It has been shown that it was less arid in the Tarim Basin during the colder periods like the Little Ice Age (Yang et al., 2004). Consequently, the challenges for combating desertification would be increasingly greater if it became warmer.

Although a precise chronology is still lacking, the available data about the Holocene climatic history in the sandy lands in the eastern part of the Asia's middle latitude desert belt broadly show that the increase in temperature was accompanied by an increase in rainfall, similar to the boundary conditions during the warmer epoch of mid-Holocene. Under this logic, the vegetation cover would become denser during the process of global warming. But as various studies have suggested (e.g., Introduction of this book by Dodson), the frequency of abrupt events including droughts will increase in the future, being disadvantageous to the ecological rehabilitation in this region. Some climate models predict that future global warming may reduce soil moisture over large areas of semi-arid grassland in Asia and North America (Manabe and Wetherald, 1986). Such a climate trend will exacerbate the degradation of semi-arid lands even if human impact could be fully controlled. In order to restore the degraded ecosystem in dryands of northern China, many grazing activities have been arranged to cease in formerly intensively used pasturelands and the herdsmen have been allocated to other sectors of employment. For example, 5,778 local inhabitants of one county in the southern Hunshandake have been moved to the newly constructed towns with the financial assistance from

Chinese government (Peng et al., 2005). In the new and centralized settlement, the former herdsmen make a living by working in tertiary industry or by feeding animals using forage collected from less degraded areas. Some areas of the Hunshandake have been protected as natural reserves by fences since 2000. The features of future climate will decide whether the rehabilitation efforts will be ultimately successful in cases where the human impact has been eliminated.

In the case of the Kalahari in southern Africa, global warming might become a key factor triggering the reworking of the stable desert dunes. The GCM (General Climate Model) forecasts that the ratio between precipitation and potential evaporation will decline and the mean wind velocity will increase in the Kalahari. By 2099 all stable dunes in the Kalahari are likely to be reactivated (Thomas et al., 2005). This kind of trend will need to be given profound attention for the aims and plans of regional developments.

Interactions Between Climate Change and Desertification

It was suggested that the global area of desert lands will increase by 17% during the climate changes expected with a doubling of atmospheric CO_2 content (Emanuel, 1987). However, the regional responses to the warming are significantly different at water-shed level, depending on timing and distribution of rainfall and soil water-holding capacity (Feddema, 1999). On one hand, the changes in precipitation, evaporation and winds could have profound impacts on the process of desertification. In the other, desertification and its related processes have a significant feedback effect in the Earth system. Through dust storms, desertification may affect the global geochemical cycles and may play a great role in climatic changes. Physically, the process of desertification is modifying the concentration and features of aerosols that absorb or scatter radiant energy. Simulations suggests that strong Asian dust emissions during glacials may have prevented the formation of permanent snow cover in northern Asia and may even have determined the location and extent of the last large ice sheets (Krinner et al., 2006; Goudie, 2009).

In principal, desertification can also transfer soil carbon into the atmosphere, directly causing increases in greenhouse gases. The quantitative assessment of carbon emission via desertification needs to be a focus of future research, as the available data differ greatly. For instance, the total release of carbon into the atmosphere from the desertified land in China during the last 40 years has been estimated to be as high as 91 M t C by Duan et al. (2001), and as low as 2.2 M t C by Feng et al. (2002). A study in the Hunshandake demonstrates that the significant portion of carbon lost on-site was re-deposited in downwind areas through deflation and re-deposition. Consequently, the carbon exchange between the atmosphere and desertified lands is much smaller than the total carbon initially contained in the deflated soil (Yang et al., 2008). More recent investigations suggest that the alkaline soil of deserts may sequester large quantities of atmospheric CO_2 in an inorganic form. This process has been detected in the Guerbantonggute Desert of western China (Xie et al., 2009) and in the Mojave Desert of USA (Wohlfahrt et al., 2008). It was found that the Mojave Desert, square meter for square meter, may absorb the same amount of CO_2 as some temperate forests. But it is assumed that the visible surge in vegetation in many deserts due to a recent increase in precipitation is the trigger for the observed CO_2 absorption. A potential decrease in the annual rainfall in those deserts would release the stored carbon again and lead to an increase in atmospheric CO_2 (Stone, 2008).

Conclusions

Although desertification was firstly given intense global attention after the severe drought that directly triggered dramatic decrease in agricultural production and massive famine in the Sahel region of northern Africa between 1968 and 1974, climatic variations were excluded as a cause of desertification in the conception of the United Nations Environment Program (UNEP) for a long time. In its earlier definitions only human impact was accepted as the driver of land degradation in drylands. Now in the United Nations Convention to Combat Desertification, both climatic variations and human activities are described as causes of this process. From the scientific point of view, natural deserts formed during past geological times should not be treated as cases of desertification. So far, the estimates of desertification extent on regional and global scales are often inconsistent and even contradictory due to both different opinions on the definition and data of varying levels of authenticity. In general, hyper-arid regions are not included in the issue of desertification, but the changes of the oases in the hyper-arid Taklamakan Desert suggest that climatic variations and possible mismanagement of water resources during historical times might have caused deterioration and even abandonment of the former settlements and their related production bases. Therefore, the deterioration of oases in hyper-arid environments should be included in the UN Convention to Combat Desertification. The case of the semi-arid Hunshandake Sandy Land raises some thoughts about the temporal framework of desertification. Should the mid-Holocene shift from soil formation to aeolian deposition in the Hunshandake, caused by the changes of East Asian monsoon system, be defined as desertification? The answer for this question tends to be supportive if it is considered that humans have been active in this region at latest since the Neolithic. The changes of the last four decades between land use (grazing and agricultural cultivation) and ecological rehabilitation are to a large degree decided by governmental policies, although the climatic impact on vegetation cover has also been obvious. In the course of global warming, both the oases in the Taklamakan and the Sandy Land of Hunshandake will face greater challenges due to the potential decrease in ice melt runoff to the rivers and the possible decline of the ratio between precipitation and evaporation. Through impacts on dust storms and global geochemical cycles, the processes of desertification may play a large role in the Earth system although many details of this require further deep and profound studies.

Acknowledgments We are very grateful to Professor John Dodson for the opportunity to participate in this IYPE project and for his more than generous help in improving the writing. Our work on drylands in recent years has been financially supported by the Chinese Academy of Sciences (Grant no.: kzcx2-yw-119) and the National Natural Science Foundation of China (Grant no.: 40930105). Sincere thanks are extended to Hongwei Li for cartographical assistance.

References

Aubréville A (1949) Climats, forêts et desertification de l'Afrique tropicale. Société d'Editions Géographiques. Paris: Maritimes et Coloniales.

Berg L (1907) Ist Zentral-Asien im Austrocknen begriffen? Geogr Z 13:568–579.

Conacher A, Gisladottir G (2006) UN year of deserts and desertification: Policy implications. Quat Sci (in Chinese) 26:1030.

Crutzen P (2002) Atmospheric chemistry in the "Anthropocene." In: W Steffen, J Jäger, D Carson et al. (Eds), Challenges of a Changing Earth: Proceedings of the Global Change Open Science Conference. Amsterdam, The Netherlands, 2001. Berlin, Heidelberg: Springer.

Dodson J, Taylor D, Ono Y, Wang P (2004) Climate, human, and natural systems of the PEP II transect. Quat Int 118–119:3–12.

Duan Z, Xiao H, Dong Z, He X, Wang G (2001) Estimate of total CO_2 output from desertified sandy land in China. Atmos Environ 35:5915–5921.

El-Baz F (2008) Let deserts be. Nature 456:30.

Emanuel K (1987) The dependence of hurricane intensity on climate. Nature 326:483–485.

Feddema J (1999) Future African water resources: Interactions between soil degradation and global warming. Clim Change 42:561–596.

Feng Q, Cheng G, Masao M (2002) The carbon cycle of sandy lands in China and its global significance. Clim Change 48:535–549.

Goudie A (2002) Great Warm Deserts of the World: Landscapes and Evolution. Oxford: Oxford University Press.

Goudie A (2009) Dust storms: Recent developments. J Environ Manage 90:89–94.

Helldén U, Tottrup C (2008) Regional desertification: A global synthesis. Glob Planet Change 64:169–176.

Huang W (1958) Records of Archaeological Investigations in the Tarim Basin (in Chinese). Beijing: Science Press.

Ji X (1985) Annotations to the Historical Book "Records from the Travel to the West During the Tang Dynasty" (in Chinese). Beijing: Zhonghua Book Company.

Jiang W, Guo Z, Sun X, Wu H, Chu G, Yuan B, Hatté C, Guiot J (2006) Reconstruction of climate and vegetation changes of Lake Bayanchagan (Inner Mongolia): Holocene variability of the East Asia monsoon. Quat Int 65: 411–420.

Krinner G, Boucher O, Balkanski Y (2006) Ice-free glacial northern Asia due to dust deposition on snow. Clim Dyn 27:613–625.

Li S, Sun J, Zhao H (2002) Optical dating of dune sands in the northeastern deserts of China. Palaeogeogr Palaeoclimatol Palaeoecol 181:419–429.

Liu T (2004) Demand of Anthropocene study in the new stage of geoscience: in honor of late geologist Huang Jiqing for his innovative spirit. Quat Sci 24:369–378 (in Chinese).

Liu J, Diamond J (2005) China's environment in a globalizing world. Nature 435:1179–1186.

Liu S, Wang T (2007) Aeolian desertification from the mid-1970s to 2005 in Otindag Sandy Land, Northern China. Environ Geol 51:1057–1064.

Lu H, Miao X, Zhou Y, Mason J, Swinehart J, Zhang J, Zhou L, Yi S (2005) Late Quaternary aeolian activity in the Mu Us and Otindag dune fields (north China) and lagged response to insolation forcing. Geophys Res Lett doi: 10.1029/2005GL024560.

Luo C, Liu W, Peng Z, Yang D, He J, Liu G, Zhang P (2008) Stable carbon isotope record of organic matter from the Lop-Nur lacustrine sediment in Xinjiang, northwest China. Quat Sci 28:621–628 (in Chinese).

Manabe S, Wetherald R (1986) Reduction in summer soil wetness induced by an increase in atmospheric carbon dioxide. Science 232:626–628.

Mensching H (1990) Desertifikation: ein weltweites Problem der ökologischen Verwüstung in den Trockengebieten der Erde. Darmstadt: Wissenschaftliche Buchgesellschaft.

Peng Y, Jiang G, Liu M, Niu S, Yu S, Biswas D, Zhang Q, Shi X, Yang Q (2005) Potentials for combating desertification in Hunshandak Sandland through nature reserve. Environ Manage 35:453–460.

Qin D (2002) The Synthesis Report on the Evaluation of Western China's Environmental Evolution (in Chinese). Beijing: Science Press.

Research Group "study on combating desertification/land degradation in China" (1998) Study on Combating Desertification/Land Degradation in China. Beijing: China Environmental Science Press.

Ruddiman W, Guo Z, Zhou X, Wu H, Yu Y (2008) Early rice farming and anomalous methane trends. Quat Sci Rev 27:1291–1295.

Seuffert O (2001) Landschafts(zer)störung – Ursachen, Prozesse, Produkte, Definition & Perspektiven. Geo-Öko 22:91–102.

Sino-French Expedition Team (1997) Outlines of the results from the archaeological excavations in the reaches of the Keriya River, Xinjiang. Xinjing Antiquities 48:1–12 (in Chinese).

Suo X, Li S (2007) Preliminary studies on epochs of Hongshan Culture in the relics of Nuiheliang. Archaeology 10:52–61 (in Chinese).

Stone R (2008) Have desert researchers discovered a hidden loop in the carbon cycle? Science 320:1409–1410.

Thomas D, Middleton N (1994) Desertification: Exploding the Myth. Chichester: Wiley.

Thomas D, Knight M, Wiggs G (2005) Remobilization of southern African desert dune systems by twenty-first century global warming. Nature 435:1218–1221.

UNCCD (1999) United Nations Convention to Combat Desertification in Those Countries Experiencing Serious Drought and/or Desertification, Particularly in Africa: Text with Annexes. Bonn: Secretariat of the Convention to Combat Desertification.

UNEP (1977) World Map of Desertification at a Scale of 1:25,000,000. Nairobi: FAO/UNEP/WMO.

Wang W, Ren B (2007) Onqin Daga Desert and Korqin Desert since the North Wei Dynasty. J Taiyuan Normal Univ (Natural Science Edition) 6:107–110 (in Chinese).

Wohlfahrt G, Fenstermaker L, Arnone J (2008) Large annual net ecosystem CO_2 uptake of a Mojave Desert ecosystem. Glob Change Biol 14:1475–1487.

Xia X, Wang F, Zhao Y (Eds) (2007) Chinese Lop Nuer (in Chinese). Beijing: Science Press.

Xiao J, Xu Q, Nakamura T, Yang X, Liang W, Inouchi Y (2004) Holocene vegetation variation in the Daihai Lake region of north-central China: A direct indication of the Asian monsoon climatic history. Quat Sci Rev 23:1669–1679.

Xie J, Li Y, Zhai C, Li C, Lan Z (2009) CO_2 absorption by alkaline soils and its implication to the global carbon cycle. Environ Geol 56:953–961.

Xu Z, Li C, Kong Z (2004) On the fossil *Pediastrum* from the Gaoximage Section, Hunshandak Sandy Land and its ecological significance since 5000 a BP. Acta Bot Sin 46:1141–1148.

Yan S, Mu G, Xu Y, Zhao Z (1998) Quaternary environmental evolution of the Lop Nur region, China. Acta Geogr Sin 53:332–340 (in Chinese).

Yang X (1998) Desertification and land use in the arid areas of central Asia. Quat Sci 18:119–127 (in Chinese).

Yang X, Rost K, Lehmkuhl F, Zhu Z, Dodson J (2004) The evolution of dry lands in northern China and in the Republic of Mongolia since the Last Glacial Maximum. Quat Int 118/119:69–85.

Yang X, Liu Z, Zhang F, White P, Wang X (2006) Hydrological changes and land degradation in the southern and eastern Tarim Basin, Xinjiang, China. Land Degrad Dev 17:381–392.

Yang X, Ding Z, Fan X, Zhou Z, Ma N (2007) Processes and mechanisms of desertification in northern China during the last 30 years, with a special reference to the Hunshandake Sandy Land, eastern Inner Mongolia. Catena 71: 2–12.

Yang X, Zhu B, Wang X, Wang X, Li C, Zhou Z, Chen J, Wang X, Yin J, Lu Y (2008) Late Quaternary environmental changes and organic carbon density in the Hunshandake Sandy Land, eastern Inner Mongolia, China. Glob Planet Change 61:70–78.

Zhu K (1972) Preliminary studies on climatic changes in China during the last 5000 years. Acta Archaeologica Sinica 1: 13–38 (in Chinese).

Zhu Z (1998) Concept, cause and control of desertification in China. Quat Sci 18:145–155 (in Chinese).

Zhu Z, Lu J (1991) A study on the formation and development of aeolian landforms and the trend of environmental changes in the lower reaches of the Keriya River, Central Taklimakan Desert. Die Erde Ergaenzungsheft 6: 89–97.

Zhu Z, Wu Z, Liu S, Di X (1980) An Outline of Chinese Deserts (in Chinese). Beijing: Science Press.

Climate Change, Societal Transitions and Changing Infectious Disease Burdens

Emily Fearnley, Philip Weinstein, and John Dodson

Abstract Human health is directly and indirectly influenced by the effects of climate change – air and sea temperatures, rainfall and more frequent and severe climate extremes. These effects do not impact on human populations uniformly, however, and this chapter looks at the interplay between climate and the different lifestyles that follow societal transitions, from hunting and gathering, through agriculture, industrialisation and globalisation. We illustrate the importance of considering such lifestyle effects by focusing on infectious diseases as a case study of disease amplification with climate change. The range and intensity of gastrointestinal and vector-borne diseases are likely to be altered as a result of climate change, and communities with different lifestyles will be affected differently. It follows that situation and societal-specific recommendations for public health interventions will be required, with our framework as a potential basis for considering these differences.

Keywords Climate change · Faecal-oral diseases · Societal transitions · Vector-borne diseases

Introduction

Human health, for individuals and populations, is influenced by a number of factors; broadly speaking three major groups of determinants can be conceptualised: social and economic environment factors, physical environment factors, and individual characteristics and behaviours. Of particular interest within this chapter is the interaction between different lifestyles that form part of human societal transitions, with respect to the current significant issue of climate change. Climate and weather have always had a powerful influence on human health, and can be conceptualised as operating directly and indirectly to affect health through these three modalities (World Health Organisation, 2003). For example, heat waves operate at the level of individual behaviours; they affect the physical environment via vector-borne disease and have socioeconomic impacts via food production and subsequent cost of food (see Fig. 1). In particular, the transmission, incidence and prevalence of infectious diseases in the world display spatial and temporal variations that reflect seasons, interannual weather fluctuations, strong single weather events and other natural disasters (Zell et al., 2008).

There is international acceptance of the science of global warming, as evidenced by rising average air and ocean temperatures, snowmelts, glacier declines and increasing sea levels. The Fourth Assessment Report of the Intergovernmental Panel on Climate Change (IPCC, 2007) states that air temperatures are expected to continue to increase at approximately 0.2°C per decade under a range of different global greenhouse gas emissions scenarios. Likely changes in climate include increased areas affected by droughts, increased frequency of heatwaves and hot days, increased heavy rainfall events, increased sea levels and higher ocean temperatures, shifts in timing of seasonal events and animal, plant and algal phenology and habitat ranges (IPCC, 2007).

E. Fearnley (✉)
School of Population Health, University of Queensland, Brisbane, Australia
e-mail: e.fearnley@uq.edu.au

J. Dodson (ed.), *Changing Climates, Earth Systems and Society*, International Year of Planet Earth, DOI 10.1007/978-90-481-8716-4_9, © Springer Science+Business Media B.V. 2010

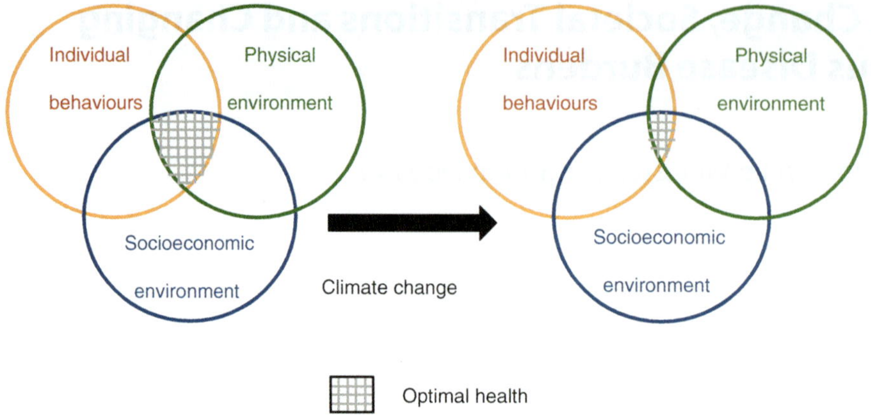

Fig. 1 The impact of climate change on the overlap of individual behaviours, physical environment and socioeconomic environment, reducing the area of optimal health

The health effects of climate change can be divided into direct and indirect impacts. The direct impacts are those caused by changes in exposure extremes (heatwaves, winter cold) and extreme weather events (floods, storms, droughts) and rise in air pollutants. The indirect changes are those including changing patterns of transmission of infectious diseases, particularly water-borne, food-borne and vector-borne diseases, and changes in regional food production (McMichael et al., 2003). More diverse impacts will also be felt by the displacement of populations if sea levels rise significantly, general environmental decline, overcrowding of populations, ecological disruption and loss of food production and conflict over resources (World Health Organisation, 2003). We aim to look at how these climate change impacts are likely to differ by lifestyle group, considering hunter-gatherers, agricultural, industrial and globalised cultures. Some of the health effects of climate change will be positive, with reduced cold related mortality and restricted distribution of some diseases where temperatures may exceed maximum thresholds for vectors or parasites; however, the majority of impacts will be negative (Confalonieri et al., 2007). The approach to minimising this negative impact requires a combination of adaptation to emerging conditions, along with the need to mitigate the factors contributing to climate change (Costello et al., 2009).

The influence of climate on human health and behaviours is not new in the history of the world; existing disease patterns are heavily influenced by climate and climate will continue to influence the health of populations in the future. Infectious diseases were a leading cause of death in past centuries, but their effect has been reduced over time with improvements in sanitation, nutrition, housing and hygiene, changes in social environments and medical advances (Avila et al., 2008). However, developing and developed countries still have a significant burden of infectious diseases, and an increasing risk with emerging infectious diseases (Weiss and McMichael, 2004). Climate variations and extreme weather events impact significantly on infectious diseases. Infectious agents, including bacteria, protozoa and viruses, and their often associated vectors of mosquitoes and ticks, are ectotherms that cannot control their internal temperatures, hence reproduction and survival rates are significantly affected by changes in environmental air or water temperatures (Lafferty, 2009a; Patz et al., 2005).

Importantly, different populations are impacted differently by climate, as a result of their individual behaviours, physical environment and socioeconomic status (see Fig. 1). An interesting integration of these drivers is provided by the natural changes in infectious disease burden that have accompanied human societal transitions (Weinstein and Cook, 2007). The original hunter-gatherer society was largely replaced with the advent of agriculture, followed by industrialisation and leading onto the stage of globalisation, all of which display differing disease patterns due to social, environmental and personal characteristics. All of these societies continue to exist within the world today, although most populations live within industrialised and globalised societies, with hunter-gatherer and agricultural communities diminishing. Human societal transitions therefore provide a unique temporal and geographic context in which to examine the possible impacts of climate change on the infectious disease burden, and we do so in this chapter by exploring two main groups of infectious diseases: Faecal-oral and vector-borne.

Faecal-Oral Infectious Diseases

In hunter-gatherer societies, populations are exposed to the elements and live in a nomadic and subsistence manner. Without access to permanent shelter and technology, food cannot be readily stored under temperature controlled conditions, such as refrigeration at or below 4°C, to limit the growth of bacteria (Savige et al., 2004). Foods of animal origin are susceptible to bacterial contamination, such as from cross-contamination during butchering of a carcass to provide meat, as pathogenic bacteria can be present in the gastrointestinal tract of animals and be transferred to meat. Once meat is contaminated with bacteria, from the gastrointestinal tract of the animal or from other environmental sources including insects and other animal pests, without adequate temperature control, pathogenic bacteria such as *Salmonella* can rapidly multiply to levels high enough to cause infection after ingestion. Numerous pathogenic bacteria are mesophilic and undergo rapid logarithmic growth under warm ambient temperatures of 20–40°C (Alcamo, 2001). With climate change increasing ambient air temperatures and the lack of cold refrigeration storage available to hunter-gatherer populations, there is a greater chance of rapid bacterial growth on foods kept in warmer environments, therefore increasing the occurrence of gastroenteritis in the exposed population. Figure 2 displays unsafe food transport and storage practices present in developing nations.

Settled horticultural (and later industrial) populations can be more vulnerable to some of the effects of climate change than are hunter-gatherers; a point well illustrated by the effects of rising sea levels in the Pacific. The 15-cm rise in sea levels since 1900 has been sufficient to cause, on some islands, a loss of arable lowlands and potable groundwater to salt water infiltration (Nunn, 2009). On atolls, sea level rise also lifts the freshwater lens closer to the land surface, increasing the likelihood of contaminating precious drinking water resources with faecal pathogens from both humans and animals, as the depth of sand to filter pathogens is reduced. Food security and health are thus adversely affected, whereas the mobile hunter-gatherers that first settled the Pacific, the Lapita people, would have been less vulnerable due to their mobility and lower population density.

The transition to agricultural societies encouraged the development of larger, more permanent communities in towns and small cities. Food supplies were more readily available with regular crop production and mass domestication of animals, allowing populations to expand due to improved nutrition and adequate access to food year round. Foods could be stored in permanent housing including cellars, but with limited technology, therefore spoilage of foods could still occur. Populations lived in closer contact with each other and with their animals, allowing infectious diseases to be more readily spread between crowded populations and between species. More permanent settlements also led to people being in closer contact with their waste and that of their animals, and essential resources such as drinking water supplies may also be located close to large animal herds, and be relatively unprotected. In the case of climate change, with increased significant rain events, flooding of unprotected drinking water sources are more likely to become contaminated with animal faeces, and can become a major cause of gastrointestinal diseases such as *Cryptosporidium* and *E. coli*. These phenomena of rain events leading to contaminated water supplies and gastroenteritis outbreaks are problems encountered with climate change in both agricultural and industrialised societies, and exacerbation in the latter. The centralisation of drinking water supplies in industrialised populations can greatly increase the scale and risk of the problem as more people can be affected by contamination of their public drinking water supplies. Flooding and high rainfall events can contaminate drinking water sources with runoff from animal grazing areas, as well as inundate and cause overflow of sewage systems. The largest recorded water-borne outbreak occurred in Milwaukee in the USA when the public water supply was contaminated with *Cryptosporidium*, leading to approximately 403,000 cases of gastroenteritis and 54 deaths (Hoxie et al., 1997; Mac Kenzie et al., 1994). The

Fig. 2 Microbiologically unsafe food transport and storage practices in developing countries. (Photos courtesy of Laura Fearnley, taken in Vietnam)

public water supply was sourced from Lake Michigan, which was contaminated from nearby cattle and other waste sources after heavy spring rains and snowmelt (Mac Kenzie et al., 1994). The risk of water-borne disease is generally considered to be at increased risk with climate change inducing more storm and heavy rainfall events (Patz and Olson, 2006), as evidenced by single outbreaks linked to such events and by longer term analyses of historical outbreaks in both England and USA (linking up to 68% of waterborne outbreaks with higher than average rainfall events; Curriero et al., 2001; Nichols et al., 2009).

The next step in human societal transitions was from agricultural societies to industrialisation and the mechanisation of processes. Industrialisation included mass production of foods in artificial and contained environments, such as large-scale production of caged poultry or pigs. High intensity farmed animals will generally experience some stress during their lives, and stress reduces the fitness of animals and can lead to lower production standards or more disease. It has also been demonstrated that stressed animals can carry and shed higher concentrations of pathogenic bacteria, leading to greater contamination of their immediate environment, which can increase the risk of contamination of the final food products. Stressful events can include starvation, long duration transportation, dehydration and heat stress. Under projected increases in air temperature with climate change, animals such as chickens in mass production batteries may be subject to greater levels of heat stress. Heat stress can increase internal bacterial load, susceptibility to pathogenic bacteria colonisation and bacteria shedding rates in poultry for organisms such as *Campylobacter* and *Salmonella* (Burkholder et al., 2008; Rostagno, 2009); therefore increased heat stress in poultry could lead to a greater risk of human disease from the consumption of contaminated poultry meat or eggs. The large-scale production, close living quarters and single food supply are further factors that can enhance the spread of disease within an animal population, which may then be transferred to the human population. Investigations into current patterns of gastroenteritis have identified statistical associations between increased incidence of disease and increased ambient air temperatures, including a study of salmonellosis in five Australian cities over 10 years (D'Souza et al., 2004) and increased risks of *Campylobacter*, *E. coli* and *Salmonella* infection in the two Canadian provinces studied (Fleury et al., 2006). A similar European study, including data from ten countries, also identified increased rates of salmonellosis with increasing temperature, and within England and Wales a stronger temperature dependent relationship was identified for *S. enteritidis*, a serotype strictly related to transmission via food, compared to *S. typhimurium*, which can be transmitted via food or environmental contact (Kovats et al., 2004).

The most recent societal transition is to a globalised lifestyle, characterised by the expansion of travel and trade throughout the world, the intensification of urbanisation and creation of dense periurban populations. Most importantly, populations continue to expand and overexploit existing environmental resources, leading to ecosystem failures and new or increasing contact between wild animals and humans, therefore increasing the chance of cross species infection. Such conditions can lead to an age of emerging infectious diseases, which are characterised as diseases with increased rates or expansion of geographic areas that they affect as well as newly emerged diseases (Weiss and McMichael, 2004). The increase of global trade and travel facilitates and enhances the spread of infectious diseases. Ballast waters in ships are possible sources of introduction of marine species into new waters, including new strains of *Vibrio cholerae* (Ruiz et al., 2000). The rapid and frequent travel of large numbers of people internationally also increases the chance of new strains of pathogenic bacteria being introduced into naive populations, such as from endemic cholera areas in developing countries. The current (seventh) pandemic of a particular strain of cholera, the El Tor strain, has spread across more continents than previous pandemics, to include Asia, Europe, Africa and the Americas originating in the 1960s; the magnitude and persistence of this pandemic is thought to be significantly influenced by the large scale of population movement between continents, the speed of shipping trade, pollution of oceans and the growth of peri-urban slums with inadequate sanitation (McMichael, 2004). Rising sea surface temperatures and more intense El Niño events with global climate change enable higher concentrations and increased growth of plankton that in turn increases concentrations of *Vibrio cholerae* in marine waters, therefore increasing the chance of cholera outbreaks in human populations (Colwell, 1996; Lipp et al., 2002). Therefore, with the combination of globalisation events increasing the chance of new species of cholera being introduced into new waters and populations, along with warmer temperatures increasing concentrations of cholera in waters, cholera could expand to increase the global burden of disease from this organism.

It is possible over time that such events will lead to a gradual globalisation of pathogens, particularly if climate change increases the number of receptive areas to pathogens that are currently temporally and/or geographically restricted by climatic conditions. From this perspective, pathogens are no different from other pest species that are capitalising on anthropogenic globalisation opportunities – a process characteristic of an era that has been dubbed the "Homogocene" (Low, 1999).

Vector-Borne Diseases

Vector-borne diseases, particularly mosquito-borne diseases, are expected to be susceptible to global climate change due to their transmission dynamics and geographic distribution being very sensitive to the conditions in the surrounding environment, including changes in air temperature, humidity, floods and ecological determinants of food sources (Patz et al., 2008). Insect vectors and free living stages of numerous infectious organisms are ectotherms; meaning that they cannot regulate their own temperature and are therefore reliant upon ambient conditions for their growth rates and survival (Lafferty, 2009b). The expansion of ranges of some vector-borne infectious diseases has been listed as a potential adverse health effect, supported by high confidence level scientific evidence by the IPCC (Confalonieri et al., 2007). The vector- or mosquito-borne diseases susceptible to climate change impacts includes malaria, dengue and Ross River virus. Figure 3 shows a female mosquito, *Aedes camptorhynchus,* a known Ross River virus vector.

The oldest known mosquito was identified in 89- to 100-million-year-old amber from Burma (Borkent and Grimaldi, 2004), and for a modern form, *Paleoculicus minutes*, identified in 70- to 84-million-year-old amber from Canada (Poinar et al., 2000). It is likely that mosquitoes have plagued hominids since well before our common ancestor diverged from the Great Apes over four million years ago because in seeking the most readily available blood meals, mosquitoes have

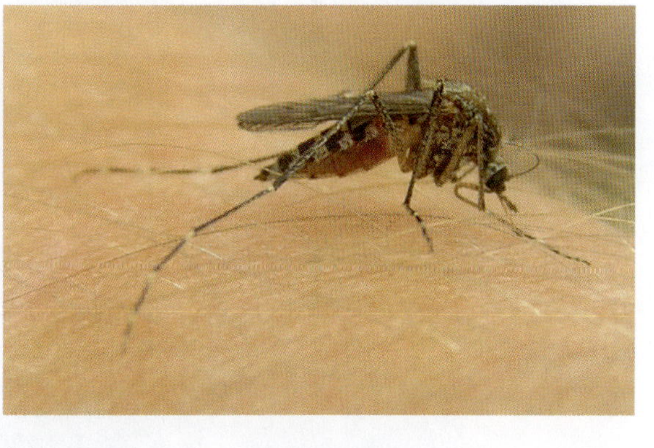

Fig. 3 Female *Aedes camptorhynchus* mosquito, a known Ross River virus vector in Australia, taking a blood meal from a human. (Photo courtesy of Scott Carver)

adapted to new hosts again and again – perhaps most obviously in shifting their diet from dinosaurs to mammals when the former became extinct in the Cretaceous period (Poinar and Poinar, 2007). Small groups of hunter-gatherer humans would however not have been the best of hosts from a mosquito's perspective: apart from their occurrence in relatively low densities, humans have blood deficient in the amino acid isoleucine, essential for the development of healthy egg batches by the female mosquito. Vector-borne diseases such as malaria therefore presented only an occasional problem to our hunter-gatherer ancestors, until human blood meals became so readily available as to compensate for their poor nutritional value (Barnes, 2005). This latter situation occurred with the advent of agriculture, when population densities increased, and settlements became permanent and often located in the vicinity of water.

The advent of agricultural societies provided increased opportunities for the transmission of disease, with larger, denser human populations living in close contact with large groups of animals, where disease could easily be transmitted between species, with the assistance of rodent and mosquito vectors. Soil salinisation as a result of agricultural clearing activities has recently been shown as a potential pathway for significantly increasing the burden of mosquito-borne disease. The replacement of deep rooted native perennial vegetation with shallow rooted annual crops and pastures is a direct result of agricultural expansion and can cause water tables to rise. In some areas water logging may result, and where salt has been deposited in the soil profile over geological time, dissolution of these salts and movement to the soil surface can result in severe salinity problems. Soil salinisation can then lead to changes in the abundance and distribution of vector mosquitoes by various mechanisms, including: directly, as the waterlogged soil provides a greater area and temporal duration of potential mosquito breeding habitat following rainfall; and/or indirectly, through a decrease in biodiversity of aquatic invertebrate predators that may be more salt sensitive, leading to a reduction in interspecific resource competition with a proliferation of salt tolerant vector species (Jardine et al., 2008; 2007) (see Figs. 4 and 5). An example of this is provided in the south-west of Western Australia, where the development of dryland salinity appears to have favoured the inland spread of the Ross River virus vector *Aedes camptorhynchus*, thereby increasing the risk to human health (Jardine et al., 2008) (See also Fig. 3).

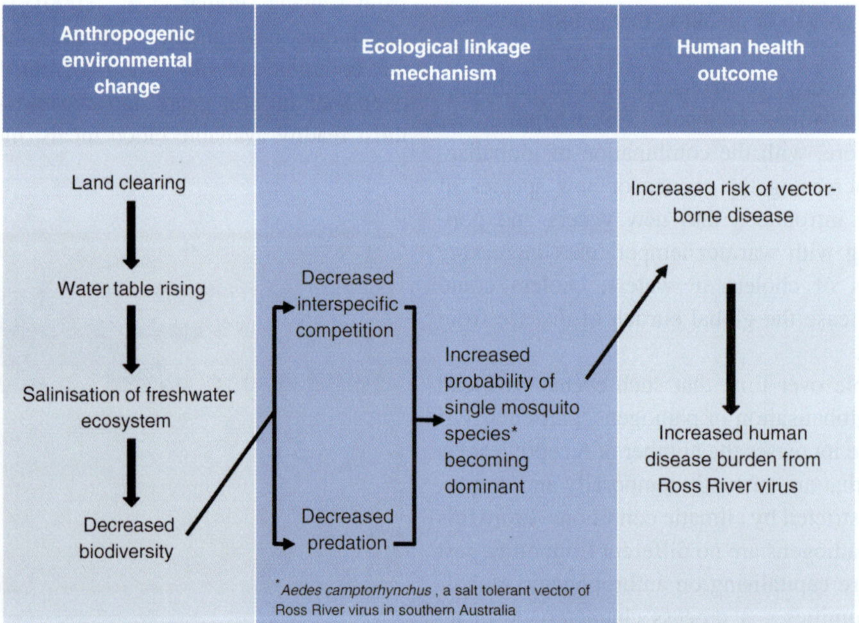

Fig. 4 Example of increased risk of Ross River virus (RRv) in Australia due to anthropogenic environmental changes associated with increasing agricultural activity (based on O'Sullivan et al., 2008)

Fig. 5 Interaction between climate change, agriculture and vector-borne disease: A mosquito trap (*centre*), monitoring the potential for Ross River virus transmission in salinised area of Western Australia, where agriculture and drought have combined to increase the prevalence of disease. (Photo courtesy of Scott Carver)

Dengue is predominantly transmitted by the *Aedes aegypti* mosquito, which has adapted well to breeding in urban environments, particularly in small containers of freshwater from local rainfall such as pot plants and roof gutters. Dengue fever cannot be treated with drugs or prevented with vaccination, rather vector control and avoiding being bitten are the only minimisation mechanisms available, making a disease of particular significance in public health as outbreaks have the potential to spread rapidly in populations (McMichael et al., 2003). The dengue mosquito is also a morning and evening biting mosquito that preferentially feeds on humans rather than other animals. Temperature, rainfall, humidity and solar radiation are important determinants of the growth of the mosquito and higher temperatures also increase virus replication rates in laboratory experiments (Watts et al., 1987). In industrial societies with large urban environments, which incorporate the production of large amounts of garbage, often including numerous containers that can be places for rainfall to pool and collect in, urban adapted mosquitoes such as *Aedes aegypti* can thrive and spread dengue fever. With increasing temperatures and more rainfall associated with global climate change, the geographical range where dengue carrying mosquitoes can breed will increase, and with population growth within susceptible areas, the number of people at risk of contracting the disease could also subsequently increase (Hales et al., 2002) (see Fig. 6). Moving to globalised societies, rapid air travel and frequent movement of populations between endemic and non-endemic countries increases the chance of dispersal of the disease through infected travelling human hosts or mosquito eggs being transported on freight vehicles (Russell et al., 2009). Coupled with the increased geographic area receptive to mosquito breeding in the midst of changing temperature and rainfall patterns worldwide due to changing climate patterns, the chances of different dengue virus serotypes being present within a given human population increases, as well as the chance for dengue outbreaks in general to occur. Humans infected with more than one serotype of dengue virus are at increased risk of dengue haemorrhagic fever and shock syndrome with potentially fatal outcomes (Heymann, 2008). The risk of this more severe disease outcome could be enhanced by the combination of climate change factors and the movement of human societies into more globalised, rapidly travelling activities.

Several examples have been discussed, illustrating the relationship between increasing temperature or rainfall and higher incidence rates of some diseases, with respect to lifestyles. For example, vector-borne disease is generally accepted as a clear illustration of endemic zone expansion with climate change. However, most predictions do not take adaptation into account. In industrialised countries it is, for example, likely that intervention measures such as mosquito screens, insecticiding and source reduction will at least partly mitigate such a possible increase in mosquito-borne disease burden; the net disease burden resulting from climate change is therefore likely in many cases to be less than predicted by models that do not take adaptation into account. Importantly, it is also possible for apparently adaptive interventions to

A

B

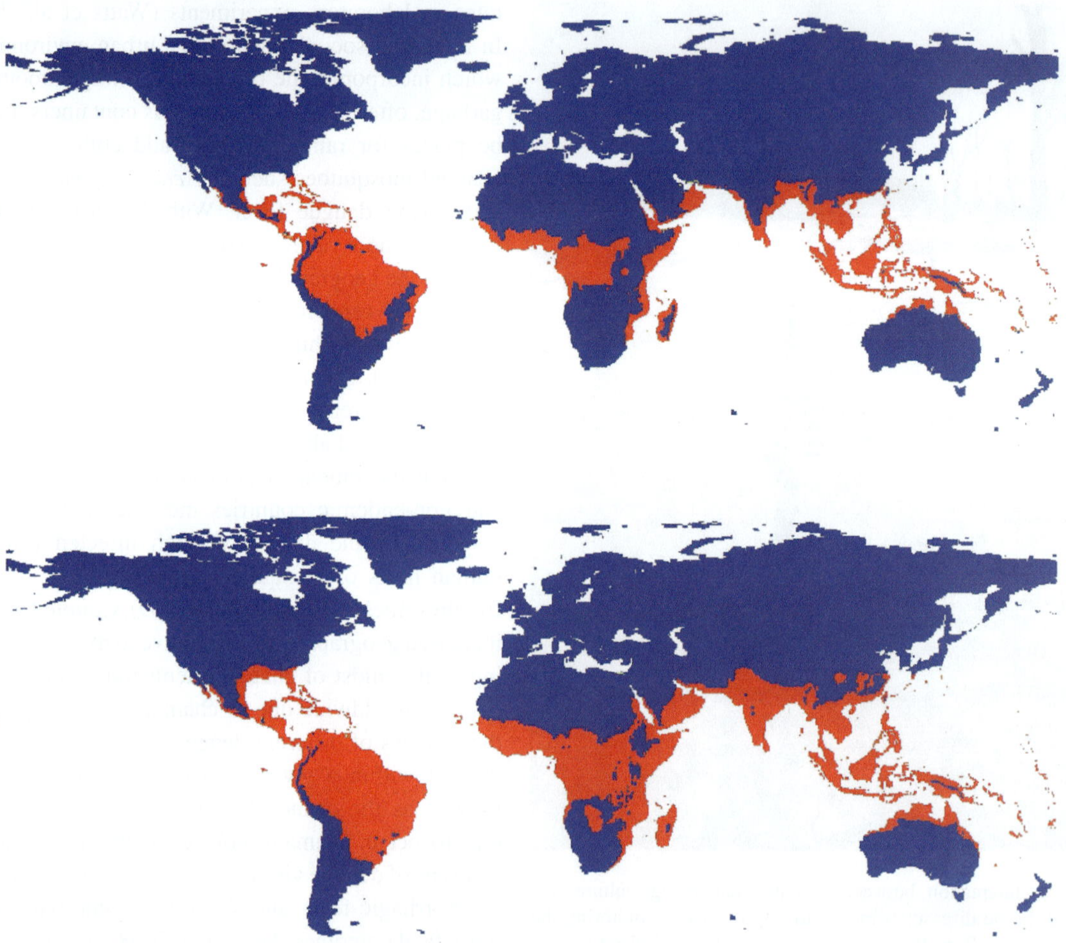

Fig. 6 Areas of 50% or greater probability of dengue fever transmission in *red*. (**a**) Estimated population at risk in 1990. (**b**) Estimated population at risk in 2085 with projected climate change influences. Figure adapted from Hales et al. (2002)

indirectly exacerbate the potential disease burden, such as when post-disaster malaria control efforts enhance the potential for development of drug resistance in plasmodium, thereby impeding long term malaria control globally (Weinstein et al., 2009). Conversely, adaptation measures, per se, may lead to infectious disease emergence. One example is the emergence of the respiratory disease, legionellosis, which we describe in detail because of its direct relevance to societal transitions that are a focus of this chapter. The background to the emergence of this pathogen as a public health problem lies in the societal transition to industrialisation, which is characterised by even greater population growth and density increase, with the production of larger cities and significant increases in the use of technologies. Many technologies improve health, but several also produce toxins or habitats for the proliferation of harmful organisms. Technology has created artificial micro-climates that can facilitate the transmission of organisms from different environments to human hosts. The artificial environment of air conditioners and cooling towers proved to be beneficial for the growth of *Legionella*, an organism found in the natural environment of ponds and other surface waters. The technology of air conditioners then provided a route of transmission for the bacteria to then infect humans via aerosols and cause legionellosis or legionnaires disease (Avila et al., 2008). With climate change inducing more heat waves and warmer weather in general, human adaptation to try and cool the environment by having more air conditioners and cooling towers is likely to increase, therefore increasing the risk of exposure to *Legionella* and subsequent legionellosis respiratory infection (Semenza and Menne, 2009). Recent increases in sporadic cases of legionellosis in Philadelphia, United States were investigated and linked to the environmental risk factors of high humidity and recent rainfall events (Fisman et al., 2005), furthering the link with environmental climate change. Also, the developing world is undergoing rapid industrialisation, such as in India and China; these countries will have a dramatic increase in cooling towers and a subsequent increase in risk of legionellosis.

In this chapter we have attempted to illustrate the complex interaction between the potential climatic intensifiers of infectious disease as found in different societal lifestyles and transitions; increasing temperature, changing rainfall patterns and more frequent extreme events will influence infectious diseases differently in hunter gatherer, agricultural, industrial and global societies. The complexities of these interactions suggest that differential exacerbation or amelioration of the infectious disease burden will occur in different places at different times. When the various exacerbators and societal lifestyles conspire to maximise the disease burden, significant outbreaks or new disease emergences can occur. As a final integrative example, consider the relationship between flooding, slums and leptospirosis.

Flooding and leptospirosis: Changing climatic factors of more frequent and intense extreme events such as floods and storms, along with rising sea levels increasing the chance of coastal flooding are expected as a result of anthropogenic environmental change. The impacts of floods vary depending upon population vulnerability, but generally include a mixture of immediate effects including mortality and injuries, together with outbreaks of infectious diseases including faecal-oral, vector-borne and rodent-borne diseases (Ahern et al., 2005). Leptospirosis is a rodent-borne disease, where human infection results from direct contact with infected rat urine or via indirect exposure from contaminated soil or water, often through skin abrasions or cuts and occasionally through ingestion or inhalation of aerosols from contaminated fluids (Heymann, 2008). In the event of floods, humans are at increased risk of leptospirosis due to rodent populations seeking refuge indoors, therefore increasing the chance of contact with rat urine, in addition to direct contact with flood waters during floods or post-flood clean-up activities. Leptospires also have increased environmental survival in warm, humid conditions (Levett, 2001). Outbreaks of leptospirosis after floods or extreme rainfall events have been identified in Brazil (Ko et al., 1999), India (Maskey et al., 2006) and Nicaragua in central America (Trevejo et al., 1998).

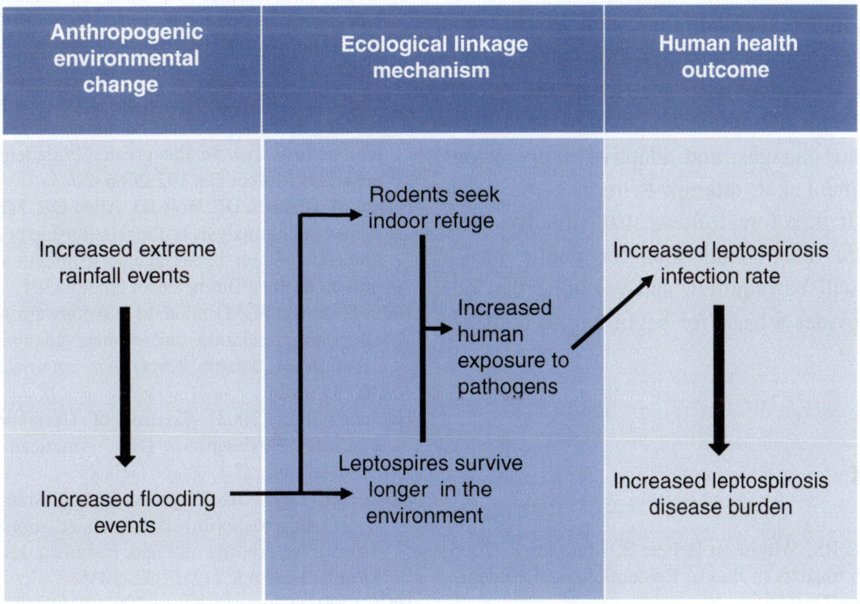

Fig. 7 Example of increased risk of leptospirosis due to anthropogenic environmental changes

Societal transitions to increasing urbanisation, and high density living in slums in the transition to globalisation, have facilitated the spread of leptospirosis, with poor sanitation facilities and high rates of garbage production that attract rodents – see Fig. 7 – for the relationship between the environmental change of more flooding and the human health outcome of increased leptospirosis. Poor drainage and sanitation in slums also favour the pooling of flood waters and therefore increase the risk of leptospirosis transmission. Leptospirosis was once a rare disease, limited to rural and tropical regions, particularly associated with occupational exposure to sewage and livestock, has become an endemic and epidemic disease in high density urban slums correlated with the wet season and extreme events such as floods and hurricanes (Ko et al., 1999; Levett, 2001).

Conclusion

Climate change is not a standalone risk factor for disease and ill-health in most circumstances; rather it is an amplifier of existing health risks, which are heavily influenced by the physical environment, socioeconomic environment and individual behaviours, as shown earlier in Fig. 1 (Costello et al., 2009). Climate change is an overarching exacerbator of other anthropogenic environmental disruptors, such as land use change and alterations of lifestyles, from hunter gatherers through to agricultural, industrial and global lifestyles. Different lifestyles will be impacted differently by climate changes, and adaptive interventions will be implemented to attempt to reduce the negative impacts. It therefore follows that situation- and societal-specific recommendations for public health interventions will be required, and we hope that our framework provides a basis for beginning to consider such differences.

References

Ahern M, Kovats RS, Wilkinson P, Few R, Matthies F (2005) Global health impacts of floods: Epidemiological evidence. Epidemiol Rev 27:36–46.

Alcamo IE (Ed) (2001) Fundamentals of Microbiology. Sudbury, MA: Jones and Bartlett Publishers.

Avila M, Said N, Ojcius DM (2008) The book reopened on infectious diseases. Microbes Infect 10(9): 942–947.

Barnes E (2005) Diseases and Human Evolution. Albuquerque, NM: University of New Mexico Press.

Borkent A, Grimaldi D (2004) The earliest fossil mosquito (Diptera: Culicidae), in Mid-Cretaceous burmese amber. Ann Entomol Soc Am 97(5):882–888.

Burkholder KM, Thompson KL, Einstein ME, Applegate TJ, Patterson JA (2008) Influence of stressors on normal intestinal microbiota, intestinal morphology, and susceptibility to *Salmonella enteritidis* colonization in broilers. Poult Sci 87:1734–1741.

Colwell R (1996) Global climate change and infectious disease: The cholera paradigm. Science 274(5295): 2025–2031.

Confalonieri U, Menne B, Akhtar R, Ebi KL, Hauengue M, Kovats RS, Revich B, Woodward A (2007) Human health. In ML Parry, OF Canziani, JP Palutikof, PJ van der Linden CE Hanson (Eds), Climate Change 2007: Impacts, Adaptation and Vulnerability (pp. 391–431). Contribution of Working Group II to the fourth assessment report of the Intergovernmental Panel on Climate Change. Cambridge: Cambridge University Press.

Costello A, Abbas M, Allen A, Ball S, Bell S, Bellamy R, Friel S, Groce N, Johnson A, Kett M, Lee M, Levy C, Maslin M, McCoy D, McGuire B, Montgomery H, Napier D, Pagel C, Patel J, Antonio J, de Oliveira P, Redclift N, Rees H, Rogger D, Scott J, Stephenson J, Twigg J, Wolff J, Patterson C (2009) Managing the health effects of climate change. Lancet 373(9676):1693–1733.

Curriero FC, Patz JA, Rose JB, Lele S (2001) The association between extreme precipitation and waterborne disease outbreaks in the United States, 1948–1994. Am J Public Health 91(8):1194–1199.

D'Souza RM, Becker N, Hall G, Moodie KBA (2004) Does ambient temperature affect foodborne disease? Epidemiology 15(1):86–92.

Fisman D, Lim S, Wellenius GA, Johnson C, Britz P, Gaskins M, Maher J, Mittleman MA, Spain V, Haas C, Newbern C (2005) It's not the heat, it's the humidity: Wet weather increases legionellosis risk in the greater Philadelphia metropolitan area. Am J Infect Dis 192:2066–2073.

Fleury M, Charron DF, Holt JD, Allen OB, Maarouf AR (2006) A time series analysis of the relationship of ambient temperature and common bacterial enteric infections in two canadian provinces. Int J Biometeorol 50:385–391.

Hales S, de Wet N, Maindonald J, Woodward A (2002) Potential effect of population and climate changes on global distribution of dengue fever: An empirical model. Lancet 360:830–834.

Heymann DL (2008) Control of Communicable Diseases Manual. Washington, DC: American Public Health Association.

Hoxie NJ, Davis JP, Vergeront JM, Nashold RD, Blair KA (1997) Cryptosporidiosis associated mortality following a massive waterborne outbreak in Milwaukee, Wisconsin. Am J Public Health 87(12):2032–2035.

IPCC (2007) Climate Change 2007: Synthesis Report Summary for Policymakers (p. 22). Valencia: Intergovernmental Panel on Climate Change.

Jardine A, Speldewinde P, Carver S, Weinstein P (2007) Dryland salinity and ecosystem distress syndrome: Human health implications. EcoHealth 4(1):10–17.

Jardine A, Lindsay M, Johansen C, Cook A, Weinstein P (2008) The impact of dryland salinity on the population dynamics of vector mosquitoes (Diptera: Culicidae) of Ross River virus in inland areas of south-west Western Australia. J Med Entomol 45(6):1011–1022.

Ko AI, Reis MG, Dourado CMR, Johnson WD, L.W. Riley and Salvador Leptospirosis Study Group (1999) Urban epidemic of severe leptospirosis in brazil. Lancet 354:820–825.

Kovats RS, Edwards S, J S, Hajat BG, Armstrong KL, Ebi B, Menne and The Collaborating Group (2004) The effect of temperature on food poisoning: A time-series analysis of salmonellosis in ten European countries. Epidemiol Infect 132:443–453.

Lafferty KD (2009a) Calling for an ecological approach to studying climate change and infectious diseases. Ecology 90(4):932–933.

Lafferty KD (2009b) The ecology of climate change and infectious diseases. Ecology 90(4):888–900.

Levett PN (2001) Leptospirosis. Clin Microbiol Rev 14(2):296–326.

Lipp EK, Huq A, Colwell RR (2002) Effects of global climate on infectious disease: The cholera model. Clin Microbiol Rev 15(4):757–770.

Low T (1999) Feral Future. Ringwood, VIC: Viking.

Mac Kenzie WR, Hoxie NJ, Proctor ME, Gradus MS, Blair KA, Peterson DE, Kazmierczak JJ, Addiss DG, Fox KR, Rose JB, Davis JP (1994) A massive outbreak in milwaukee of cryptosporidium infection transmitted through the public water supply. N Engl J Med 331(3):161–167.

Maskey M, Shastri JS, Saraswathi K, Surpam R, Vaidya N (2006) Leptospirosis in Mumbai: Post-deluge outbreak 2005. Indian J Med Microbiol 24(4):337–338.

McMichael AJ, Woodruff RE, Whetton P, Hennessy K, Nicholls N, Hales S, Woodward A, Kjellstrom T (2003) Human Health and Climate Change in Oceania: A Risk Assessment 2002 (p. 128). Department of Health and Ageing. Canberra, ACT: Commonwealth of Australia.

McMichael AJ (2004) Environmental and social influences on emerging infectious diseases: Past, present and future. Philos Trans R Soc Lond Ser B 359:1049–1058.

Nichols G, Lane C, Asgari N, Verlander NQ, Charlett A (2009) Rainfall and outbreaks of drinking water related disease and in England and Wales. J Water Health 7(1): 1–8.

Nunn PD (2009) Vanished islands and Hidden Continents of the Pacific. Honolulu, HI: University of Hawaii Press.

O'Sullivan L, Jardine A, Cook A, Weinstein P (2008) Deforestation, mosquitoes, and ancient rome: Lessons for today. BioScience 58(8):756–760.

Patz JA, Campbell-Lendrum D, Holloway T, Foley JA (2005) Impact of regional climate change on human health. Nature 438(7066):310–317.

Patz JA, Olson SH (2006) Climate change and health: Global to local influences on disease risk. Ann Trop Med Parasitol 100(5–6):535–549.

Patz JA, Olson SH, Uejio CK, Gibbs HK (2008) Disease emergence from global climate and land use change. Med Clin North Am 92(6):1473–1491.

Poinar G, Zavortink T, Pike T, Johnston P (2000) Paleoculicis minutus (Diptera: Culicidae) n. gen., n. sp., from Cretaceous Canadian amber, with a summary of described fossil mosquitoes. Acta Geol Hisp 35(1–2):119–128.

Poinar G, Poinar R (2007) What Bugged the Dinosaurs? Insects, Disease and Death in the Cretaceous. Princeton, NJ: Princeton University Press.

Rostagno MH (2009) Can stress in farm animals increase food safety risk? Foodborne Pathog Dis 6(7):767–776.

Ruiz GM, Rawlings TK, Dobbs FC, Drake LA, Mullady T, Huq A, Colwell R (2000) Global spread of microorganisms by ships. Nature 408:49.

Russell RC, Currie BJ, Lindsay MD, Mackenzie JS, Ritchie SA, Whelan PI (2009) Dengue and climate change in Australia: Predictions for the future should incorporate knowledge from the past. Med J Aust 190(5):265–268.

Savige G, Dalais F, Wahlqvist M, Coveney J, Fallowfield H (2004) Food safety. In N Cromar, S Cameron, H Fallowfield (Eds), Environmental Health in Australia and New Zealand (pp. 229–250). Melbourne, VIC: Oxford.

Semenza JC, Menne B (2009) Climate change and infectious diseases in Europe. Lancet Infect Dis 9(6):365–375.

Trevejo RT, Rigau-Perez JG, Ashford DA, McClure EM, Jarquin-Gonzalez C, Amador JJ, de los Reyes JO, Gonzalez A, Zaki SR, Shieh WJ, McLean RG, Nasci RS, Weyant RS, Bolin CA, Bragg SL, Perkins BA, Speigel RA (1998) Epidemic leptospirosis associated with pulmonary hemorrhage – Nicaragua, 1995. J Infect Dis 178: 1457–1463.

Watts DM, Burke DS, Harrison BA, Whitmire RE, Nisalak A (1987) Effect of temperature on the vector efficiency of Aedes aegypti for Dengue 2 virus. Am J Trop Med Hyg 36(1):143–152.

Weinstein P, Cook A (2007) Epidemiological transitions and the changing face of medical geology. Ambio 36(1): 67–69.

Weinstein P, Goff J, Skelly C (2009) Do post-disaster public health interventions impede malaria eradication? Med Hypotheses. ePub Available online November 7, 2009. doi: 10.1016/j.mehy.2009.09.055.

Weiss RA, McMichael AJ (2004) Social and environmental risk factors in the emergence of infectious diseases. Nat Med 10(12):S70–S76.

World Health Organisation (2003) Climate Change and Human Health: Risks and Responses – Summary. Geneva: World Health Organisation.

Zell R, Krumbholz A, Wutzler P (2008) Impact of global warming on viral diseases: What is the evidence? Curr Opin Biotechnol 19(6):652–660.

Don't We All Want Good Weather and Cheap Food?

Elisabeth Simelton

Abstract Today the Earth produces enough food for everyone. Yet one of seven world citizens is malnourished or starving as the access to food or money to buy food is inequitably distributed across countries, societies and families. This chapter considers two aspects of food security for a growing world population. First, the demand for food is growing in two fundamental ways that suggests access to food is becoming increasingly tied to food prices. Second, the means to supply the amounts needed in a food secure future are exposed to two unprecedented stresses that will require stronger multi-level institutional collaboration on managerial and technical solutions. This food-security equation has a chance to balance if consumers and producers alike account more holistically for their environmental and social impacts – and act responsibly.

Keywords Food security · Food supply · Food demand · Climate impacts · Adaptation

Introduction

Food Security – A Human Right Since 1948

Living in a comfortable climate and producing food is an on-going human challenge. Our predecessors, the Neanderthals, are believed to have died from hunger due to climate change some 24,000 years ago (Rincon, 2006). Since then achieving sufficient food production has been a fundamental ingredient of religious practices, power manifestations and scientific research. Despite that, food security was not defined until 1974, but this was quickly followed up by hundreds of definitions (Maxwell, 1996). The first UN World Food Conference was prompted by global inflation and soaring oil prices that greatly raised prices in the early 1970s. In addition, food supply had been reduced by large scale droughts across four continents. The Soviet Union bought so much US wheat that it threatened the stability of world food supplies. The UN talks therefore highlighted quantitative, temporal and commercial aspects of food security at the global scale:

> availability at all times of adequate world food supplies of basic foodstuffs to sustain a steady expansion of food consumption and to offset fluctuations in production and prices (UN, 1975)

The conference also noted that few of the least developed countries were taking part in the global trade. During the following decade famines remained in Africa while they became rare in the rest of the world. Famine, a Latin word for hunger, denotes severe forms of starvation that are related with malnutrition and impaired bodily functions. Nobel laureate Amartya Sen argued that hunger and poverty were part of the same problem (Sen, 1981). He and other scholars meant that the African famines illustrated that food security is not simply about availability in the function of supply and growth at the global scale, the food must be within reach and the people be able to buy it:

> [...] ensuring that all people at all times have both physical and economic access to the basic food that they need (FAO, 1983)

E. Simelton (✉)
Centre for Climate Change Economics and Policy,
University of Leeds, Leeds LS2 9JT, UK
e-mail: E.S.Simelton@leeds.ac.uk

J. Dodson (ed.), *Changing Climates, Earth Systems and Society*, International Year of Planet Earth,
DOI 10.1007/978-90-481-8716-4_10,

The expression "basic food" was soon revised from a minimum amount for survival to the purpose of "an active, healthy life" (WB, 1986). Bothered with the slow progress, the next World Food Summit in 1996 agreed to halve the number of hungry and malnourished people by 2015 (Millennium Development Goals). Differences between hunger (quantity of food) and malnourishment (quality of food) were underlined. The summit published one of the most widely used definitions:

> Food security, at the individual, household, national, regional and global levels [is achieved] when all people, at all times, have physical and economic access to sufficient, safe and nutritious food to meet their dietary needs and food preferences for an active and healthy life (FAO, 1996)

This sentence expresses a more complex understanding of the matter. Food security at one spatial scale is not necessarily a guarantee for food security at finer scales, not even within a household. The redefinitions allow us to re-evaluate food security. However, the small print appears robust in comparison with Article 25.1 in the Human Rights declarations from 1948, which states that

> Everyone has the right to a standard of living adequate for the health and well-being of himself and of his family, including food [. . .] (UN, 1948)

Although a number of countries have ratified the declaration and are legally bound to the subsequent Covenants (National Coordinating Committee for UDHR50, 2001), 30,000 people die every day from hunger, malnutrition and hunger-related diseases (UNDP, 2007). Paradoxically, other parts of the world have become food "oversecure" by passing the standards for "adequate for the health and well-being" to suffer from overweight-related illnesses.

While I am writing this (September 2009), a photo of hundreds of dying and dead cattle cover the front page of a British newspaper as East Africa suffers severe droughts for the fourth year in a row. Why do the most severe forms of food insecurity, famines, persist in Africa and why are they not prevented? The answer is generally to be found at political levels, notably governance and donors (Devereux, 2009). Having seen examples from West Africa in 2007, the executive director of United Nations World Food Program, Josette Sheeran, warned that the triple challenges of climate change, rising food prices and population growth "threatened to unleash a perfect storm on the world's 850 million hungry people" (WFP, 2007). Then clouds gathered up and darkened the global financial sky. In 2009 United Kingdom's Chief Research Scientist, Professor John Beddington, said that within 20 years the whole world will face "[a]. . . perfect storm of food shortages, scarce water and insufficient energy resources [that] threaten to unleash public unrest, cross-border conflicts and mass migration as people flee from the worst-affected regions." (Sample, 2009). Such dramatic exclamations take food insecurity beyond being simplistic causes of population growth or climate change. What makes people food insecure at less severe rates than the persisting famines vary fundamentally in different parts of the world (Smith et al., 2005) because food systems are a complex web of historic, human, economic and environmental systems from the global to individual level (Ericksen, 2008; Gregory et al., 2005). Bearing in mind the complex nature of food systems, the purpose of this chapter is to highlight six key nodes of the food-security web.

This chapter is organised around two simplified aspects of food security: changes in demand and in supply. Imbalances between supply and demand can cause global food crises, similar to those in the early 1970s and late 2000s (Table 1). One factor that is expected to increasingly affect this balance is climate change. The first section of the chapter deals with what people (want to) eat as a major driver of food production. Three key changes on the demand side are illustrated: population growth, purchasing power and access to food through markets and distribution networks. The second section asks the logical follow-up question: can we produce enough of it? The focus shifts to the capacity to supply food in the light of climate change, environmental restrictions, and technological solutions.

Changes in Demand – Drivers to What We Want

Over-population has traditionally been seen as a major threat to global food security. This view overlooks the fact that rising living standards changes eating

Table 1 Major contributors to global food price inflation in 2007–2008

	Demand	Supply
Long-term (1980s–2000s)	Living standards improve	Grain stock levels decline Subsidies to European and US producers/Structural adjustment programs in developing countries Investments in agricultural research decline Yields decline
Intermediate (2000s)	Biofuel demand increase	Fuel and fertiliser cost increase Production decline
Short-term (2007–2008)	Exchange rate changes US dollar versus other currencies Consumer subsidies/price controls maintain demand	Weather-related crop failures Consumer subsidies/price controls (remove incentives to produce) Export restrictions (trigger speculation and price, not necessarily production)

Source: Abbott et al. (2008), FAO (2008c), UNCTAD (2008), and von Braun et al. (2008).

habits, in particular the quantities of food that we buy. Distribution networks, the means to enhance people's access to food through, e.g. more shops, longer opening-hours and wider selections, are closely tied with purchasing power.

Population Growth

In 1960 there were 3 billion children, women and men on our planet. By 2000 this number had doubled and around 2050 we are expected to reach, and plateau at, 9 billion. Until then it means 75 million new mouths to feed every year (Table 2). The total population increase currently is the highest in Asia, while the fastest population growth rates occur in a band across the mid-latitudes: Central America, central Africa, parts of Middle East and Oceania. Food production in most of these regions is already constrained by high pressures on land and water resources, and facing serious exposure and vulnerability to climate change (IPCC, 2007a).

While high fertility rates are not necessarily related with poverty, high child mortality rates certainly are (Fig. 1). There are strong feedbacks between reducing the need to reproduce to secure the survival of a few, with improved health and raised living standards. Such changes may not be evident immediately in the national statistics, as the population continues to grow for at least two generations after fertility rates start to decline before the demographic transition eventually finds a new balance (e.g. Ethiopia, Table 2). Halting the number of births can be achieved without policy interventions and perhaps, irrespective of religious belief. Russia's and China's fertility rates are down to 1.3 and

Table 2 The world's most populated countries in 2009 and projected population (total population and percentage change relative to 2009) in 2030 and 2050. The top ten 2009 countries account for more than two-thirds of the world's total population and produced twothirds of the world's cereals: China produced 20%, the US 16%, India 11%, Brazil, Japan and Russia 3% each, Bangladesh 2%, Japan, Nigeria and Pakistan 1% each, all based on the average production for 1995–2007

Rank	2009 (million people)		2030 (million people)		% change	2050 (million people)		% change
1	China	1,339	China	1,462	9	India	1,657	13
2	India	1,157	India	1,461	26	China	1,424	−3
3	US	307	US	374	22	US	439	18
4	Indonesia	240	Indonesia	289	20	Indonesia	313	8
5	Brazil	199	Brazil	240	21	Ethiopia	278	71
6	Pakistan	175	Pakistan	232	33	Pakistan	276	19
7	Bangladesh	156	Nigeria	212	42	Nigeria	264	25
8	Nigeria	149	Bangladesh	204	31	Brazil	261	9
9	Russia	140	Ethiopia	162	91	Bangladesh	234	14
10	Japan	127	Philippines	138	41	Congo (Kinshasa)	189	52

Source: US Census Bureau (2009) and FAOSTAT (2009).

Fig. 1 The value of health as an asset for coping with climate change is invaluable. Reducing infant mortality rates improves not only women's and children's health but also living standards for the household as a whole. Sanitation and access to safe water are two preconditions for reduced infant mortality rates that bring long-term benefits to the overall society. Photo: the author (2009)

1.8 births per couple, which is below the replacement rate – China with a strict one-child policy and Russia without. The average couple in Iran and Italy, predominantly Muslim and Catholic, respectively, have 2 and 1.3 births, respectively (US Census Bureau, 2009). While UN and non-governmental organisations repeatedly recommend that educating (empowering) girls will improve the lives of many, it is worth noting that they rarely propose family health and human well-being in boys' education.

So approaching 9 billion people – is there going to be enough food for all? We need only basic arithmetic to show that the present 960 million hungry people would not need to feel their stomachs rumbling or having their body functions distorted. In 2004 the total global meat production reached 260 billion kg (FAO, 2008a). Thus, in theory there was 40 kg of meat for each of the 6.4 billion persons. That equals 100 g/day, which interestingly is the size of "one portion" recommended by various western National Food Agencies from post-war times. Meat, or rather protein, production thus needs to increase to meet future health requirements of 9 billion people. But here is a drawback. Livestock contributes to climate change (see "Climate Change" section). Policy makers (in the west) therefore recommend that meat consumption should be gradually reduced in favour of milk, egg and vegetarian protein sources (Tukker et al., 2009). How about carbohydrates? A common estimate for per capita annual consumption of grain seeds range from 350 to 375 kg, which translates to a yearly demand of 3,150–3,375 million tonnes for 9 billion people (excluding the need for animal feed and biofuels). The global production of cereals, roots and tubers in 2007 was 2,400 million tonnes (Fig. 2), which is 355 kg/person. If the per capita consumption and the global average productivity (3.4 t/ha) were to remain constant, another 230–300 million hectares (about the size of Argentina) would be required for 9 billion people. For future global food security it is therefore going to be essential to increase productivity (see "Technological fix" section below). However, these back-of-the-envelope calculations also demonstrate that the current food security is not altogether about quantity. Indeed, the most food insecure countries in the 1990s shared the same characteristics: enough food but high poverty (Smith et al., 2005). Food security is about having the means to access the food.

Access to Food

People can access food legally by producing it, by buying it (including by exchange) or by receiving it as gift or aid (transfer). Famines occur when all three access routes fail (Sen, 1981), and it worsens if governments and other agencies fail to intervene appropriately. For this reason, it is quite rare that weather-related crop failures alone result in famines. Instead, famines quite often originate with small declines in crop or livestock production (possibly related to weather) that have been exacerbated by price inflations that increase the cost for food and reduce the value of people's assets (Devereux, 2009).

Economic growth and improved living standards are expected to have far more impact on food demand than that of population growth. For example, in the beginning of the twentieth century the average Swedish

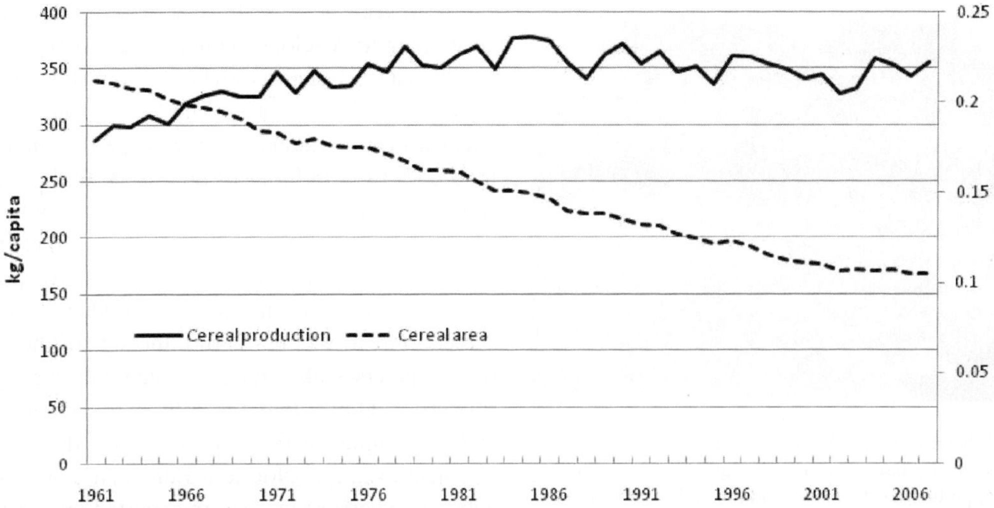

Fig. 2 Global average per capita cereal production (*left axis*) and per capita land area allocated to cereal production (*right axis*) between 1961 and 2007. Source: FAOSTAT (2009)

person ate about 20 kg of meat per year (Neset Schmid and Lohm, 2004). In 1995 this amount increased to 55 and in 2005 to 65 kg (EurActiv, 2009), less than a quarter of the meat in a full grown beef. However, in reality Swedes eat more than one cow in 4 years. Fairly unchanged food prices since 1980 (Hazell and Wood, 2008) have made it possible to buy more food and many cuts that people used to eat now appear on the menu of pets. The transition from an agricultural economy, where a majority of people produced their own food, to a service-based economy, where a majority of people buy their food, comes with ironic twists. First, more food is binned. For example, British consumers throw away some 6,700 million kg of food in a year, of which 160 million kg is meat (WRAP, 2008) (equal to 2 kg meat per person). Second, while some can afford to throw away edible food, others become more vulnerable to food price fluctuations. During the global price inflation that peaked in 2007–2008, about 75–100 million people went hungry – not because there was not enough food but simply because prices went up (Tukker et al., 2009). The economic crisis gradually undermined the transfer route to access food. Thus the World Food Program had to prioritise who to leave hungry as procurement for food aid became too expensive to serve all the needy. Then, remittances from relatives declined. Private capital flows to developing countries, which contributes by up to 40% of GDP in some countries, halved between 2007 and 2008 from 1,158 to 707 billion US dollars, and this is forecasted to halve again to 363 billion dollar for 2009 (World Bank, 2009).

Food prices are unlikely to return to pre-2007 levels, and particularly urban poor households without facilities to grow some subsistence food are at risk of becoming food insecure (FAO, 2008b). Having large groups of food insecure people or mass unemployment rates places pressure on the political stability. Food riots that broke out in 2007–2008, such as in Egypt or Haiti, generally reflected urban food insecurity. In China, millions of migrant workers lost their jobs in the economic crisis. Many were fortunate to be the first generation to try their luck in generating off-farm incomes in the cities, which meant that many had farms to return to (Fig. 3). Many will be tempted to rent or sell their land as urban salaries generally increase the living standard faster than in rural areas. As more people move to cities and become permanently landless, they turn increasingly to dependence on purchasing (through private savings/the market/the state) rather than producing (in rural areas) for food security.

Markets and Distribution

Are the market and distribution systems improving the access to food for everyone? The globalisation of agricultural markets over the past three decades has had

Fig. 3 Post-harvest losses. As much as climate change and air pollution (note the hazy sky) reduce agricultural production, there are opportunities to make traditional labour and land-intensive rice production more efficient in the post-harvest stage alone. Such changes are likely to reduce the need of agricultural labour and the generation of the young boy on the photo is about to become landless urban citizens. However, villages like this one, near Hangzhou in southeast China, are absorbing family members who lost their jobs in the cities, such as during the recent financial crisis. These family members now, at least, have a chance to produce their own food. Photo: the author (2008)

four major consequences: more competition, higher demand for quality, fewer actors and thus longer transportation times. Globalisation has closed some traditional trade routes, particularly of coffee and tea from Africa, while enabling new trade patterns of higher value goods, such as flowers, wine and fish, from developing to developed countries and between developing countries, such as soybeans from Brazil to China (Hazell and Wood, 2008). The food chains in OECD countries can be likened to the shape of a time-glass; food goes from millions of producers, to thousands of suppliers and manufacturers, passes through hundreds of brokers and via thousands of supermarkets to millions of consumers. Although the effective power is with the few actors that link the producers and consumers (Gregory et al., 2005), the size of the number of consumers makes them a potentially strong pressure group. Quite often it is difficult to combine ethical and environmental-friendly interests – distribution is a typical example. Northern European consumers are debating whether tomatoes transported on trucks from southern Europe are a more efficient use of energy than growing them locally in greenhouses. The same goes for apples shipped from New Zealand versus transport by truck from France, and whether to support a developing country and buy air-freighted Kenyan peas and flowers.

The enduring failure to incorporate Sub-Saharan Africa and Central America into the global market (Fig. 4) and to build up domestic markets are essentially ascribed to be unfair market policies: structural adjustment programs in the 1980s to early 1990s and agricultural subsidies and trade agreements still favour European and American producers and consumers

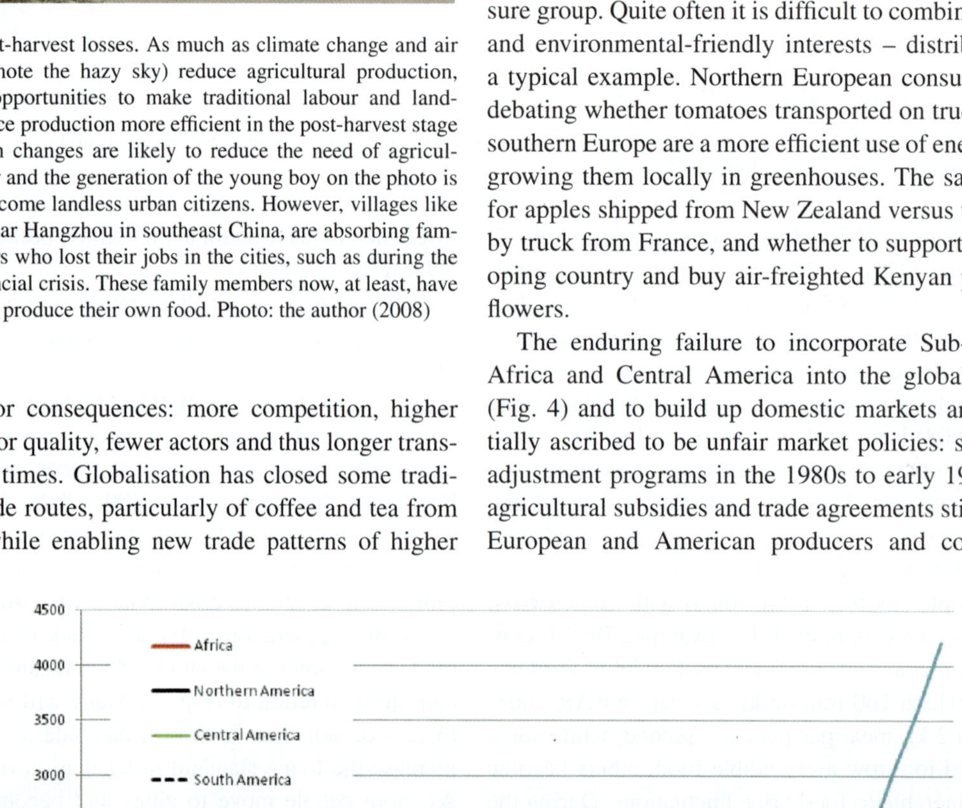

Fig. 4 Export value of total agricultural merchandise trade. China dominates Asia's strong trend from mid-1980s. Oceania includes Australia and New Zealand. Source: FAOSTAT/Pricestat (2009)

(Hazell and Wood, 2008). There are further institutional differences. The Green Revolution, although implemented on three continents, did a better job of making Asian people more food secure, healthier and wealthier than Africans and Latin Americans. East and Southeast Asian successes are often explained to be due to the more stable political situations than is the case for Africa and Latin America, who were caught in structural adjustments in exchange for loans from IMF/WB. For the Asian situations this arguably resulted in a positive spiral of upgraded infrastructure, transport networks and foreign investment. Productivity in the former Asian communist countries boomed from around 1980, when small-scale farmers could sell their own produce at local markets. Larger African farms are predominantly export and cash-crop oriented with a large share owned by foreign cash-crop enterprises that reinvest little in the host country. In contrast, Asian agriculture was based on a broad variety of traditional staple crops and tropical cash crops supported by advanced irrigation systems. The export was run by state-owned enterprises that were not necessarily managed efficiently; however, this meant that at the time for transition to market economy there were established, and growing, domestic markets that generated purchasing power and demand before entering the international arena. State-funded research and development in Asia buffered for companies' potential losses during periods of trial-and-error. The African and Latin American farmers did not have such back-up. Over the past 40 years maize had the strongest production increase of the Sub-Saharan grain crops, with a steady increase in Eastern Africa and progress in Western Africa in the 1990s. Nevertheless, Africa's cereal imports have continued to increase while exports remain constant. Trade is further restricted since many African countries are land-locked (World Bank, 2009). In extension, this means that where local food markets function it can be more beneficial to provide cash transfer or jobs to those who are in risk instead of food aid when food security is threatened (Barrett et al., 2009). Such interventions are less distorting on the local production and have emerged in recent years thanks to improved early warning systems.

The future of the global agricultural market holds two main pathways. One is free trade, the other holds to various degrees of regulation. A common argument for deregulation is that subsidies typically disincentivise producers and favour consumers, which oftentimes leads to (economically) inefficient land use. Two examples can be given here. New Zealand and Brazil minimised government interference in agricultural production systems in response to the economic crises in the 1980s and the 1990s, respectively. In these cases farmers specialised into sectors where rising international prices compensated for the removed subsidies, such as meat and energy-rich crops (Chaddad and Jank, 2006). Those who favour regulated agriculture argue that consumers and producers need to be protected from price fluctuations, particularly those that are not driven by agriculture. Due to their sizes in population and agricultural production, China and India are occasionally accused of triggering food price rises. Both countries have, however, been self-sufficient in grain and meat for decades and have minimal trade of grains. Admittedly, China's weak points are its lack of animal feed and vegetable oils, which influences the global food trade flows of non-staples, especially through imported soybeans from Brazil, and it is thus a major contributing agent to Amazonian deforestation (Abbott et al., 2008; Chen, 2009; O'Brien and Leichenko, 2000). However, China's and India's latent roles should not be underestimated. China's oil demand has influenced world oil prices and hence indirectly food price inflation (Abbott et al., 2008).

The price of oil plays two important roles: for the cost of agricultural inputs such as fuel and fertilisers and, in determining whether crops are grown for food or energy (Table 1). It is estimated that 5% of global cereal production in 2007–2008 went to biofuel production (FAO, 2008b). Political incentives to reduce the use of fossil fuels, such as through carbon taxes, could similarly make bioenergy plantations on cropland more competitive and hence help drive food price directions (Johansson and Azar, 2007). This has raised concern as to whether agricultural land use needs to be regulated to guarantee a certain amount for food production (von Braun et al., 2008). China's government has decided to set aside 120 million hectares for permanent agricultural use.[1] Consequently land use needs to be intensified, small

[1] 120 million hectares correspond to the area currently denoted as "Total cultivated area," according to the National Bureau of Statistics of China. About 105 million hectares of this

Fig. 5 Land use changes on the Loess plateau, northwest China. With a highly erosive soil and an unpredictable annual total rainfall ranging from 300 to 800 mm, the options for rain-fed agricultural diversification are limited. Cultivation and grazing on steep slopes has been banned and the land is replanted with grass or perennial trees in one of the world's largest reforestation projects ever. Agricultural production is cramped onto floodplains in strong competition with housing, industrial and infrastructure development (note two roads on each side of the river and a railway is in the planning). For climatic and economic reasons farmers now prefer cash crops in greenhouses to land-intensive cereal production. Photo: the author (2005)

fields merged, new housing provided and off-farm jobs created (Chen, 2009). Modernisation and globalisation provides another dilemma to farmers (Figs. 3 and 5). As supermarket chains control more of the links between farmers and consumers, the demand for standardisation, quality and deliverability means that small producers cannot afford to supply their products. To stay in business they trade-off between increasing mono-cropping to gain economy of scale or diversifying the production, which would make them less vulnerable to price shocks (Gregory et al., 2005).

The debate on free market brings up another aspect of the global food distribution. Individual countries' responses during the food crisis (Table 1) provide an unprecedented study of behaviour during critical times. A survey of 77 countries across the world showed that more than half used some measure to reduce the costs for consumers while only one quarter took action to increase the supply, for example, export restrictions. The Middle East, North Africa, Europe and Central Asia prioritised for consumer costs while parts of Africa and South Asia focussed on supply increases. Only 16% showed no policy action at all, and the majority of those countries were in Latin America and Africa (FAO, 2008c). Among the countries that introduced export bans, several were self-sufficient grain producers like China and India (Table 2), but also significant grain exporters like Vietnam, Argentina and Egypt. This shows that trade-agreements such as those for WTO-members can be rebuffed leaving food net importing nations particularly vulnerable.

Changes in Supply – Can We Get It?

Distribution channels are an essential link to make people food secure by taking food from where it can be produced to where it can be sold. However, two factors that limit the quantities of food that can be produced are climate and land. Since both are likely to continue to restrict food supply, agro-technological innovations are expected to bring food to all.

Climate Change

Media is an important (sometimes objective) source of information for the general public. However, a noteworthy share of English-speaking media do not seem to reflect the scientific consensus on causes to human-induced climate change, as, for example, the impact that our food systems have (Boykoff, 2008; Boykoff and Mansfield, 2008; Neff, 2008). There are at least three basic features of climate change with respect to food production where news media often present an ad hoc simplified angle that is bound to end up in contradictions. When talking of impacts of climate change on agriculture, it is paramount to

(1) distinguish between effects of changes in the average climate (e.g. mean temperatures or mean rainfall) from changes in climate variability (e.g. range

land were used for cereals, tubers and soybeans in 2008 (www.stats.gov.cn).

and frequency of hot spells or rainfall events). While mean climate change is comparatively easier to adapt to, changes in the variability are difficult to attribute and brings along more unpredictability in timing, amounts, impacts and adaptive options. The impacts of changes in variability are further linked with local variations such as variations in topography and land use. For example, aerosols can alter the distribution of rainfall as they make clouds too heavy to ascend further, which causes them to precipitate around the sources of the pollution such as around industrial zones instead of more regionally, such as over agricultural plains. On larger scales pollution can reallocate rain, e.g. creating floods in southern parts of India and China and droughts in their northern regions (Menon et al., 2002).

(2) To holistically understand what positive *and* negative effects environmental change, including climate change, has on agricultural production and human livelihoods and vice versa. Global warming prolongs growing seasons and can open up new farming opportunities in high latitudes while temperature and water stress is likely to reduce productivity for semi-arid and tropical agriculture. Areas that already are dry, such as the Sahel and southern Europe, are likely to become even drier (IPCC, 2007b, p. 47), and the areas for rain-fed agriculture become more restricted.

In terms of climate change, land use contributes to one-third of the anthropogenic driven greenhouse gas emissions. Half of this comes from agriculture, of which the majority are nitrous oxides from inorganic fertilisers and methane-emissions from digesting ruminants. Other sources are paddy fields, biomass burning and manure. The other half are carbon dioxide emissions via the carbon sink that is lost through deforestation for farmland expansion (IPCC, 2007b, p. 36; Scherr and Sthapit, 2009).

As for the greenhouse gases, carbon dioxide may speed up the photosynthesis rate and improve water uptake (carbon dioxide fertilisation). Although difficult to prove, an additional 10 parts per million of carbon dioxide approximates to a 1% increase in yields; some argue that this has contributed to the increased productivity of crops in the past century (Lobell and Field, 2008). The long-term effects of atmospheric carbon dioxide, as the concentration increases, on crops, are also uncertain. In combination with warmer temperatures crops may mature early. The potential benefits are minimal in combination with other greenhouse gases, such as ground level ozone which destroys plant cells at a more rapid pace (Slingo et al., 2005) and puts near-urban agriculture at particular risk (Fig. 3). Some ways to reduce the environmental impact from agriculture, including mitigating greenhouse gas emissions, are presented in the "Technological Fix" section below.

(3) To learn to live with environmental changes, including climate change, even though this is difficult to define (in contrast to black carbon pollution or the ozone hole). But define them we must if we are to adapt or mitigate the consequences of them.

> We do not look at the future climate change. We adapt to the climate that is changing now. (Blessings Mwale, FAO Malawi)

> Before we used to plant when the first rain fell in October. Now the rain has changed. Sometimes it starts in November, sometimes in December and we don't know if it is the first rain or just a sprinkle [. . .] so we don't know when to plant [. . .]. And we don't know how long the rains will last so we don't know what to plant. (Farmers in Novu village, Malawi)

Everybody, rich or poor, who do not invest in adaptation are losers: rich because they are economically vulnerable and poor because they are more exposed both socioeconomically and climatically (Diffenbaugh et al., 2007, Fig. 1). While it is easier to prepare for long-term mean changes, e.g. by adjusting planting windows, using stress tolerant seeds, training and experimenting, our past experiences have no common characteristics with new extreme situations (Smit and Wandel, 2006). In fact, there are physical limitations too. Daily temperatures above 35°C reduce productivity of crops and livestock as well as threatening human lives (Slingo et al., 2005). In contrast to prevailing trends towards larger farm units, research shows that when factoring in climate larger farms in Europe are more vulnerable to higher temperatures than smaller ones (Reidsma et al., 2009). More variable climate and extreme events means that agricultural production, i.e. the food supply, is likely to become more variable and that prices are likely to follow accordingly. Let me illustrate with two recent examples of visits into near-future weather forecasts. In 1972 the summer temperatures in south Soviet Union were 4°C above the long-term mean. This resulted in 13% lower grain production compared to the previous year and the Soviet Union had to purchase American grain. This was in the middle of the oil crisis and thus contributed to food price inflation. In 2003 European summer temperatures went to 3.6°C above normal. Agricultural production declined 20–30% from the year before and it was estimated that 52,000 people died of heat stress (Battisti and Naylor, 2008).

To study future climates, scientists use a number of global or regional climate model scenarios. The climate scenarios are then used to simulate future crop production. However, climate scenarios do not predict future weather. At best they can guide in locating potential hotspots where, for example, food production may become limited (Diffenbaugh et al., 2007). Such results provide the foundation for many political decisions despite the fact that the conclusions that can be drawn from these studies are sometimes limited for the degree of certainty that is required for policy making. This depends among other things on understanding the links between biophysical and socio-economic limitations to adaptation (Challinor et al., 2009). Although farmers and plants can adapt to climate change, just as they have done over millennia, perfect adaptation is impossible (the closest we can get is to produce food in controlled environments, such as greenhouses or piggeries). Biological adaptation occurs without permission, while social adaptation may be limited by cultural and individual factors, such as that stakeholders fail to agree on the ultimate goals of adaptation (Adger et al., 2009). Furthermore, adaptive capacities vary over time (Smit and Wandel, 2006). Farmers' capacity to deal with droughts can be related to the availability of three categories of assets: land, labour and capital (Figs. 1, 3 and 5). By analysing and incorporating such incompletely understood information in crop models, places at risk and types of interventions that can be identified to strengthen the capacity to adapt to a drier climate (Simelton et al., 2009). An important next step is to deliver information in a useful format to the farmers (Patt and Schröter, 2008; Snapp et al., 2003).

Land Use Change

Land, as is water, constitutes a limiting factor for food security as it is unlikely to increase. Between 1700 and 1990 the global land area for agriculture expanded fivefold, from 300–400 to 1,500–1,800 million hectares (Goldewijk and Ramankutty, 2004). In the previous section it was estimated that unless productivity increases, an additional couple of hundred million hectares are needed to produce enough cereals for 9 billion people. However, productive cultivable land is lost as cities and industries expand and traditional farmland is switched from staples to cash crops, such as biofuels or grain for biofuel. Although the increases in cereal production since the 1960s have been accomplished on fairly constant areas of land so that each person needs a smaller plot for his/her cereal intake (Figs. 3 and 5), the areas planted for vegetables and fruits have skyrocketed since 1980s. At the turn of the millennium at least 16% of the world's arable land was degraded and two-thirds of the land in developing countries constrained by other factors, e.g. topography and low soil fertility (FAO and UNEP, 1999).

Climate change will fundamentally change the current land use by, for example, submerging coastal areas as sea-levels rise, and shifting of ecological zones as temperatures increase. Warming introduces particular problems in tropical and polar regions as they are at the respective ends of the eco-climatic spectrum. As the agro-ecological zones migrate polewards there is no existing agro-ecological zone to succeed what is currently equatorial (tropical) – and the arctic and polar regions have nowhere to go. Consequently, many groups of people are at risk of becoming marginalised when their chances to make their livelihoods become increasingly restricted by impacts of climate change. Studies show that the previous food crises made winners and losers of farmers and that access to land played an important role in this. Of households that make more than 75% of their income from farming, the richest group had the biggest improvement in welfare and the landless and poor lost welfare as staple prices increased (FAO, 2008c). Taking the effects of globalisation and climate change together, and there will be doubling of winners and losers (O'Brien and Leichenko, 2000).

Is there land to grow food for 9 billion people, feed for our livestock, and energy for our homes and industries? The countries with potential agricultural land are likely to be found in the northern latitudes as what used to be limited by temperature may benefit from warming, but they still lack sunshine hours. In Asia most land is already allocated. Many populous countries, such as India and China, still cultivate on plots smaller than one hectare (Figs. 3 and 5). Chinese policies, in particular, now endorse land mergers that can intensify land use by increasing urban populations (Chen, 2009). Those who are concerned that this land area is not enough take evidence in the accelerating land grab. At least 15–20 million hectares of farmland (the size of Cambodia) have been involved in international transactions since 2006. In contrast to the Europeans' land grab some hundred years ago, this land grab is predominantly carried out by Asian and Middle-East countries. The target remains the same: poorer less economically developed countries, but this time focussed on Sub-Saharan Africa and South East Asia. Land acquisitions abroad, both by private and state-owned companies, are expected to increase not only because of increasing food/feed/fuel demand but also from reduced ability to ensure domestic food security (supply) as a result of climatic marginalisation (Grain, 2008; von Braun and Meinzen-Dick, 2009). The Human Rights Commissioner has condemned this, concerned that these types of agreements will have little benefit to the local population or contribute to their food security (UNSR, 2009).

Technological Fix

Agriculture must increasingly serve the double purposes of having to feed an increasing number of people while mitigating its environmental side-effects, in particular from enhancing greenhouse gas emissions. Optimists stress that global grain supply in the recent past has largely been achieved by increasing productivity (without increasing vast areas of arable land), so there are several ways to fix both food security and climate change by seeds and by management. In fact, agriculture originated when farmers started practising natural selection, e.g. save the best seeds for planting the next season. This idea then moved to grafting and cross-breeding to select preferred characteristics. Technology now allows grain crops with more seeds and shorter stems that do not break when the grains fill, crops that are resistant to climate, insects, fungus and viruses and tolerate herbicides, and crops that mature faster or contain additional vitamins (Le Page, 2009). While research on GM-crops is largely taking place in the labs of privately funded research institutes, public research institutes are instead predominantly oriented to conventional agriculture. Perhaps the isolation helps in creating public scepticism. For example, questions have been raised about the cost, both of the research itself and for farmers as they no longer can multiply the seeds themselves, and of various environmental side-effects. Adding to the confusion is different national regulations for testing and releasing new seeds, hence the technology is advancing rapidly where legislations are more liberal (Cohen, 2005). Nevertheless, by 2008 over 13 million farmers cultivated 125 million ha with transgenic crops. The majority of the farmers were in developing countries while more than two-thirds of the land area was in the United States (Marshall, 2009). Another similarly value-laden debate covers what organic/ecological agriculture can do for developing countries. This would be one way to minimise misuse of agro-chemicals, chemicals that often are banned in western countries, improve the environment and empower small-scale farmers. Those in opposition claim that ecological agriculture is less productive and cannot produce the amounts of food that are required, particularly of staple crops.

Most scientists agree there are prospects to double the food productivity in Sub-Saharan countries and by relatively cheap measures. This requires appropriate advice and technology for small-scale farmers, and committed governments. For example, in 1998–1999 Malawian farmers received a "Starter Pack" of seeds and fertilisers to prove and spread hope that their yields can double. They did. In 2000 farmers were told to plant the seeds in a way that required more fertilisers, which few could afford. Yields plunged. However, rains were good in 1998 and 1999 in contrast to 2000, so typical for short experimental periods it has been difficult to attribute and quantify the direct causes of the observations. As the quotes above suggest, there are difficulties to match fertiliser recommendations with the timing of rainfall and the fertilisers are expensive so the adoption levels are limited (Snapp et al., 2003). The farmers in Novu village (Malawi) mixed local and hybrid maize seeds to increase their probability of at least getting some harvest. The local seeds were low-yielding but more drought-resistant than the expensive hybrid seeds, which would only succeed if the rains were good and the fertilisers applied in a timely fashion. Other technological advances occur outside the agricultural field. In some cases it may be to reduce losses after harvest (Fig. 3). In other cases farmers and pastoralists use mobile phones to get official prices and become stronger negotiators.

Incidentally, many of the changes in management that can mitigate climate change are also having other environmental benefits as well as increasing harvests (ADAS, 2009; Scherr and Sthapit, 2009). To reduce methane and nitrous oxide emissions the following can be applied:

(1) Change paddy rice varieties or timing for flooding the rice fields.
(2) Natural grazing is preferred to animals relying on hard-digestive pellets (that are manufactured of imported grains or soybeans).

To increase carbon sinks:

(1) No-tillage reduces fuel for tractors, prevents soil erosion and fertiliser leakage, and improves soil organic matter and soil water-holding capacity.
(2) Intercropping enhances synergy effects between plants, increases habitats for natural pollinators and biodiversity.
(3) Mulch reduces water-stress in crops, prevents soil erosion and reduces the need for synthetic fertilisers.

Among the drivers for agricultural production at the global, national and local scales, the implications of climate change on supply and demand are different at each level (Hazell and Wood, 2008): At the global scale the main drivers of agricultural production are related to trade and thus the predictability of impacts of climate change. This is the appropriate scale for setting emission targets to mitigate climate change. At the country scale the drivers are related to policies that favour economic growth, in the light of demographic changes, and back up producers/consumers/business against negative impacts of climate, such as water and energy scarcity, or pandemics. The national scale is ultimately responsible for achieving the mitigation targets as well as coming up with adaptation strategies. At the local scale the drivers are immediately related to basic human needs for survival and beyond – access to technology to produce food/markets to sell or buy food/jobs to improve the purchasing power of householders, and status of the environment in order to cope with weather and adapt to climate change. This requires responsibility at all stages – from producers, middlemen and consumers to policymakers and interest groups. Citizens in democratic countries have a particular responsibility (Fig. 6).

Fig. 6 Much of western culture and economy is organised around consuming (and wasting) agricultural products. Those habits have changed enormously since these houses in Manchester, UK, were built. It is tempting to imagine the scenery in 2050. Photo: the author (2009)

Concluding Remarks

This chapter has viewed food security across six aspects of food demand and supply. The demand for food will no doubt continue to increase. *Population growth* is likely to increase food demand where the vulnerability to climate change and environmental degradation already is high. *Increased purchasing power* across the whole of society eventually makes more people food secure; however, above a certain threshold consumers are likely to waste food. The supply of food may well increase; the question is will it reach all people, at all times? *Climate change* is likely to make food production unpredictable (perhaps until greenhouse gases eventually stabilise) because climate variability will increase, the combined environmental impacts are blurred and exceed the human experience for adaptation. *Land use changes* have largely been pointing towards maximising the economic rather than the ecological return, and away from less-paid staple crops. *Technical solutions* to increase food production exist but do not necessarily change underlying behaviour in a way that contributes to food insecurity, malnutrition or unequal access to food, such as irresponsible governance and corruption, insufficient water and sanitation and waste generation. *Markets* can be understood as a means to distribute food from producers to consumers so that people can access food. However, if food prices increase faster than salaries, which can happen for a number of reasons including unexpected weather, pandemics, severe pollution accidents, civil unrest or economic inflation, then access to food is bound to become more unequal. For some, food security drifts into food insecurity.

Food security, as a definition, has travelled a journey from national supply to one of household purchase capacity – but it remains words without responsibility, and charges for those who fail to ensure safeguards that everybody has enough food to live an active and healthy life. The lack of coordinated action during the financial crisis in 2007–2008 clearly illustrates that agricultural trade agreements need restructuring. Several countries acted against the treaties and still failed to secure their people from hunger and poverty. Such behaviour furthermore demonstrates that poverty and instability are big obstacles for long-term investments. Until basic needs are secured, such as in housing, food, regular income and physical safety, it is difficult to be proactive with a dollar a day. Imagine the possible outcomes if the 365 richest people in the world gave one dollar each to everyone who earns less than one dollar a day.

One advantage with the global food crisis was that it greatly illuminated the complexity of secure food systems. This will result in more holistic analyses of direct and indirect impacts of production and consumption patterns, hopefully in science, media as well as policy implementation. Climate change is likely to soon increase the costs of living; as water resources become scarcer the cost to produce and transport food increase and more energy will be needed for air conditioning and refrigeration. Consequently, there is scope to explore more efficient trade-offs between energy for feed, food and fuel. This requires, among other things, a more flexible and multi-purpose type of agricultural land use and production. The first purpose of such agriculture is that it contributes to food security, by improving the management from pre-sowing to the final leftovers and by increasing productivity without compromising the nutritional content. The second purpose is that it contributes to ecosystem services, including mitigating climate change and neutralising environmental impacts. Life cycle analyses of various types of agricultural production can help clarifying certain products' environmental impacts – and introduce fairer payment for environmental degradation. Producers and consumers need clearer labels, not more of them.

Finally, I would like to encourage the reader to engage in discussions about our daily food – what does food security mean to you? Who is responsible for food security? Should failing to achieve food security be penalised? What impacts on the environment are acceptable for everybody's food security? If the environmental costs are not included in the price of the product, how should it be paid for?

Find Out More About...

... detailed trade flows. Online map tool that allows you to explore who imports and exports what agricultural products from where http://faostat.fao.org/DesktopModules/Faostat/WATFDetailed2/watf.aspx?PageID=536

... Global Information and Early Warning System on Food and Agriculture. A tool that allows you to monitor food security situation around the world http://www.fao.org/giews/english/index.htm

... your environmental impact. World Wildlife Fund gives you a quick estimate of how many planet earths your lifestyle requires http://footprint.wwf.org.uk/

... food and climate research. The Food Climate Research Network reviews and publishes the most recent reports from across the world. It also enables you to get in touch with scientists and practitioners in the field http://www.fcrn.org.uk/

... or run your own climate scenarios with EdGCM (Educational Global Climate Modelling) http://edgcm.columbia.edu/spotlight/

Acknowledgements I am grateful to Evan Fraser for discussing the outline of this chapter, Tiep Quang Le for collecting data and discussing economic aspects of food and to Andrew Challinor for valuable comments on the supply side. The work was funded by the UK Natural Environment Research Council's Quantifying and Understanding the Earth System's Global Scale Impacts project (QUEST-GSI) and the UK Economic and Social Research Council.

References

Abbott PC, Hurt C, Tyner WE (2008) What's driving food prices? Farm Found http://ageconsearch.umn.edu/bitstream/37951/2/FINAL%20WDFP%20REPORT%207-28-08.pdf

ADAS (2009) RMP/5142 Analysis of Policy Instruments for Reducing Greenhouse Gas Emissions from Agriculture, Forestry and Land Management, Defra.

Adger NW, et al. (2009) Are there social limits to adaption to climate change? Clim Change 93:335–354.

Barrett CB, Bell R, Lentz EC, Maxwell D (2009) Market information and food insecurity response analysis. Food Secur 1:151–168.

Battisti DS, Naylor RL (2008) Historical warnings of future food insecurity with unprecedented seasonal heat. Science 323:240–244.

Boykoff MT (2008) Lost in translation? United States television news coverage of anthropogenic climate change, 1995–2004. Clim Change 86:1–11.

Boykoff MT, Mansfield M (2008) 'Ye olde hot aire': Reporting on human contributions to climate change in the UK tabloid press. Environ Res Lett 3. doi: 10.1088/1748-9326/3/2/024002.

Chaddad FR, Jank MS (2006) The evolution of agricultural policies and agribusiness development in Brazil. Choices 21(2):85–90.

Challinor AJ, Ewert F, Arnold S, Simelton E, Fraser E (2009) Crops and climate change: Progress, trends and challenges in simulating impacts and adaptation. J Exp Bot doi: 10.1093/jxb/erp062

Chen X (2009) Review of China's agricultural and rural development: policy changes and current issues. China Agr Econ Rev 1(2):121–135.

Cohen J (2005) Poorer nations turn to publicly developed GM crops. Nat Biotechnol 23(1):27–33.

Devereux S (2009) Why does famine persist in Africa? Food Secur 1:25–35. doi: 10.1007/s12571-008-0005-8.

Diffenbaugh NS, Giorgi F, Raymond L, Bi X (2007) Indicators of 21st century socioclimatic exposure. PNAS 104(51):20195–20198.

Ericksen PJ (2008) Conceptualizing food systems for global environmental change research. Glob Environ Change 18:234–245.

EurActiv (2009) Sweden promotes climate-friendly food choices. http://www.euractiv.com/en/cap/sweden-promotes-climate-friendly-food-choices/article-183349

FAO (1983) World food security: a reappraisal of the concepts and approaches. Director-General's Report, Rome.

FAO (1996) Declaration on world food security. World Food Summit, FAO, Rome.

FAO (2008a) FAO statistical yearbook 2007–08. http://www.fao.org/economic/ess/publications-studies/statistical-yearbook/fao-statistical-yearbook-2007-2008/en/

FAO (2008b) High food prices and food security – Threats and opportunities. The state of food insecurity in the world 2008. ftp://ftp.fao.org/docrep/fao/011/i0291e/i0291e00.pdf

FAO (2008c) Soaring food prices: Facts, perspectives, impacts and actions required. High-level conference on world food security: the challenges of climate change and bioenergy. Rome 3–5 June 2008. http://www.fao.org/fileadmin/user_upload/foodclimate/HLCdocs/HLC08-inf-1-E.pdf

FAO (2009) FAOSTAT Online Database. FAO, Rome. http://faostat.fao.org/default.aspx

FAO and UNEP (1999) The future of our land – Facing the challenge. Guidelines for Integrated Planning for Sustainable Management of Land Resources. http://www.fao.org/DOCREP/004/X3810E/x3810e00.htm#TopOfPage

Goldewijk KK, Ramankutty N (2004) Land cover change over the last three centuries due to human activities: The availability of new data sets. GeoJournal 61:335–344.

Grain (2008) Seized! The 2008 land grab for food and financial security. Spain: Girona.

Gregory PJ, Ingram JSI, Brklacich M (2005) Climate change and food security. Philos Trans R Soc B 360(1463):2139–2148.

Hazell P, Wood S (2008) Drivers of change in global agriculture. Philos Trans R Soc B 363:495–515.

IPCC (2007a) Climate Change 2007: Impacts, Adaptation and Vulnerability. Contribution of Working Group II to the fourth assessment report of the Intergovernmental Panel on Climate Change. Cambridge: Cambridge University Press.

IPCC (2007b) Climate Change 2007: Synthesis Report, Intergovernmental Panel on Climate Change.

Johansson DJA, Azar C (2007) A scenario based analysis of land competition between food and bioenergy production in the US. Clim Change 82(3–4):267–291.

Le Page M (2009) Learn to love genetic engineering. New Sci 203:37.

Lobell BD, Field CB (2008) Estimation of the carbon dioxide (CO2) fertilization effect using growth rate anomalies of CO_2 and crop yields since 1961. Glob Chang Biol 14:39–45.

Marshall A (2009) 13.5 million farmers cultivate GM crops. Nat Biotechnol 27(3):221.

Maxwell S (1996) Food security: A post-modern perspective. Food Policy 2:155–170.

Menon S, Hansen JW, Nazarenko L, Luo Y (2002) Climate effects of black carbon aerosols in China and India. Science 297:2250–2253.

National Coordinating Committee for UDHR50 (2001) The Universal Declaration of Human Rights. Questions and Answers. New York, NY: Franklin and Eleanor Roosevelt Institute.

Neff (2008) Yesterday's dinner, tomorrow's weather, today's news? US newspaper coverage of food system contributions to climate change. Public Health Nutr doi: 10.1017/S1368980008003480.

Neset Schmid TS, Lohm U (2004) Spatial imprint of food consumption: A historical analysis for Sweden, 1870–2000. Hum Ecol 33(4):565–580.

O'Brien K, Leichenko RM (2000) Double exposure: assessing the impacts of climate change within the context of economic globalization. Glob Environ Change 10:221–232.

Patt AG, Schröter D (2008) Perceptions of climate risk in Mozambique: Implications for the success of adaptation strategies. Glob Environ Change doi: 10.1016/j.gloenvcha.2008.04.002

Reidsma P, Lansink AO, Ewert F (2009) Economic impacts of climatic variability and subsidies on European agriculture and observed adaptation strategies. Mitig Adap Strat Glob Change 14(1):35–59.

Rincon P (2006) Neanderthals' 'last rock refuge'. BBC News http://news.bbc.co.uk/1/hi/sci/tech/5343266.stm

Sample I (2009) World faces 'perfect storm' of problems by 2030, chief scientist to warn. The Guardan http://www.guardian.co.uk/science/2009/mar/18/perfect-storm-john-beddington-energy-food-climate.

Sherr SJ, Sthapit S (2009) Faming and land use to cool the planet. In: Worldwatch Institute. State of the world (pp. 30–49). Washington DC: W.W. Norton & Co.

Sen A (1981) Poverty and Famines. An Essay on Entitlement and Deprivation. Oxford: Clarendon Press.

Simelton E, Fraser EDG, Termansen M, Forster PM, Dougill AJ (2009) Typologies of crop-drought vulnerability: An empirical analysis of the socio-economic factors that influence the sensitivity and resilience to drought of three major food crops in China (1961–2001). Environ Sci Policy 12(4):438–452.

Slingo JM, Challinor AJ, Hoskins BJ, Wheeler TR (2005) Introduction: Food crops in a changing climate. Philos Trans R Soc B 360:1983–1989.

Smit B, Wandel J (2006) Adaptation, adaptive capacity and vulnerability. Glob Environ Change 16(3):282–292.

Smith L, El Obeid AE, Jensen HH (2005) The geography and causes of food insecurity in developing countries. Agric Econ 22(2):199–215.

Snapp SS, Blackie MJ, Donovan C (2003) Realigning research and extension to focus on farmers' constraints and opportunities. Food Policy 28:349–363.

Tukker A, et al. (2009) Environmental impacts of diet changes in the EU. Annex Report, European Commission Joint Research Centre, Institute for Prospective Technological Studies.

UN (1948) The universal declaration of human rights. In General Assembly of the United Nations (Editor). http://www.un.org/en/documents/udhr/index.shtml#a24

UN (1975) Report of the World Food Conference, Rome: New York 5–16 November 1974.

UNCTAD (2008) Addressing the global food crisis: Key trade, investment and commodity policies in ensuring sustainable food security and alleviating poverty.

UNDP (2007) MDGMonitor. Tracking the millennium development goals. Goal 1: Eradicate Extreme Poverty and Hunger. http://www.mdgmonitor.org/goal1.cfm

UNSR (2009) Press release: The UN Special Rapporteur on the Right to food recommends principles and measures to discipline "land grabbing" In: OD Schutter (Ed), www2.ohchr.org/english/issues/food/index.htm

US Census Bureau (2009) International Data Base. US Census Bureau, Population Division.

von Braun J, et al. (2008) High food prices: The what, who, and how of proposed policy actions. IFPRI Policy Brief May International Food Policy Research Institute.

von Braun J, Meinzen-Dick R (2009) "Land Grabbing" by foreign investors in developing countries risks and opportunities. IFRPI Policy Brief No. 13.

WFP (2007) News: WFP chief calls for support to combat "perfect storm" over Africa's rural poor. World Food Programme. http://www.wfp.org/news/news-release/wfp-chief-calls-support-combat-perfect-storm-over-africas-rural-poor

World Bank (1986) Poverty and Hunger: Issues and Options for Food Security in Developing Countries. World Bank Policy Study, Washington, DC.

World Bank (2009) World Development Report. Reshaping Economic Geography. Washington, DC: The International Bank for Reconstruction and Development/The World Bank.

WRAP (2008) The food we waste. http://www.wrap.org.uk/downloads/The_Food_We_Waste_v2__2_.97caf659.5635.pdf

Building Capacity to Cope with Climate Change in the Least Developed Countries

Hannah Reid, David Dodman, Rod Janssen, and Saleemul Huq

Abstract The least developed countries (LDCs) are particularly vulnerable to the impacts of climate change, due to a combination of physical vulnerability and limited coping capacity. The poor in these countries rely heavily on natural resources, are severely affected by climate variability such as floods and droughts and have limited savings and few alternative livelihood opportunities to respond to shocks and stresses resulting from climate change. This chapter outlines these challenges and describes the Capacity Strengthening in the LDCs for Adaptation to Climate Change (CLACC) programme, an initiative to build civil society capacity to address these. It evaluates the role of a range of activities – including supporting civil society participation at key meetings, conducting relevant research and facilitating outreach activities – that can help low-income nations and their most vulnerable citizens respond to the challenges of climate change.

Keywords Capacity strengthening · Adaptation · Least developed countries · Outreach · Evaluation

Introduction

Poor countries and poor communities are widely acknowledged as being particularly vulnerable to climate change. The Intergovernmental Panel on Climate Change (IPCC) and the United Nations Framework Convention on Climate Change (UNFCCC) acknowledge three groups of developing countries as being particularly vulnerable: the least developed countries (LDCs), Small Island Developing States (SIDS) and countries in Africa (IPCC, 2001, 2007). There is considerable overlap between these three groups, for example, Guinea-Bissau is a SIDS, and LDC and in Africa, and 33 out of the total 49 LDCs are in Africa, but together, these three groups number roughly 100 countries and are home to well over a billion people.

The fact that these countries are so vulnerable to climate change is ironic given that they have contributed least to the rising concentrations of greenhouse gas emissions in our atmosphere. Indeed, their carbon dioxide emissions (excluding South Africa's) account for only 3.2% of the global total, compared to 23.3% for the United States, 24.7% for the European Union, 15.3% for China and 4.5% for India (see Table 1) (Huq and Ayers, 2007).

These countries are amongst the most vulnerable to the impacts of climate change for several reasons. Many are located in some of the most geographically vulnerable parts of the world, such as drought prone Sahelian countries, cyclone-prone Bangladesh or low-lying small island states likely to suffer from sea-level rise. Poor communities within countries are also located in vulnerable areas. For example, in cities, informal illegal settlements commonly occupy land on floodplains or at the foot of unstable hillsides. Industry or wealthier city residents do not want to occupy this land because of the risks, but poor families have no alternative if they want to reduce their commute and travel costs (Satterthwaite et al., 2007). This perpetuates cycles of poverty for these people, which can be hard to escape.

H. Reid (✉)
Climate Change Group, International Institute for Environment and Development, London WC1H 0DD, UK
e-mail: hannah.reid@iied.org

J. Dodson (ed.), *Changing Climates, Earth Systems and Society*, International Year of Planet Earth,
DOI 10.1007/978-90-481-8716-4_11,

Table 1 Highest and lowest: carbon dioxide emissions, 2002

Country/region	Total emissions (1,000 t)	Total of global emissions (%)	Per capita emissions (t)
LDC, SIDS and Africa (excluding South Africa)	791,456	3.2	2.2
LDC, SIDS and Africa	1,155,363	4.67	2.3
India	1,105,595	4.5	1.1
China	3,783,231	15.3	2.9
US	5,773,401	23.3	19.9
EU	6,117,989.5	24.7	8.4
Global	24,756,694	100	4

Source: Huq and Ayers (2007).

The poor also tend to rely more on natural resources, such as forests, fish stocks and grazing land for their food and livelihoods, than the wealthy, and yet these natural resources are very vulnerable to climate change. The impacts of climate change look set to reduce Namibia's Gross National Product by at least 5% over the next 20 years due to the nation's heavy reliance on natural resources, for example (Reid et al., 2008).

Poor nations and communities also have a very low capacity to cope with drought, floods, cyclones and other climate change impacts when they strike (Huq et al., 2004). Low capacity to cope with climate change impacts is apparent at several levels, from the community through to local and national government and non-government organizations, and right up to those representing their countries at the United Nations climate change negotiations.

At the community level, people have few savings, few alternative livelihood opportunities, no insurance and are already close to or even below the poverty line. For many, resilience to climate change impacts has already been eroded by entrenched poverty. When a climate change related disaster strikes, there is seldom a safety net to fall back on. Harsher or more frequent disasters could therefore tip them over the edge into chronic famine or forced migration.

At the level of local or municipal governments, knowledge on climate change is limited. And even where it is clear that certain sites are at risk from climate change impacts, there are very few examples of local or municipal governments taking climate change into account in their planning, policies and activities. Budgets are usually tight and there is already a huge infrastructure deficit to consider before people can think about incorporating climate change considerations into their work.

National governments also lack knowledge and capacity on climate change. Often it is seen as a low priority when compared to pressing and more clearly definable issues such as health, employment, housing and education. Climate change is usually situated in environment ministries with relatively little power and influence compared to, for example, finance ministries. Too often climate change is seen as an issue for the future rather than the present day – something beyond the short-term thinking associated with election cycles and strategic political maneuvering.

In the international climate change negotiations conducted under the auspices of the UNFCCC, LDCs also suffer serious capacity constraints. Few LDCs can send more than a handful of delegates to these negotiations, and without the resources – financial and technical – negotiating against the large well-paid teams of lawyers and experts brought in from wealthier nations who see taking action on climate change as a threat to their economic development poses a huge challenge. Some can only send one delegate. Despite this, the least developed countries have won some important and significant concessions in these negotiations. In 2001, the seventh session of the Conference of the Parties (COP7) to the UNFCCC decided that all LDCs under the UNFCCC would prepare a National Adaptation Programme of Action (NAPA). This would identify immediate and urgent climate change adaptation needs. NAPAs recognized the need for action-oriented country-driven activities, which should be country-driven, flexible and based on national circumstances (Alam, 2007; Adger et al., 2003).

The UNFCCC has long recognized the need for capacity building to help Parties, especially developing countries, respond to climate change. Under the Convention, the Subsidiary Body on Implementation is charged with providing advice on 'ways and means of supporting endogenous capacity building in developing countries' in Article 9, while the Kyoto Protocol commits Parties to cooperating in, and promoting, 'the strengthening of national capacity building' in Article 10(e). Capacity-building cuts across many of the issues under consideration in the climate change process, including technology transfer, national communications and funding. Capacity building was first considered as a separate agenda item at COP5, resulting in frameworks for capacity building for developing countries and countries with economies in transition being agreed at COP7 in 2001. These frameworks were intended to guide the climate change capacity-building activities of the Global Environment Facility and other funding bodies.[1]

In recognition of the poor capacity amongst the least developed countries to cope with climate change, the Climate Change Group at the International Institute for Environment and Development (IIED) began a capacity-building programme in 2003. The programme is called CLACC – Capacity Strengthening of LDCs for Adaptation to Climate Change. It decided to focus on non-government organizations (NGOs), very few of which at the time had any significant knowledge or programmes of work on climate change. This was because NGOs were viewed as being able to effectively get involved with capacity building at both government and local levels, and the institutions selected already had strong links with both government policy makers and communities in their respective countries, as well as convening power and research capacity.

Building capacity to cope with climate change at all societal levels is essential in order to lessen the overall effects that climate change will have in the LDCs. But effective capacity building must be an ongoing process with long-term commitment. The CLACC programme aims to tackle this problem by building capacity within civil society within participating LDCs. It began with a fellowship programme and has since conducted research, dissemination workshops and other outreach activities to build the capacity of CLACC fellows and partner non-government organizations in selected LDCs, and hence civil society and government at large. This chapter describes CLACC programme activities conducted to date. In recent years, the focus of much CLACC research and outreach activities has been on urban areas, where large populations of vulnerable communities reside, so this is described in more detail. The chapter also describes the results of a recent external evaluation of CLACC's key successes as well as areas which require improvement. The chapter concludes with a description of the proposed direction of future CLACC activities in a world where the impact that climate change will have on the world's poor will dramatically increase.

Responding to the Challenges of Climate Change and Development: The CLACC Programme

The International Institute for Environment and Development (IIED) conceived the CLACC programme in 2003 in recognition of the low capacity to tackle climate change within civil society in LDCs (see Box 1 for CLACC aims). IIED already had a strong long-term working relationship with a group of Southern NGOs known as the RING – the Regional and International Networking Group. Indeed, long-term collaboration with institutions such as those in the RING is one of the key differences between IIED and similar NGOs working at the nexus of environment and development issues. Regional CLACC partners were, therefore, chosen from the RING network, which has a proven track record of working on environment, national policy and poverty reduction issues. These regional CLACC partners then looked in their region for LDC partner organizations with the following characteristics: some research, policy analysis and convening capacity; links with both policy makers and communities; proven track record of knowledge on

[1] See http://unfccc.int/cooperation_and_support/capacity_building/items/1033.php for more information on capacity building under the UNFCCC.

development and environment issues, not necessarily including climate change; and an interest in working on climate change and a willingness to provide a member of staff available to get involved with the CLACC programme of work.

Box 1 CLACC programme aims

- Strengthening the capacity of civil society in the LDCs to adapt to climate change and fostering adaptive capacity amongst the most vulnerable groups.
- Establishing an information and knowledge system to help countries to deal with the adverse impacts of climate change.
- Integrating adaptation to climate change into the work of key non-government institutions, and mainstreaming the NAPA process with these institutions.

The CLACC programme incorporates a variety of activities: fellowships; attending key meetings; strengthening NAPAs; research; regional workshops; and outreach activities. The following sections discuss these activities and offer a critical analysis of the programme's strengths and weaknesses.

The CLACC Fellowship Programme

Given the low levels of knowledge on climate change issues that typified most NGOs in poor nations at the time, IIED felt the first step was to raise awareness on climate change issues amongst CLACC partner institutions before any wider capacity building could occur. The first CLACC activity undertaken was, therefore, a fellowship programme. In early 2004, CLACC fellows from four regional CLACC partner organizations spent 2 months at the following Northern host institutions: IIED, the Stockholm Environment Institute (SEI), the Potsdam Institute for Climate Impacts Research (PIK) and the Centre for International Climate and Environmental Research (CICERO). CLACC focused on these four regions because they are home to many communities that will be disproportionately and negatively affected by climate change. Following these fellowships, in 2005, CLACC fellows from 12 LDCs spent 2 months with their respective regional CLACC partner organizations (Table 2 shows all regional and LDC CLACC partner

Table 2 CLACC organizations

Region	Regional CLACC partner and country	LDC partner and country
South Asia	Bangladesh Centre for Advanced Studies (BCAS), Bangladesh	CARITAS Bangladesh, and Rangpur Dinajpur Rural Service (RDRS) Bangladesh
		Royal Society for Protection of Nature (RSPN), Bhutan
		Local Initiatives for Biodiversity, Research and Development (LI-BIRD), Nepal
West Africa	Environmental Development Action In The Third World (ENDA), Senegal	Organisation des Femmes pour la Gestion de l'Energie, de l'Environnement et la promotion du Développement Intégré (OFEDI), Benin
		Amade-Pelcode, Mali TENMIYA, Mauritania
East Africa	African Centre for Technology Studies (ACTS), Kenya	Environmental Protection and Management Services (EPMS), Tanzania
		Development Network for Indigenous Voluntary Association (DENIVA), Uganda
		Sudanese Environment Conservation Society (SECS), Sudan
Southern Africa	ZERO Regional Environment Organisation, Zimbabwe	Coordination Unit for the Rehabilitation of the Environment (CURE), Malawi
		Action Group for Renewable Energies and Sustainable Development (GED), Mozambique
		Energy and Environmental Concerns for Zambia (EECZ), Zambia

organizations). CLACC fellowships have been held intermittently since, for example, when a CLACC fellow has found work elsewhere and the individual replacing them needs some exposure to CLACC activities and climate change issues to continue the CLACC programme of work.

Attending Key Meetings

The first four regional CLACC fellows attended their first UNFCCC meeting in June 2004. Following this, the final selection of LDC country partners was made and CLACC programme methodologies and activities finalized during COP10 in December 2004. Since then, CLACC fellows and regional partners have had a strong presence at all COPs and some UNFCCC subsidiary body meetings too. In addition, most CLACC fellows attended the second and third international workshops on community-based adaptation held in Dhaka, Bangladesh, in February 2007 and 2009, respectively.

Attending these meetings provides networking opportunities for CLACC fellows, introduces them to their country negotiators, builds their knowledge and capacity on climate change issues, and strengthens the voices of LDCs, both in the negotiations themselves (when CLACC fellows are able to support their national delegations and in some cases become official delegates themselves), but also through media outputs and in parallel NGO meetings and events. At each COP, fellows get involved with the Climate Action Network (CAN) – a global network of NGOs interested in climate change issues. The meetings also allow the whole CLACC team to come together at least once a year to discuss the CLACC work programme for the coming year.

Strengthening NAPAs

The first major CLACC activity was to strengthen the existing NAPA process. CLACC fellows worked in their respective countries to engage with this process. In some instances this involved attending NAPA meetings, bringing experts to those meetings to ensure key national issues were not forgotten and sharing information about the NAPA process with other in-country stakeholders. In other countries, such as Sudan and Bangladesh, the CLACC fellow became an integral part of the NAPA team, helping to lead and organize the process with a view to producing a stronger final NAPA report.

Research

The CLACC research programme began in 2005. The first research topic focused on climate change and health in the LDCs. CLACC partners engaged with in-country health experts, and with support from the London School of Hygiene and Tropical Medicine (LSHTM) conducted an assessment of key climate change and health issues in their country. This information fed into national adaptation planning (including NAPAs) and international decision-making processes. For example, a CLACC side event on health and climate change was held at COP13 in Bali in 2007. Following the research on health, the CLACC programme embarked on a research programme on climate change and urban areas. This acknowledged the fact that the effects of climate change will be felt especially strongly in LDC towns and cities. These cities often have high concentrations of people and economic activities located in vulnerable physical settings.

Subsequently, the CLACC network began an ongoing programme of work on the economics of climate change in selected LDCs. Other research programmes, both within IIED and elsewhere, have used the CLACC network, or part of it, on which to base their activities. For example, The Community Based Adaptation in Africa (CBAA) project led by ACTS in Kenya, and funded by the United Kingdom Department of International Development (DFID) and Canada's International Development Research Centre (IDRC) as part of their Climate Change Adaptation in Africa Programme (CCAA), is based largely on a CLACC group of project partners.

Regional Workshops

Building on the experience and knowledge gained from the fellowships and research projects, the CLACC programme then organized several regional workshops on adaptation to climate change. These brought together groups of interested people from civil society in the four regions to discuss, brainstorm and plan next steps on the adaptation agenda in their respective regions and countries. The workshops aimed to build capacity within each region in this way and help establish links between interested actors thus helping them work together and speak for their nations and regions more effectively. Three workshops were held in early 2008: in West Africa, East Africa and Asia. These workshops have also served to raise awareness within the CLACC institutions themselves, especially amongst institution heads, such that capacity building on adaptation to climate change extends to the CLACC partner institution rather than the individual alone.

Outreach Activities

A number of information sharing activities occurred under the CLACC programme. First, each CLACC partner institution has established a climate change and development resource library, which is open to the public and contains information on climate change and development issues. At the very minimum, this is a shelf in the CLACC partner organization's office, but some CLACC partners have made considerably more effort than this.

Each CLACC partner has established a small national group of NGOs that meet on a regular basis to discuss climate change and development issues. This is based on the model of the Up in Smoke coalition in the UK which has been tremendously successful at raising awareness on climate change and development issues, both within major development NGOs and in the policy making arena.[2]

Forums have also been established to provide feedback on the *Tiempo* bulletin on climate change and development, of which IIED is a leading editor.[3] This CLACC activity had three goals: to raise awareness about climate change and development issues within LDCs; to inform people in LDCs about *Tiempo* and encourage them to submit articles thus sharing their knowledge; and, to solicit feedback on *Tiempo* to improve its usefulness within LDCs.

A regular quarterly CLACC newsletter serves to share information about CLACC activities to all CLACC programme partners, and also donors and those interested in CLACC activities. Once a year, the newsletter produced before the COP is designed with a more general COP audience in mind, and printed copies are brought to the COP for distribution. This newsletter is coordinated by ACTS on behalf of the programme.

The CLACC website also provides information on CLACC partners and activities. This is regularly updated and managed by BCAS.[4]

Training on use of videos has also been provided to nearly all CLACC fellows. Video cameras were provided to selected fellows who were tasked with producing a short film on community based adaptation. Many of these were shown at a film festival held during COP13 in 2007.

Climate Change and Urban Development in Low- and Middle-Income Countries

The physical vulnerability of urban areas in low- and middle-income countries is frequently exacerbated by poverty, the lack of appropriate infrastructure, and weak or inefficient systems of urban management (Dodman and Satterthwaite, 2008; Bicknell et al., 2009). Yet this dense concentration can also provide the potential for effective adaptation, improved resilience and the opportunity to meet broader development needs.

In response to this particular set of challenges, the CLACC Fellows developed a programme of work to describe and assess climate vulnerability in urban centres, and to identify proposed interventions for adaptation and building resilience. The work involved a mapping exercise to assess where vulnerable areas are and illustrate this on maps, a particularly important task as official city maps do not provide such information. It also required meeting with a range of stakeholders in order to assess where key vulnerabilities lay – a process that simultaneously built knowledge and awareness among these actors.

[2] Publications by the Up in Smoke coalition include Reid and Simms (2007), Simms and Reid (2006), Simms and Reid (2005), Simms et al. (2004).

[3] See http://www.cru.uea.ac.uk/tiempo/newswatch/latest.htm for current and past issues of the *Tiempo* bulletin.

[4] See www.clacc.net

Table 3 CLACC countries and cities

Asia	Eastern Africa	Southern Africa	West Africa
Bangladesh (Khulna)	Kenya (Mombasa)	Malawi (Blantyre)	Benin (Cotonou)
Bhutan (Thimphu)	Sudan (Khartoum)	Mozambique (Maputo)	Mali (Bamako)
Nepal (Kathmandu)	Tanzania (Dar es Salaam)	Zambia (Lusaka)	Mauritania (Nouakchott)
	Uganda (Kampala)	Zimbabwe (Harare)	Senegal (Diourbel)

In 2008, CLACC fellows and experts prepared review documents highlighting climate related problems (current stressors and potential climate change impacts) and their impacts for fifteen cities (see Table 3). These were deliberately selected to incorporate cities in a variety of physical settings – from low-lying coastal cities, to inland water scarce cities, to high altitude cities – and to include secondary urban centres as well as major cities and capitals as a means of demonstrating key vulnerabilities of cities in different positions in the urban hierarchy. This process was intended to identify key threats and major institutional stakeholders, and to provide the necessary information for communities and local governments to begin the process of climate change adaptation planning.

The CLACC project identified several key climate-related issues affecting urban areas in low-income countries, many of which will be exacerbated by anthropogenic climate change. LDCs are among the countries most affected by recent droughts. Drought is predicted to become more frequent and severe as a result of climate change, and many LDC cities are already badly affected. Zimbabwe has seen a decline in average rainfall of nearly 5% since 1900, with Harare and Bulawayo both affected by water stress. The Kariba hydropower plant that serves Harare has also been impacted by water scarcity, resulting in load-shedding by electricity providers. In Mali, Bamako is seeing widespread difficulties in accessing water throughout the city: although 90% of families in the city have their own wells, the availability of water in these is declining as groundwater levels fall.

But even in towns and cities where overall rainfall totals are declining, precipitation is tending to occur in shorter, more intense bursts that can overwhelm urban drainage systems and lead to flooding. Frequent flooding has been affecting the congested slums of Kampala, particularly Kawempe, where almost half the houses are built on wetland. Heavy rainfall and flooding may also lead to landslides. In Kathmandu, Nepal, 207 mm of rainfall in a single day caused a landslide in nearby Matatirtha that killed 16 people.

Sea-level rise will affect towns and cities in the LDCs particularly severely because a relatively large proportion of their populations live in the Low Elevation Coastal Zone – the continuous area along the coast lying less than 10 m above sea level. Already, coastal erosion has damaged infrastructure (including houses and roads) in Cotonou (Benin) and necessitated heavy investment in coastal protection in Dar es Salaam (Tanzania).

Yet even within these cities, exposure to risk is distributed unevenly. In Mombasa (Kenya), low-lying areas vulnerable to coastal flooding are inhabited by low-income groups, for example, in the coastal settlement of Tudor. In Khulna (Bangladesh), a mapping exercise showed substantial overlaps between slum settlements and areas that frequently suffer from water-logging (Fig. 1). And these spatial distributions are compounded by a variety of social phenomena: low-income groups are less able to move away from vulnerable sites; whilst the very young and very old are at greater risk from heat stress and vector-borne and water-spread diseases.

Tudor Settlement, Mombasa

Photo: David Dodman

CLACC research clearly demonstrated that building urban resilience in the LDCs is important because

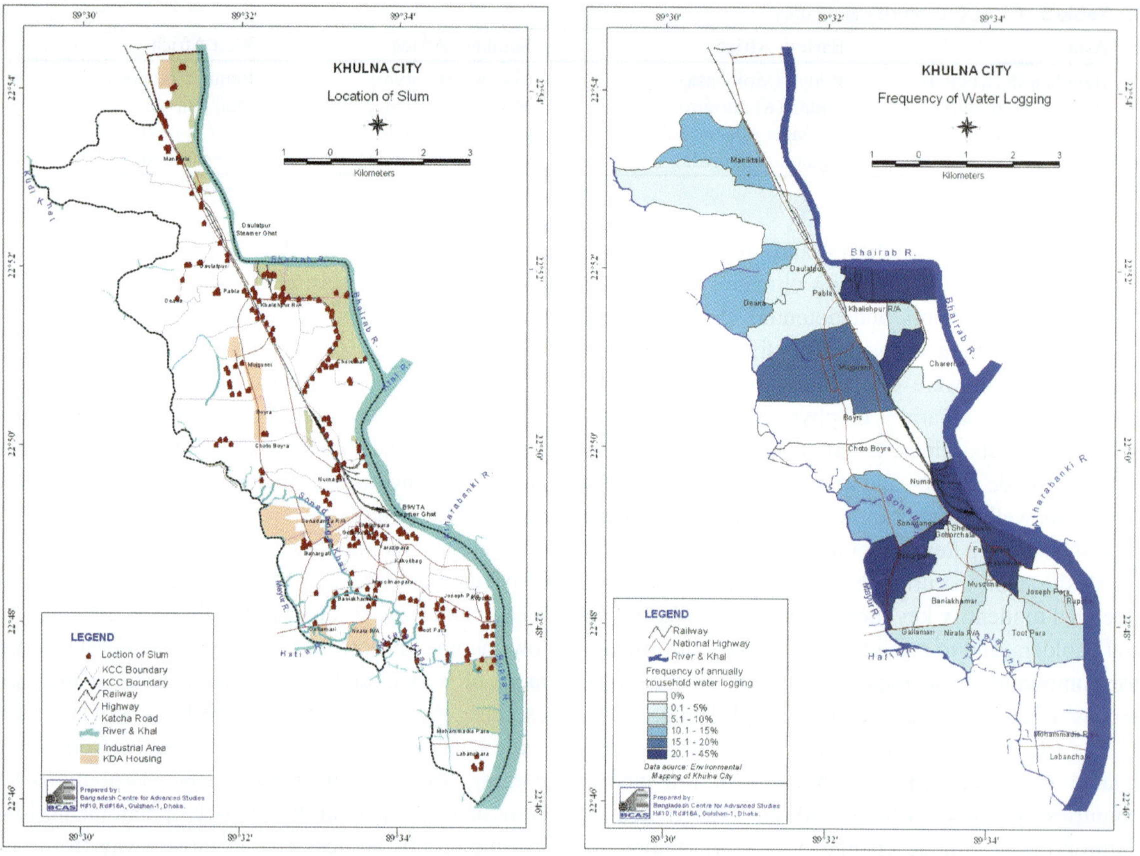

Fig. 1 Slums and Waterlogging in Khulna, Bangladesh. Source: Bangladesh Centre for Advanced Studies

of the vulnerability of large and growing urban populations to the hazards described above. But it is also important because of the potential economic costs without effective adaptation strategies: successful national economies depend on well-functioning and resilient urban centres. Building urban resilience will require not only improving urban infrastructure, but also creating more effective and pro-poor structures of governance and increasing the capacity of individuals and communities to address these new challenges.

Local and national governments have an important role to play: they influence the quality of provision of infrastructure for all areas within a city, the quality of provision for disaster-preparedness and response, and the extent to which low-income groups can acquire safe housing in safe sites. In addition, local governments strongly influence the environment within which local civil society action can take place. Community-Based Adaptation (CBA) is one such form of response and is based on the premise that local communities have the skills, experience, local knowledge and networks to undertake locally appropriate vulnerability reduction activities that increase resilience to a range of factors, including climate change.

This CLACC activity demonstrates the potential for locally based CBOs and NGOs to engage constructively in key development challenges facing their countries, and it shows how facilitating and supporting this kind of response is likely to be key to building resilience in low-income urban centres.

Managing and Evaluating CLACC

IIED manages the CLACC programme on behalf of the group and conducts the majority of fund-raising activities as well as intellectual and policy guidance. The regional partners play an important coordination and management role, particularly with the CLACC fellowships, overseeing the research and organizing the workshops. IIED is also supported by other organizations that provide guidance and leadership on certain issues, such as SEI and the LSHTM.

Funding for CLACC programme activities comes from a number of sources, namely the Royal Norwegian Ministry of Foreign Affairs, DFID, the Dexter Trust, the Embassy of the Federal Republic of Germany, the Norwegian Agency for Development Cooperation (NORAD) and the Swedish International Development Cooperation Agency (Sida).

It was decided in late 2006 that an independent ongoing monitoring and evaluation system should be put in place to follow and assess the progress being made. HELIO International, based in Paris, heads the evaluation process, and the next section of this chapter is based on results from HELIO evaluation reports. Activities involve evaluation of

1. programme management,
2. in-country outcomes as a result of CLACC activities,
3. individual CLACC fellows, and
4. results from CLACC programme meetings.

Evaluation of in-country outcomes as a result of CLACC activities is conducted using an 'in-country' monitor in each CLACC country. These monitors operate under the supervision of HELIO International and are completely independent from CLACC. The CLACC programme is evaluated in terms of its relevance, effectiveness, impact, efficiency and sustainability (Adger et al., 2005). These terms, and how the CLACC programme measures up to them, are explained in the sections below.

Relevance

Assessing whether a programme is relevant means examining the extent to which the programme objectives match the needs of the target groups, the policies of the partner countries and institutions, global development goals and the programme manager's orientation on basic development policy.

CLACC began well before climate change adaptation became a major consideration in climate debates. COP13 in 2007 saw a relative explosion of interest in adaptation, and as realization of the need to strengthen capacity in civil society and elsewhere to prepare for a 'ramping up' of adaptation activities due to the availability of large-scale funding grows (Huq, 2002; Beg et al., 2002), CLACC objectives are highly relevant. Civil society has an important role to play and CLACC, therefore, fills a key void. LDCs often have a well-developed civil society, but not one that is oriented towards climate change, and in particular adaptation. Broadening awareness and capacity more widely is also important. This includes affected sectors such as health, local government and communities. HELIO in-country monitors confirm that awareness levels in such sectors are still a major concern.

The NAPA process is also important (Ayers, 2008) and CLACC fellows are well placed to participate in and strengthen this process. NAPAs are also plans of action and many CLACC organizations are action-oriented. Following capacity building, CLACC partners now have a level of understanding that is highly relevant to the NAPA process.

Effectiveness

Evaluating effectiveness means assessing the extent to which a programme achieves its objectives and any additional direct results. Most of the CLACC work programme is directed towards achieving its three main objectives, and overall, the programme is achieving these objectives and its overall effectiveness is good. Many HELIO in-country monitors believe, however, that the fellows could be more effective. Milestones regarding what should be achieved and by when were not set for the objectives, but this is understandable since it is difficult to put boundaries around and quantify capacity strengthening of civil society.

Developing capacity at the national level starts with developing the capacity of individuals who will be key role players in the national context. CLACC has shown much success in this regard. Many of the CLACC fellows had no climate change experience prior to CLACC. Through the programme, however, the fellows have become confident and highly motivated and feel empowered by the knowledge they have gained. They are proud of the network they belong to, the international experience they have gained and the national and international recognition they are receiving. They are gaining visibility and respect in their countries and are looked upon, more and more, as authoritative voices on the subject. Most are not too senior but they have demonstrated the ability to be leaders amongst the next generation of climate change and adaptation experts.

CLACC fellow participation in the COPs has grown. At COP13, some made high caliber presentations, while others made first rate interventions from the floor in various side events. Many CLACC fellows were also active in the video presentations on community-based adaptation and three were also members of their country delegations. Most CLACC fellows belong to the adaptation group within CAN and the CLACC fellow from Senegal organized the West African CAN group. Involvement with CAN is important because it allows fellows to get involved in a global network of NGOs, to follow the climate change negotiations through the lens of this NGO community and to raise awareness of LDC perspectives amongst the NGO community. More experienced CLACC fellows have, over the years, mentored younger fellows new to the COPs.

The fellows are working with civil society in their own countries, but strengthening capacity through developing networks and establishing *Tiempo* focus groups will take time. The resource libraries are available and could be important for other NGOs, but to date, they have been considerably underutilized, often poorly publicized, and in the francophone countries, handicapped because almost all material is in English. The NGO networks and *Tiempo* focus groups need to play a stronger role. Understandably, this takes time, but the initial energy must be there. Some NGO groups have taken off and even produced their own publications,[5] but others have been slower to get off the ground. Several of the fellows have also been very active in the NAPA process, which is very positive.

The health research, for the most part, has been done well. LSHTM had concerns about the quality of some of the reports but the work has made a strong contribution to national and international discussions. National responses to the research were quite favourable, although more stakeholders could have been brought into the consultation process and there is potential for increasing awareness further. National level dissemination of the research needs improving in many countries as awareness of the major issues researched remains weak. The event at COP13 on the health studies was well attended and well received.

The capacity development of CLACC partner organizations is less obvious in some cases where climate change or adaptation is not a main organization objective. Some organizations felt that more emphasis should be placed on them and slightly less on the fellows themselves. The capacity of these institutions also varied according to region and country. West African francophone organizations, for example, are less strong on climate change than those in East and Southern Africa. A weaker command of English further inhibits these organizations from engaging in global processes and accessing information. Regional CLACC fellows successfully acted as mentors to help them complete their assignments and alleviate this problem.

Some feel that CLACC could be much more effective nationally if there was a longer-term perspective on strategy and activities. Despite this, the wide range of CLACC activities can be seen as a set of building blocks, each one contributing piece by piece to strengthening LDC capacity, first with the fellows and their organizations and then the country as a whole.

Impact

Evaluating impact means assessing the extent to which the programme helps achieve the desired overarching development objectives as well as other indirect development benefits. It is still relatively early to consider the impact of the CLACC programme, but some indications do exist.

In acknowledgement of their growing expertise, several CLACC fellows have been invited to participate in prestigious events and activities. Three have been included in their national delegations to the climate change conference, which is a strong recognition of their public stature and visibility. Understandably, this is not entirely due to CLACC, but the programme has certainly strengthened their positions.

CLACC has also been important for the NAPA process in some countries. For example, in the Sudan, the CLACC fellow actively participated in the

[5] For example, Joshi (2007).

development of the Sudanese NAPA. As a member of the Technical Committee, she was involved in the drafting process and providing basic guidance to sectoral working groups. In Bangladesh the CLACC fellow was seconded from his institution to work for the government on coordinating the NAPA process.

The CLACC newsletter and website help disseminate information and raise awareness about CLACC and climate change adaptation. Many CLACC fellows have been interviewed by the media, thus raising awareness about LDC issues more widely. It is, however, too early to say how much impact CLACC is having on raising the awareness of climate change and adaptation in LDCs and getting civil society mobilized to take action. The signs are positive but there are also signs that the full potential of CLACC for this has not been tapped.

Efficiency

Efficiency measures the relationship between the resources invested (funds, expertise, time, etc.) and the results achieved. The fellows receive a relatively small annual sum for quite an extensive work programme. For some fellows, CLACC activities fit neatly into their own work, but others are over-burdened as CLACC work is generally outside the mainstream activities of their organizations. In some cases, they must do it out of normal office hours. There is concern that the CLACC fellows are required to deliver more for CLACC than is realistic, given their work priorities. And it is uncertain whether all fellows realized what being a CLACC fellow entailed and how much effort it would require. CLACC is evolving and expanding. For those working in organizations where they can devote most of their time to CLACC, this is very good. But some fellows have many other responsibilities and this can be a problem. They are keen to get involved but juggling priorities is problematic. Because of this there must be a solid agreement between IIED and each partner organization. There is also a need for a common vision that goes beyond 1 year.

IIED has relatively good systems in place for CLACC management. Senior management at IIED is kept informed of progress, as are other non-CLACC partners. Communication can sometimes be problematic when working with LDCs, but this does not seem to have had a major impact on CLACC work. The regional fellows play an important support role for the CLACC fellows, as do organizations such as SEI Oxford and LSHTM.

There are often delays in many of the deliverables but they are generally not serious enough to hamper overall programme performance. The health research, for example, fell behind schedule, but draft reports were used for local consultations. Reports from the health studies were also slow to be available on the CLACC website.

Sustainability

Assessing sustainability means evaluating whether programme impacts are durable, and benefits are ensured beyond the end of the programme. The sustainability of capacity building is fundamental because LDCs need permanent capacity to articulate and address their own vulnerabilities and concerns. Some help from outside is inevitable, but the LDCs need to be much more aware of the implications of climate change on countries and to be able to mobilize resources both within governments and civil society to tackle them. This will take time.

There has been a relatively large turnover of CLACC fellows. This is to be expected, and is, in one sense, a positive outcome of the CLACC programme because CLACC fellow capacity has clearly been built to the extent that they become valuable recruits for other organizations searching for climate change expertise such as the European Community and IDRC, OXFAM and CARE. This can cause problems, however, for their parent CLACC organizations, which need to be seen as centres of adaptation knowledge.

The turnover of CLACC fellows has brought about a new category of alumni. These are former fellows who want to stay in regular contact with CLACC even though they have changed jobs. Such alumni can, and usually do, stay in touch with CLACC, attending CLACC meetings where possible and remaining as ambassadors for CLACC. The continuing interest of these alumni in CLACC, climate change and adaptation is an important aspect of sustainability.

Capacity building in the LDCs must be long-term in order to ensure sustainability. IIED has forged long-term links with RING and CLACC partner organizations over time. Choosing CLACC regional partners from the RING network in Zimbabwe and Kenya, which are not LDCs, confused some early observers of the CLACC programme, but this is typical of IIED's strong commitment to long-term working relationships with developing country partners as opposed to the shorter contractual/consultancy type relationships commonly characterizing the way that similar organizations work. CLACC is now in its seventh year and so it is also relatively long-lived for a development programme of its type. Funding has come from many sources to sustain this longevity.

The long-term commitment from IIED is matched by the willingness of CLACC partner organizations to continue with CLACC activities even when fellows leave the organizations or there are gaps in funding. All the CLACC partner organizations, without exception, have demonstrated strong loyalty and willingness to work as part of the CLACC team, and this has also contributed to the sustainability of the programme.

Unfortunately, because funding has been piecemeal and from numerous sources, the contracts with CLACC partner organizations have been renewed on an annual basis. This has meant that CLACC partner organizations have not been able to commit to the CLACC programme perhaps as strongly as they might have done if secure long-term financial commitments were in place. It does, however, serve to prevent complacency from settling in, and provides the CLACC partners with the opportunity to shape research and programme activities each year as these have not been set in stone.

What Next for CLACC?

The capacity of CLACC fellows to engage with climate change issues of national and international importance is now such that they are starting to share the knowledge and skills they have gained with others. CLACC activities during 2009 are therefore focusing on maintaining or scaling up the outreach activities described earlier, with particular focus on networking with other NGOs and community-based organizations in their country with an interest in climate change. Media outreach activities are continuing, and in some cases CLACC fellows are engaged in training activities to help raise awareness about climate change both within their own organizations and more broadly amongst other interested individuals and organizations in their country. Activities at COPs and other international climate change meetings will also continue, as will involvement in the national NAPA processes.

The workplan for post-2009 is currently under development. More than ever before CLACC fellows and their organizations are being given the opportunity to directly determine what climate change activities they will get involved with at the national level based on an assessment of national need, and institutional and individual priorities and capacity. IIED has committed to try and raise funds for these activities in addition to the regular CLACC activities. CLACC is, therefore, moving further away from a 'one size fits all' approach to its activities, and is becoming increasingly demand-led with tailor-made activities undertaken according to individual country circumstances.

Lessons Learned for Capacity-Building Programmes Elsewhere

The UNFCCC process has emphasized the need for capacity building on climate change, especially amongst developing countries. It has even compiled a list of capacity-building needs and activities in developing countries, based on national communications and submissions from Parties and organizations.[6] IIED has chosen to focus on building climate change capacity in civil society in selected LDCs, but lessons from its CLACC programme have broader relevance. While CLACC is unique, it could prove to be a good model for other capacity-building initiatives. There is no doubt that more awareness creation and capacity strengthening are needed in addressing global climate change.

Some of the key lessons learned from CLACC are listed below.

1. Long-term commitment to partnerships and trust between partner organizations is essential. The CLACC programme has been operating since 2003, but it is based on the RING, which was

[6] See http://unfccc.int/cooperation_and_support/capacity_building/items/4093.php

founded in 1991. Levels of trust between partner organizations are high.

2. Flexibility is important. Smaller partners need more time and assistance to fulfil tasks, and when fellows leave the replacements need time to catch up. In some cases, activities for certain CLACC fellows and organizations are altered to accommodate national, organizational and individual circumstances, thus building on existing strengths and making the work more relevant to the local context.
3. Bottom-up inputs into the programme of work are essential. Whilst IIED raises most funding, all partners agree the CLACC programme of work at the annual planning meeting. Attending the COP allows CLACC fellows to make their own contacts, spend time with each other and follow their particular specialties and interests.
4. Working as a team is important. Much of the CLACC work programme is devolved to programme partners. For example, BCAS manages the CLACC website and ACTS produces the CLACC newsletter.
5. Opportunism can be important. For example, NAPAs provided an opportunity to engage with governments at the national level after contacts had been made at the international level.
6. Capacity building is needed both within government, for example, through the NAPA process, and civil society, for example, through NGO networking.
7. Connecting local, national and global processes is important. For example, participation in the second and third international community-based adaptation conferences held in Bangladesh in 2007 and 2009 provided a local focus, the NAPA process was national, and attending COPs provides international experience. Climate change cuts across all these levels and it is important to understand the links between them.
8. Cross-sectoral work needs to be encouraged. Whilst some sectors, such as water management in Bangladesh or agriculture in Mali, are starting to consider climate change issues, most sectors in most LDCs have not even begun to incorporate climate change into their policies and planning.
9. Adopting a learning-by-doing approach is important given that knowledge on climate change adaptation, particularly in the LDCs, is still a relatively new and fast evolving arena.
10. Long-term funding is beneficial but not essential as long as there is trust amongst partners.

References

Adger WN, Huq S, Brown K, Conway D, Hulme M (2003) Adaptation to climate change in the developing world. Prog Dev Stud 3(3):179–195. DOI: 10.1191/1464993403ps060o.

Adger WN, Arnell NW, Tompkins EL (2005) Successful adaptation to climate change across scales. Glob Environ Change A 15(2):77–86.

Alam M (2007) The NAPA process. Tiempo Bull Clim Dev 65:3–6.

Ayers J (2008) Progress implementing NAPAs. Tiempo Bull Clim Dev 69:15–19.

Beg N, Morlot JC, Davidson O, Afrane-Okesse Y, Tyani L, Denton F, Sokona Y, Thomas JP, La Rovere EL, Parikh JK, Parikh K, Rahman AA (2002) Linkages between climate change and sustainable development. Clim Policy 2(2/3): 129–144.

Bicknell J, Dodman D, Satterthwaite D (Eds) (2009) Adapting Cities to Climate Change: Understanding and Addressing the Development Challenges. London: Earthscan.

Dodman D, Satterthwaite D (2008) Institutional capacity, climate change adaptation and the urban poor. Inst Dev Stud Bull 39(4):67–74.

Huq S (2002) The Bonn-Marrakech agreements on funding. Clim Policy 2(2002):243–246.

Huq S, Reid H, Konate M, Rahman A, Sokona Y, Crick F (2004) Mainstreaming adaptation to climate change in Least Developed Countries (LDCs). Clim Policy 4(1):25–43.

Huq S, Ayers J (2007) Critical List: The 100 Nations Most Vulnerable to Climate Change. Sustainable Development Opinion Paper. London: International Institute for Environment and Development (IIED).

IPCC (2001) Climate Change 2001: Impacts, Adaptation, and Vulnerability (p. 1005). Contribution of Working Group II to the third assessment report of the Intergovernmental Panel on Climate Change. Cambridge: Cambridge University Press.

IPCC (2007) Climate Change 2007: Impacts, Adaptation and Vulnerability. Contribution of Working Group II to the fourth assessment report of the Intergovernmental Panel on Climate Change. Geneva: IPCC.

Joshi GR (ed) (2007) Nepal Up in Smoke: Time for Action on Climate Change. Nepal: LI-BIRD.

Reid H, Simms A (2007) Up in Smoke? Asia and the Pacific. Fifth report of the Working Group on Climate Change and Development. London: New Economics Foundation.

Reid H, Sahlén L, Stage J, MacGregor J (2008) Climate change impacts on Namibia's natural resources and economy. Clim Policy 8(5):452–466.

Satterthwaite D, Huq S, Reid H, Pelling M, Romero-Lankao P (2007) Adapting to climate change in urban areas: The

possibilities and constraints in low- and middle-income nations. Human Settlements Discussion Paper Series. Climate change and Cities 1. London: IIED.
Simms A, Magrath J, Reid H (2004) Up in Smoke. Threats From, and Responses to, the Impact of Global Warming on Human Development. London: New Economics Foundation.
Simms A, Reid H (2005) Africa – Up in smoke? The second report from the Working Group on Climate Change and Development. London: New Economics Foundation.
Simms A, Reid H (2006) Up in smoke? Latin America and the Caribbean. The threat from Climate Change to the Environment and Human Development. The third report from the Working Group on Climate Change and Development. London: New Economics Foundation.

Climate Change Mitigation Policy: An Overview of Opportunities and Challenges

Timothy J. Foxon

Abstract This chapter provides an overview of the opportunities and challenges associated with climate change mitigation policy, in order to achieve the target of limiting global average temperature rise to 2°C above pre-industrial levels. Scenarios from the International Energy Agency and others show that a large number of mitigation 'wedges' from the large-scale deployment of a range of low-carbon technologies, including a range of renewable energy technologies, nuclear power and carbon capture and storage from coal- and gas-fired electricity generation, are likely to be needed. The Stern Review set out the economic case for climate change mitigation, arguing that the costs of mitigation are likely to be much smaller than the costs and risks of the impacts of climate change. Governments are now attempting to implement policies to deliver effective and efficient mitigation in three complementary areas: carbon pricing, through taxes or tradable permit schemes; increasing support for R&D, demonstration and early commercialisation of low-carbon technologies; and measures to overcome non-market barriers to the deployment of energy efficiency and low-carbon technologies. However, the rationales for these different types of mitigation policy come from different areas of economic theory, and it is argued that a more holistic framework may be needed to stimulate a transition to a low-carbon economy, drawing on new economic thinking, such as a 'green fiscal stimulus' or a global 'green new deal'.

T.J. Foxon (✉)
Sustainability Research Institute, Centre for Climate Change Economics and Policy, School of Earth and Environment, University of Leeds, Leeds LS2 9JT, UK
e-mail: t.j.foxon@leeds.ac.uk

Keywords Climate change mitigation policy · Carbon pricing · Low-carbon innovation · Green fiscal stimulus · Green new deal

Introduction

The increasing scientific evidence of the likely severe impacts and consequences of human-induced climate change is set out in the Fourth Assessment Report of the Intergovernmental Panel on Climate Change (IPCC, 2007), as well as in the other chapters of this book. However, the response by national and international policy-makers is struggling to address the global and inter-connected nature of the challenge of climate change mitigation, with (at the time of writing) hopes hanging in the balance of achieving a comprehensive and stringent agreement at the Copenhagen Climate Change Conference in December 2009. This chapter sets out the main technological, economic and policy responses so far, arguing that, though progress has been made, this has been constrained by the actions of players with vested interests in maintaining current systems, and by ideological commitments to free-market based solutions. The severe challenge of setting the world on a pathway to a low carbon transition, whilst enabling economic development in developing countries, suggests that more radical approaches may be needed to overcome these difficulties.

Climate Change Mitigation Options

In summary, rising atmospheric concentrations of carbon dioxide (CO_2) and other greenhouse gases (GHG)

J. Dodson (ed.), *Changing Climates, Earth Systems and Society*, International Year of Planet Earth,
DOI 10.1007/978-90-481-8716-4_12, © Springer Science+Business Media B.V. 2010

have resulted from human-induced emissions, particularly from the burning of fossil fuels (coal, oil and gas) for energy, transport and manufacturing uses, and from deforestation and other land use changes. These rising concentrations trap heat in the atmosphere and so alter the energy balance at the earth's surface, leading to rising temperatures, rising sea levels and other physical impacts. These changes lead to feedbacks affecting the energy balance, e.g. cloud cover, and the further emissions of greenhouse gases, such as methane from warming permafrost, that make exact predictions of future levels of warming difficult. Nevertheless, advanced climate systems models predict mean global temperature rises of 1.7–6.4°C by 2100 if past emissions trends continue. This is supported by evidence of a 0.7°C mean temperature rise over the last century, and discernible human influences on temperature increases, melting polar ice and glaciers, and increased frequency of extreme weather events (IPCC, 2007).

The challenge of mitigating the impacts of climate change thus focuses on efforts to reduce human emissions of GHGs from fossil fuel burning, deforestation and land use change. In this chapter, we focus on reducing emissions from activities that have largely relied on fossil fuel burning. This is so challenging because these activities have contributed to the large increase in human wellbeing in industrialised countries whilst, because of increasing returns to the adoption of technologies and associated institutional rule systems, human societies are now 'locked-in' to these high carbon systems (Unruh, 2000). To avoid the worst impacts of climate change whilst maintaining and enhancing human wellbeing requires a transition to a low carbon development path.

A low carbon path will require the adoption of a range of energy efficiency improvements in delivery of energy services for households, businesses and transport, and the further development and adoption of a range of low carbon energy supply technologies. These are likely to include a number of renewable sources of electricity and heat generation, such as wind power, solar photovoltaics, solar thermal electric power, sustainably sourced biomass, wave and tidal power, as well as next generation nuclear power, and carbon capture and storage (CCS) from coal and gas-fired electricity generation plants. Rather than a single 'silver bullet' technological solution, a large number of mitigation 'wedges' from the large-scale deployment of many or all of these technologies are likely to be needed to achieve significant levels of emissions reductions by 2050 (Pacala and Socolow, 2004). The International Energy Agency has examined scenarios for reducing global CO_2 emissions by 50% to below 20 Gt CO_2 by 2050, identifying the wedges needed relative to a 'business-as-usual' emissions trajectory (IEA, 2008). They confirm that significant mitigation wedges would be needed from the power generation, transport, industry and buildings sectors, each of which would require contributions from a number of low-carbon technology options (see Fig. 1). The implementation of many of these options will give rise to huge technical and political challenges, with advocates both in favour and against particular technological solutions (e.g. Romm, 2006; Mackay, 2009; Giddens, 2009).

Economic Case for Climate Change Mitigation

To inform action on climate change, in 2006, the UK Government commissioned Sir Nicholas Stern, former Chief Economist at the World Bank, to review the economics of climate change. The resulting Stern Review (Stern, 2007) laid out the economic case for government action, arguing that climate change represents the 'greatest market failure the world has ever seen'.

However, as Stern recognised, the economic case on its own is unlikely to stimulate action to mitigate climate change, as there are social and ethical issues that also need to be addressed. Climate change results from the *externality* associated with greenhouse gas (GHG) emissions, i.e. costs that are not paid by those who create the emissions. It has a number of features that distinguish it from other social and environmental problems:

- Mitigating climate change is a global public good (i.e. the benefits of mitigation accrue to everyone on the planet and can not be bought by any individual at the expense of others);
- Impacts are long-term and persistent;
- Uncertainties and risks are pervasive; and
- Risk of major, irreversible change with non-marginal economic effects.

Hence, questions arise of equitable sharing of the responsibility for past emissions and the costs of

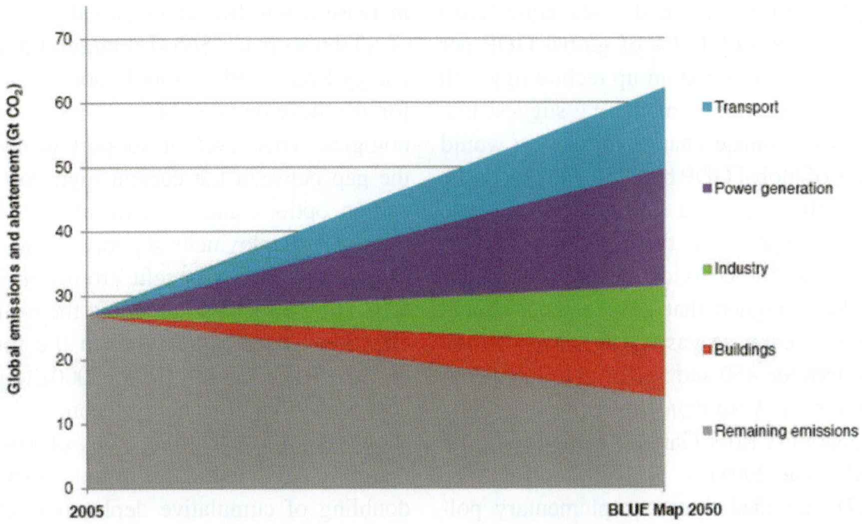

Fig. 1 Mitigation wedges for IEA 'Blue Map' mitigation scenario (based on IEA, 2008)

achieving a low-carbon transition, both between richer and poorer countries and between present and future generations. The above features also imply that some standard tools of economic analysis, such as cost-benefit analysis, are limited in their usefulness, since they assume only marginal changes. The issue of the appropriate rate for discounting future costs and benefits also proved to be controversial, with Stern and colleagues arguing that standard treatments of discounting are inappropriate for comparing potential future mitigation pathways (Dietz et al., 2007).

Whilst not undertaking a single global cost-benefit analysis to calculate the optimal level of climate change mitigation, Stern used a number of different economic tools to perform separate calculation of costs and benefits. This approach has been criticised both by neo-classical economists for a lack of rigour (Nordhaus, 2008) and by ecological economists for still retaining a number of debatable assumptions (Spash, 2007). To calculate the economic costs of climate change impacts, Stern used simple 'integrated assessment models', giving equal weight to impacts in poorer countries and on future generations. These models include impacts for which there is a 'market' value, such as agriculture and food supplies, due to changes in crop patterns, energy use, due to additional cooling requirements, and coastal zones, e.g. impacts on fisheries; 'non-market' impacts, including impacts on human health, e.g. increased prevalence of diseases and impacts on natural ecosystems; and 'system change' impacts, such as higher levels of conflict and migration. Stern argued that the likely annual social and economics costs of the impacts of climate change would be in the range 5–20% of global GDP, now and forever.

To calculate the economic costs of climate change mitigation, Stern used both 'bottom-up' and 'top-down' models. Bottom-up economic analyses incorporate a range of low-carbon technological options, and assume that the costs of these options fall with their implementation as a result of learning effects and economies of scale (IEA, 2000). They then calculate the costs of a low-carbon pathway, compared to the costs of a 'business-as-usual' pathway, with little or no consideration of macroeconomic factors. Bottom-up modelling by Professor Dennis Anderson for the Stern Review calculated that to reduce global emissions by 33% to 18 $GtCO_2e$ by 2050, the addition annual cost of following this low-carbon pathway would rise from \$134bn in 2015 to \$930bn in 2050. Assuming continuing high levels of economic growth over this period, this would imply that the costs of mitigation would be equal to 1% of global GDP by 2050. Top-down economic analysis uses macroeconomic models of the global economy, with a relatively small number of regions and economic sectors. These models incorporate the implications of changes in investment patterns on wider economic activity and most, but not all, models assume a 'general equilibrium' framework. They calculate that stabilisation of GHG concentrations at 450–550 ppm CO_2e (carbon dioxide equivalent) implies mitigation costs of 1–2% of global GDP per year by 2050. Hence, both bottom-up technology-rich and top-down macroeconomic modelling suggest that the annual costs of climate change mitigation would be around 1–2% of global GDP by 2050. On this basis, Stern concluded that there is a strong economic case for undertaking mitigation, as the costs are likely to be much lower than the costs of the impacts of climate change. Stern argued that governments should aim to stabilise greenhouse gas (GHG) levels in the atmosphere at between 450 and 550 ppm CO_2 equivalent. A review for the Australian Government in 2007 by eminent economist Ross Garnaut came to similar conclusions (Garnaut, 2008).

Stern (2007) identified three complementary policy areas as necessary to deliver timely, effective and economically efficient climate change mitigation:

- carbon pricing, through taxes or tradable permit schemes;
- increasing support for R&D, demonstration projects and early stage commercialisation of clean technologies; and
- measures to overcome institutional and other non-market barriers to deployment of energy efficiency and low carbon measures.

Based on the scale of low carbon R&D and deployment needed, Stern recommended that deployment incentives for low-emission technologies should increase two to five times globally from current levels of $33bn to reach $65–150bn, and that global public energy R&D funding should double, to around $20bn, for the development of a diverse portfolio of technologies. This level of support is needed to bridge the gap between the current high costs of many low carbon options and the current high carbon alternatives. The deployment support would enable the low carbon options to benefit from learning, scale and adaptation effects, so reducing their unit costs. These effects are usually analysed in the form of learning or experience curves (IEA, 2000). Analysis of past cost reductions for energy technologies has typically shown empirical learning rates of 10–25%, meaning that a 10–25% reduction in unit costs results from a doubling of cumulative deployment (McDonald and Schrattenholzer, 2001). Hence, as shown in Fig. 2, early deployment support would be expected to reduce the cost of low carbon options, so that they would become cost competitive with current technologies, under the general support of a carbon price, provided by a carbon tax or trading scheme.

Implementation of Climate Policy Measures

The first internationally agreed targets for climate change mitigation were those set by the 1997 Kyoto Protocol to the United Nations Framework Convention

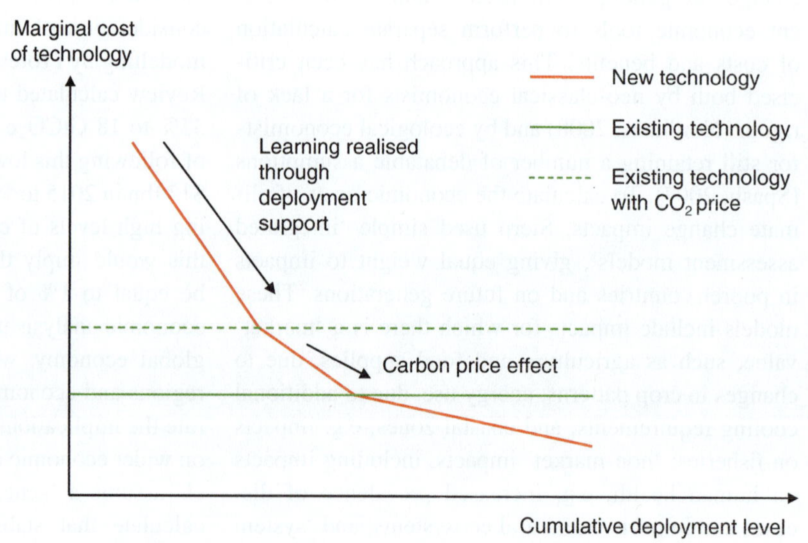

Fig. 2 Interaction between carbon pricing and deployment support (based on Stern, 2007)

on Climate Change (UNFCCC). The Kyoto Protocol set targets for industrialised countries only to achieve an average 5% reduction in GHG emissions by 2008–2012, relative to 1990 levels. The Protocol came into force in 2005, with ratification by the Russian government, following that of all other industrialised countries, except the USA and Australia.[1] Some countries, including the UK and Germany, are on track to meet their Kyoto targets, at least partly due to the side effects of non-climate-related policies, such as the dash for gas-fired electricity generation in the UK following liberalisation of electricity markets. However, many other countries, including Canada and Spain, are highly unlikely to achieve their Kyoto targets, and, overall, worldwide GHG emissions grew by 38% between 1992 and 2007.

The Kyoto Protocol also provided for carbon emissions trading between governments and by firms, designed to stimulate the lowest-cost emissions reductions which could then be traded within overall caps. The European Emissions Trading Scheme, which began operation in 2005, followed this cap-and-trade approach in order to help the EU meet its target of 8% reduction in emissions by 2008–2012 from 1990 levels. This covers emissions from power generation and energy-intensive industries totalling around 40% of total EU emissions. The Waxman-Markey Bill, which was passed by the US House of Representatives in June 2009, but (at the time of writing) had yet to be agreed by the US Senate, would set up a similar cap-and-trade for the US to achieve emissions reductions of 17% by 2020 and 83% by 2050, relative to 2005 levels.

The policy areas identified by Stern are beginning to be reflected in policies of the European Union member states. These are seen as forming part of a 'new, green industrial revolution' (Miliband, 2007; Barroso, 2007). Such calls have recently been echoed by the new US Secretary of Energy (Chu, 2009). In December 2008, the European Council of Ministers agreed on an Energy and Climate Policy package, both for domestic action and as a basis for negotiation at the Copenhagen meeting. This package aims to address environmental targets, whilst, at the same time, contributing to ensuring security of energy supply for EU countries. The main aims of this package are to achieve by 2020:

- a 20% reduction in carbon emissions, with a promise of a 30% reduction in carbon emissions by 2020, if there is an international agreement at the COP15 meeting in Copenhagen;
- 20% of final energy from renewables; and
- a 20% improvement in energy efficiency.

In July 2009, the Major Economies Forum, including China, India, Brazil and the G8 countries, recognised that, to avoid the most serious impacts, the increase in global average temperature above pre-industrial levels ought not to exceed 2°C. This is likely to require atmospheric concentrations of greenhouse gases to stabilise below 400 ppm CO_2 equiv (Meinshausen, 2006; Anderson and Bows, 2008). Even for a very small chance of exceeding 4°C rise by 2100, global emissions need to peak by 2016 and then reduce by around 3% per year to 2100 (Committee on Climate Change, 2008). This implies global GHG emissions will need to be reduced by 50% by 2050, from current 40 $GtCO_2e$ to 20–24 $GtCO_2e$ by 2050. Assuming fair allocation of these emissions amongst 9 billion people by 2050, this implies a share of 2.1–2.6 tCO_2e per person. Hence, if these emissions are shared equitably between countries, this implies that at least 80% reductions are required by industrialised countries. Emissions reductions of these orders of magnitude will require dramatic transformation of systems of production and consumption, involving different possible transition pathways to low carbon power generation, transport and energy use systems (Foxon et al., 2010). The UK government has committed itself to a legally binding target of reducing its GHG emissions by 80% by 2050, from 1990 levels, with a new institution, an expert-led Committee on Climate Change to recommend 5-yearly budgets to put the UK on track to achieving this target (Committee on Climate Change, 2008, 2009). Other members of the G20 group of major economies, including the US, are considering bringing in similar legally binding targets.

The main policy instruments to achieve these targets being applied in European and other industrialised countries are Emissions Trading Schemes (ETS), and R&D and price-support measures for deployment of new renewable and other low-carbon energy technologies. As well as stimulating the deployment of existing low-carbon technologies and processes, these measures are intended to promote innovation and rapid take-up of low-carbon alternatives, such as wind

[1] Australia later ratified the protocol in 2007.

power, solar photovoltaics, nuclear power, and carbon capture and storage (CCS), which are currently more expensive than the dominant alternatives of using coal and gas for electricity generation. Complementary measures to improve the building stock, in order to reduce the heat demand by homes and businesses, have received more attention in some countries, such as Germany, than others, such as the UK.

Critique of Current Policy Approaches

Unfortunately, as described above, current policy measures have so far had relatively little success in stimulating emissions reductions. Indeed, it is notable that the 2008–2009 economic recession has led to more rapid, though temporary, emissions reductions (Bowen et al., 2009b). We argue that this lack of success is at least partially the result of the theoretical framing of current climate policies, which draw on a range of theoretical approaches. This leads to potential inconsistencies and arguably forms an inadequate basis for addressing long-term environmental problems, in which actors need to make decisions in the face of high levels of risk and uncertainty, both in relation to outcomes of current actions and the potential for the development of alternatives. Table 1 outlines the main policy measures and their underlying framing.

As can be seen from the Table, these measures focus largely on the important issue of reducing emissions from energy supply and energy-intensive industries, which are within the scope of the EU Emissions Trading Scheme. There is relatively little focus on wider systems of production and consumption, to stimulate a more general greening of these systems by creating space and incentives for eco-innovators and greening of markets. Whilst markets in tradable carbon permits are likely to have an important role to play, it may be argued that the wider greening of markets and other incentives to promote wide and deep low-carbon innovation are more important to achieving a long-run transition to a low carbon economy (Andersen and Foxon, 2009).

The rationale for carbon pricing comes from environmental economics. Here, a 'market failure' is identified in relation to the existence of negative *environmental externalities,* i.e. environmental by-products of consuming or producing activities that affect third parties but are not reflected in market transactions and prices (Pigou, 1932). In this case, the emissions of CO_2 and other greenhouse gases from energy production and other industrial activities has historically been the unpriced externality. Arguably the simplest way to price carbon emissions would be to impose a carbon tax on activities leading to emissions. However, taxes tend to be politically highly unpopular and there are concerns about the threat to international competitiveness of industries faced with taxes, especially if there is the potential for firms to relocate production to countries without a carbon tax (so-called 'carbon leakage'). Efforts to introduce a European carbon tax in the early 1990s failed to gather enough political support to be enacted.

The alternative pricing mechanism is to impose an emissions trading scheme. Following the success of an emissions trading scheme in the US to reduce emissions of sulphur dioxide from coal-fired power stations and other industrial firms, the legal basis for emissions trading was agreed in the 1997 Kyoto

Table 1 Framing of current climate change mitigation policies

Policy area	Theoretical framing	Example policy	Example target
Carbon pricing	Neo-classical economic theory	EU Emissions Trading Scheme	21% reduction in EU ETS sector emissions by 2020 (compared to 2005)
Support for R&D, demonstration projects and early commercialisation of clean technologies	(Neo-classical) innovation theory/(evolutionary) innovation systems theory	Feed-in tariffs for renewable energy technologies in many EU countries	20% share of final energy from renewables by 2020
Overcoming institutional and non-market barriers to deployment	Institutional and behavioural economics	Fiscal, regulatory or information incentives to take-up (cost effective) energy efficiency improvements	20% reduction in energy consumption by 2020

Protocol, which set carbon emissions reduction targets for industrialised countries. The European Union subsequently agreed a carbon Emissions Trading Scheme (ETS), which began with a first period in 2005–2007, to enable learning, followed by a second period covering 2008–2012, to coincide with the commitment period under the Kyoto Protocol. As part of the Climate Policy package agreed in December 2008, the EU agreed that the third phase of the ETS would run from 2013 to 2020. The ETS is a 'cap-and-trade' system, setting an overall emissions cap for firms included in the scheme, with permits tradable between firms that are likely to exceed their allocation and firms that have spare credits. The economic rationale is that this enables the system to find the least-cost reduction opportunities. The political and competitiveness concerns were also eased by having a high proportion of free allocation of permits to firms. However, the environmental effectiveness of the scheme is determined by the level of the overall cap. In Phase One of the EU ETS, the cap set by the sum of the national allocations was too high to require emissions reductions much beyond business-as-usual, leading to a collapse in the price of permits in early 2007. The free allocation of permits also led to 'windfall' profits for electricity companies, who were given permits for free, but received the benefits of the permit prices being included in electricity costs to consumers. For Phase Two, the European Commission required countries to impose stronger caps, though the permit price began to fall again in late 2008, as a result of reductions in energy demand due to the economic recession. In Phase Three, the allocation of permits will increasingly move from free allocation to auctioning, though one of the concessions to the East European coal industry was that it should continue to receive a proportion of free allowances. Though the new agreement for Phase Three from 2013 to 2020 gives a certain level of certainty to firms that there will be a carbon price in this period, no mechanism was agreed for setting a floor or ceiling to the carbon price. This uncertainty in the level of the carbon price means that, on its own, it is unlikely to stimulate significant levels of investment leading to innovation in low-carbon technologies and processes. Thus, a carbon price is a necessary but not sufficient driver of low-carbon innovation (since without a carbon price, the economic 'benefit' of unpriced emissions to firms would be likely to override any positive incentives for low-carbon innovation). The plan to introduce a similar carbon trading scheme in the US under the Waxman-Markey Bill looks set to reproduce many of the same advantages and drawbacks of the EU scheme.

The rationale for increasing support for R&D, demonstration projects and early stage commercialisation of clean technologies comes from innovation theory (Foxon, 2003). The economic rationale is that since new knowledge is often easy to copy, innovators cannot always appropriate the full benefits of their investment in knowledge creation, and so private firms may lack the incentives necessary to undertake socially efficient levels of innovative activity. In addition, historical evidence shows that the costs of new technologies typically reduce along learning curves as they are introduced into the market.

The rationale for measures to overcome institutional and other non-market barriers to deployment comes mainly from institutional and behavioural economics. It is observed that firms and consumers do not act as purely rational economic agents, but their behaviour is influenced by the social and institutional context in which they act. The factors that prevent purely rational behaviour are often referred to as barriers, but they reflect these more complex drivers of behavioural change. For example, it is observed that many energy efficiency opportunities, such as installing wall or loft insulation, are not taken up, despite the fact that the initial capital costs would be quickly paid back by reduced energy bills, implying that they would be taken up by economically rational actors. In this case, the barriers could relate to the persistence of individuals' 'habits' preventing change, such as the behavioural predisposition to consider capital and running costs separately (Marechal, 2009), or to the fact that culturally embedded patterns of behaviour are slow to change (Nye et al., 2010). Similarly, firms may not invest in potentially economic low-carbon innovation opportunities because these conflict with existing routines that firms follow based on their historical experiences (Unruh, 2000).

Thus, it is noticeable that the rationales for the different types of mitigation policy instrument come from different areas of economic theory. In particular, there is an absence of a holistic framework for understanding how these different areas could come together to achieve a transition to a low-carbon economy. Of course, it could be argued that such a piecemeal approach is the only pragmatic possibility. However, the relatively slow pace of mitigation achieved so far, and the difficulties in reaching agreement at the 2009 Copenhagen Climate Conference on anything like the emissions reduction levels that the climate science says would be necessary to limit global temperature rises to the 2°C target, suggest that a more radical re-framing of the problem may be necessary.

New Economic Thinking for Climate Change Mitigation

Whilst most observers agree that some mechanism to price carbon emissions, either through a carbon tax or a cap-and-trade emissions scheme, is a necessary component of a climate change mitigation policy package, some have argued recently that more radical measures are likely to be needed to stimulate a global transition to a low-carbon economy. This would require global GHG emissions to peak within the next 10 years and then reduce by around 4% per annum. To enable economic development in developing countries, these countries would be required to decouple their rate of emissions increase from continued economic growth, so that rich and poor countries would equalise their emissions at around 2 tonnes of CO_2 equivalent per person by 2050. Debates centre on whether this can be achieved by a strengthening of the three existing types of policy instrument, or whether a more radical re-framing is necessary to achieve a more coherent policy mix. Amongst the approaches being discussed are a 'green fiscal stimulus' or 'green new deal', a focus on reducing emissions upstream at the production end rather than downstream at the consumption end, and challenging the accepted economic growth paradigm.

The 2008–2009 global economic crisis has created additional difficulties in moving towards a low carbon economy, particularly in relation to whether the levels of private investment funding needed will be available. However, it has also been argued by some that this represents an opportunity for simultaneously addressing economic and environmental concerns. Most industrialised and rapidly developing countries have adopted a public fiscal stimulus package, parts of which are focussed on investment in green technologies and infrastructure. For example, South Korea is focussing around 80% of its overall fiscal stimulus on green technology and manufacturing, China is investing heavily in the installation of large wind farms and solar photovoltaics, and the US is supporting the renewal of its electricity grid and moves towards a 'smart grid' that would enable more intelligent management of demand and integration of intermittent renewable energy sources. However, these and other countries are also investing in support of old, high-carbon industries, such as car manufacturing, without necessarily requiring firms to move more rapidly towards developing low carbon vehicles. Nicholas Stern and colleagues proposed that a 'green fiscal stimulus' (Bowen et al., 2009a) of the order of 0.8% of global GDP, or $400 billion of extra public spending worldwide on 'green' measures over the next 2 years would be appropriate.

Others have suggested that a green fiscal stimulus needs to be complemented by a wider range of institutional and regulatory changes to promote a more rapid low carbon transition. This is referred to as a 'Green New Deal', after President Roosevelt's New Deal in the 1930s, which created millions of jobs and helped the US to recover from the Great Depression. As part of its Green Economy Initiative, UNEP (the United Nations Environment Programme), in collaboration with a wide range of international partners and experts, is examining the conditions and requirements for a 'Global Green New Deal' (UNEP, 2009). This builds on a report that it commissioned from the respected environmental economist Edward Barbier which sets out the economic case for action (Barbier, 2009). The three broad objectives proposed in the March 2009 UNEP policy brief are as follows:

(1) make a major contribution to reviving the world economy, saving and creating jobs, and protecting vulnerable groups;
(2) reduce carbon dependency and ecosystem degradation, putting economies on a path to clean and stable development; and
(3) further sustainable and inclusive economic growth, achievement of the Millenium Development Goals, and end extreme poverty by 2015.

UNEP argues that this will require co-ordinated government action in three areas:

(a) a 'green' fiscal stimulus of the order of 1% of global GDP ($750 billion) over the next 2 years, or around a quarter of the total size of the fiscal stimulus packages currently proposed by the G20 countries;

(b) domestic policy reforms to enable the success of green investment within domestic economies; and
(c) reforms to international policy architecture and international co-ordination to enable and support national initiatives.

The 'green' stimulus would cover investment in energy efficiency of buildings, greener vehicles and transport infrastructure, renewable energy projects, 'smarter' electricity grids, and more sustainable agriculture and freshwater systems. A range of domestic policy interventions would aim to ensure a 'level playing field' to enable the investments in green sectors to take hold and flourish as commercially viable businesses. Reforms to the international policy architecture would aim to provide the framework for a transition towards a more sustainable economic system, including action in the areas of international trade, international aid, a global carbon market, global markets for ecosystem services, development and transfer of technology, and further international co-ordination to enable the participation of both industrialised and developing countries in the global Green New Deal initiative.

In July 2008, a group of leading UK economists and environmentalists independently proposed a 'Green New Deal' to tackle the financial, energy and climate crises (Green New Deal Group, 2008). Their programme aims to combine stabilisation in the short term with longer term restructuring of the financial, taxation and energy systems. They set out an even more radical programme of action, including the following:

- executing a bold new vision for a low-carbon energy system making 'every building a power station' through a $80 billion programme of investment in energy efficiency and local renewable electricity generation;
- creating and training a 'carbon army' of workers to create the human resources for a vast environmental reconstruction programme;
- ensuring that fossil fuel prices are high enough to create the economic incentive to drive efficiency and bring alternative fuels to the market, and establishment of an Oil Legacy Fund, paid for by a windfall tax on the profits of oil and gas companies;
- minimising corporate tax evasion by clamping down on tax havens and corporate financial reporting;
- re-regulating the domestic financial system to ensure the creation of money at low rates of interest, combined with tighter controls on lending and the generation of credit; and
- the breaking-up of large banks and other financial institutions seen as being 'too big to be allowed to fail' in the current economic crisis.

These types of reforms would obviously be opposed by those firms and institutions that perceived them as a threat to their strategic position or interests. Some observers have expressed concerns that, with the immediate financial crisis having been averted by government and central bank actions, such as higher public deficits and the creation of additional money supply through 'quantitative easing', the incentives for governments to undertake these types of wider institutional and regulatory reforms has weakened. The likelihood of their adoption will depend on continued public pressure and on how the state of the global and national economies evolve over the coming years.

Finally, the question has been raised of whether a sustainable, low-carbon economy is compatible with current patterns of ever-increasing material consumption and the focus on achieving economic growth as the primary policy objective in industrialised countries (Victor, 2008; Jackson, 2009). The recent report of the Stiglitz Commission to the President of France, chaired by Economics Nobel Laureate Joseph Stiglitz, noted the evidence that, after basic needs have been met, further increases in consumption do not bring systematic improvement in people's reported happiness (Stiglitz et al., 2009). The Stiglitz Commission argued that policy in industrialised countries should focus more on supporting the achievement of desired goals such as high levels of employment, reducing social inequalities and personal wellbeing through fulfilling social interactions, using a wider range of indicators than just GDP growth.

Towards Achieving Global Climate Change Mitigation

This chapter has examined the opportunities and challenges of moving to a low carbon economy to mitigate the severe threat posed by human-induced climate change. Whilst there is some agreement on the outline of the steps needed – to put a price on carbon emissions; to promote innovation and deployment of low carbon technologies; and to overcome institutional and non-market barriers to adoption of energy efficiency and low carbon measures – there is much less agreement on the technical and economic feasibility and political acceptability of these steps.

The mainstream political position has been to work in incremental steps to implement changes that will gradually reorientate economic decisions of firms and individuals in low carbon directions, whilst setting stronger long-term emissions reduction targets. Thus, the broad outlines of the proposed deal at the UN Climate Change Conference in Copenhagen in December 2009 are expected to be

(1) goals for reductions of GHG emissions by industrialised countries in absolute terms for 2020 and 2050;
(2) goals for reduction in GHG emissions by developing countries relative to their expected increases in GDP, for 2020 and 2050;
(3) funds provided by industrialised countries for transfer of low carbon technologies to developing countries;
(4) funds provided by industrialised and rapidly developing countries for adaptation to impacts of climate change in poorer countries; and
(5) regulatory and/or financial incentives for avoiding deforestation in developing countries.

The details, though, of the levels of emissions reduction commitments by individual countries and of the financial transfers between industrialised and developing countries to promote mitigation and adaptation are highly contentious.

Among the main instruments for achieving these targets are expected to be increases in the scope and coverage of national and international carbon markets, such as the European Emissions Trading Scheme, and further regulatory and financial incentives for the innovation and deployment of low-carbon technologies, such as renewable energy sources, electric vehicles and carbon capture and storage (CCS) of emissions from coal- and gas-fired electricity generation. However, as we have seen, the initial implementation of carbon markets in the European Union has not been without problems. Whilst some argue that these initial difficulties will be overcome as more experience with carbon markets is gained and stronger caps are imposed in future phases, others argue that this type of 'downstream' trading system is not likely to be the most effective mechanism for achieving high levels of emissions reductions. This is because there are a very large number of sources covered by this type of scheme. Hence an 'upstream' carbon trading system, based on a smaller number of sources at or closer to primary energy production, such as oil refining and electricity generation has been argued for, as part of a more streamlined policy approach (Tickell, 2008).

The scale of the transformation of systems of production and consumption to achieve a transition to low carbon economies at national and global levels has led some to argue that more radical approaches are needed. Proponents of a 'green new deal' have argued that both a larger green economic stimulus and significant institutional and regulatory changes are needed to address the inter-locking challenges of climate change, ecosystem degradation and economic credit crunch. Recently, the question has been raised of how compatible a sustainable, low carbon economy can be with current patterns of ever-increasing material consumption and focus on economic growth as the prime aim of policy. It is important that all these ideas are subject to public debate and scrutiny, as whether a low carbon future is possible depends on people's willingness to accept significant changes to current socio-economic systems and support for an alternative vision of a more equitable and maybe less materialistic world. In any case, it is clear that high levels of political will, technological innovation, institutional change, business leadership and citizen engagement will be needed to deliver and successfully implement the current and further global agreements to put the world on a pathway to a sustainable and prosperous low-carbon future.

Acknowledgements The author is a Research Councils UK Academic Fellow in the Sustainability Research Institute, University of Leeds and a member of the ESRC Centre for Climate Change Economics and Policy. He would like to thank John Dodson, Andy Gouldson and Elisabeth Simelton for helpful comments on an earlier draft.

References

Anderson K, Bows A (2008) Reframing the climate change challenge in light of post-2000 emissions trends. Philos Trans R Soc A 366:3863–3882.

Andersen MM, Foxon TJ (2009) The greening of innovation systems for eco-innovation – Towards an evolutionary climate mitigation policy. Paper presented at DRUID summer conference on Innovation, Strategy and Knowledge. Denmark: Copenhagen Business School. http://www2.druid.dk/conferences/papers.php?cf=32

Barbier E (2009) A global green new deal: Final report. Report prepared for UNEP. http://www.unep.org/greeneconomy/

Barroso JM (2007) 'Europe's energy policy and the third industrial revolution', speech by President of the European Commission at Loyola de Palacio energy conference, Madrid, 1 October 2007

Bowen A, Fankhauser S, Stern N, Zenghelis D (2009a) An outline of the case for a 'green' stimulus. Policy Brief, Centre for Climate Change Economics and Policy, LSE. Available at http://www.cccep.ac.uk/pdf/AnOutlineOfTheCaseForAGreenStimulus.pdf

Bowen A, Forster PM, Gouldson A, Hubacek K, Martin R, O'Neill D, Rap A, Rydge J (2009b) The implications of the economic slowdown for greenhouse gas emissions and targets. Centre for climate change economics and policy (CCCEP). Working Paper No. 11, London School of Economics and University of Leeds.

Chu S (2009) Secretary Chu's Keynote Address at the EIA 2009 Energy Conference. Available at http://www.eia.doe.gov/conference/2009/plenary/chu_transcript.pdf

Committee on Climate Change (2008) Meeting carbon budgets – The need for a step change, 1st Annual Report to Parliament, October 2009. http://www.theccc.org.uk/reports

Committee on Climate Change (2009) Building a low carbon economy – The UK's contribution to tackling climate change, 1st Report, December 2008. http://www.theccc.org.uk/reports

Dietz S, Anderson D, Stern N, Taylor C, Zenghelis D (2007) Right for the right reasons: A final rejoinder on the Stern Review. World Econ 8(2):229–258.

Foxon TJ (2003) Inducing Innovation for a Low Carbon Future: Drivers, Barriers and Policies. London: The Carbon Trust. Also available at http://www.carbontrust.co.uk/Publications/publicationdetail.htm?productid=CT-2003-07&metaNoCache=1

Foxon TJ, Hammond GP, Pearson PJ (2010) Developing transition pathways for a low carbon electricity system in the UK. Technol Forecast Soc Change (in press).

Garnaut R (2008) The Garnaut Climate Change Review. Cambridge: Cambridge University Press.

Giddens A (2009) The Politics of Climate Change. London: Polity Press.

Green New Deal Group (2008) A Green New Deal. London: New Economics Foundation. http://www.greennewdealgroup.org/

Intergovernmental Panel on Climate Change (IPCC) (2007) Climate Change 2007: Synthesis Report, Available at http://www.ipcc.ch/pdf/assessment-report/ar4/syr/ar4_syr.pdf

International Energy Agency (IEA) (2000) Experience Curves for Energy Technology Policy. Paris: OECD/IEA.

International Energy Agency (IEA) (2008) Energy Technology Perspective 2008. Paris: OECD/IEA.

Jackson T (2009) Prosperity Without Growth: Economics for a Finite Planet. London: Earthscan.

Mackay D (2009) Sustainable Energy – Without the Hot Air. Cambridge: UIT Press.

Marechal K (2009) An evolutionary perspective on the economics of energy consumption: The crucial role of habits. J Econ Issues XLIII(1):69–88.

McDonald A, Schrattenholzer L (2001) Learning rates for energy technologies. Energy Policy 29: 255–261.

Meinshausen M (2006) What does a 2C target mean for greenhouse gas concentrations? A brief analysis based on multi-gas emission pathways and several climate sensitivity uncertainty estimates. In HJ Schellnhuber, W Cramer, N Nakicenovic, T Wigley, G Yohe (Eds), Avoiding Dangerous Climate Change (pp. 253–279). Cambridge: Cambridge University Press.

Miliband D (2007) 'Time for a green industrial revolution', speech by Environment Secretary at University of Cambridge, 5 March 2007.

Nordhaus W (2008) A Question of Balance: Weighing the Options on Global Warming Policies. New Haven, CT: Yale University Press.

Nye M, Whitmarsh L, Foxon TJ (2010) Socio-psychological perspectives on the active roles of domestic actors in transition to a lower carbon electricity economy. Environ Plan A 42:697–714.

Pacala S, Socolow R (2004) Stabilization wedges: Solving the climate problem for the next 50 years with current technologies. Science 305(5686):968–972.

Pigou A (1932) The Economics of Welfare (4th ed). London: Macmillan.

Romm J (2006) Hell and High Water: Global Warming – The Solution and the Politics -and What We Should Do. New York, NY: William Morrow.

Spash C (2007) The economics of climate change à la Stern: Novel and nuanced or rhetorically restricted? Ecol Econ 63:706–713.

Stern N (2007) The Economics of Climate Change – The Stern Review. Cambridge: Cambridge University Press.

Stiglitz J, Sen A, Fitoussi J-P (2009) Report by the commission on the measurement of economic performance and social progress. Report to President Sarkozy, Paris, France. http://www.stiglitz-sen-fitoussi.fr/

Tickell O (2008) Kyoto2. London and New York, NY: Zed Books.

United Nations Environment Programme (UNEP) (2009) A global green new deal: Policy brief. http://www.unep.org/greeneconomy/

Unruh GC (2000) Understanding carbon lock in. Energy Policy 28:817–830.

Victor P (2008) Managing without growth – Slower by design not disaster. Cheltenham, UK and Northampton, MA: Edward Elgar.

Index

J. Dodson (ed.), *Changing Climates, Earth Systems and Society,* International Year of Planet Earth,
DOI 10.1007/978-90-481-8716-4, © Springer Science+Business Media B.V. 2010